Understanding Tourism: Concepts and Theories

Bob McKercher and Bruce Prideaux

(G) Goodfellow Publishers Ltd

(G) Published by Goodfellow Publishers Limited,
26 Home Close, Wolvercote, Oxford OX2 8PS
http://www.goodfellowpublishers.com

British Library Cataloguing in Publication Data: a catalogue record for this title is available from the British Library.

Library of Congress Catalog Card Number: on file.

ISBN: 978-1-915097-66-8

DOI: 10.23912/9781911635352-4280.

Design and typesetting by P.K. McBride, www.macbride.org.uk

Cover design by Cylinder

Printed by Lightning Source

Contents

Acknowledgements

This book has been 30 years in the making, since the two authors began their journeys as tourism academics. And so, there are almost too many people to thank by name for making this project possible. However, many need to be acknowledged directly and indirectly.

We started at a time when tourism scholarship was undergoing unprecedented growth. Most of the new programmes were situated in business schools or commerce departments, while most of the tourism scholars came from non-business, social science backgrounds. Our knowledge was based on the first generation of scholars who began to explore this area of study beginning in the 1970s. Their work provided the foundation for how we think about tourism the way we think about it today. The works of those pioneering scholars form the basis of this book. We, and indeed, all of tourism academics, owe a debt of gratitude to them.

In addition, we have to thank the many thousands of students we have taught over the years for challenging us, forcing us to rethink much of what was accepted to be true about tourism, yet had little real supporting evidence, and for being sounding boards for us to bounce ideas off them.

While many people can share rightfully in any success the book may have, the two authors take full responsibility for any of its deficiencies.

Part 1 – Setting the Stage

The first two chapters set the stage for the study of how our understanding of tourism has emerged over the past 75 years. Students need to be aware of how terms are conflated, misused and misunderstood. So, in these first chapters, we start to cut through the mist of what tourism is and how it is structured. Most importantly, it is essential to be able to appreciate how each discipline defines tourism and uses it for its own narrow needs.

No text examining the theories and concepts of tourism would be complete without first defining its terms. Chapter 1 sets the stage by exploring the issues of what is a theory, concept and model. Special reference is made to work by Stephen Smith and his colleagues about how the term theory is used, and mostly misused, in seven different ways to explore this phenomenon we call tourism. The chapter then moves on to a discussion of the differences between theory, concept and models.

Chapter 2 begins the exploration of tourism by examining the contribution that different disciplines and fields of study have made to the development of this field (or is it a discipline?). It begins by entering into the almost impossible debate about trying to define tourism and concludes that any comprehensive definition must include both heuristic and technical definitions, which in many ways are mutually exclusive. The chapter then recognizes that pleasure tourism grew out of the leisure paradigm. We then ask if we really know the truth about tourism, even after so many decades of enquiry and conclude that much of our mistaken beliefs are the function of persistent myths about what tourism is and how it operates.

1 Theory, Concepts and Models

By the end of this chapter, the reader will be able to:

- Know what theory is and distinguish between good and bad theory
- Be able to identify and analyse seven types of theory used in tourism
- Define the differences between theory, concepts and models

Introduction

Academics have begun to think about theory in tourism in much more detail than ever before. As a relatively young field, there have been many complaints that tourism lacks distinct theory that has held back its development. However, Stergiou and Airey (2018:165) note tourism theory does not exist independently of reality; rather it has pragmatic relevance, facilitating researchers and others to make sense of and explain the so-called real world in the form of data or other phenomena.

We, as academics seem focussed on producing theory and developing new ideas as 'theory'. Indeed, many journals now insist on 'theory' development as a core contribution before a paper can be accepted. This situation is a shame, for the risk is increased that solid, empirical and topical papers that explain how tourism works will not see the light of day because they lack 'theory'. Yet, most of us know very little about what theory is and what constitutes good theory. A key challenge, and one that must be addressed right from the outset, is the meaning of theory, in general terms and is specifically how it applies to tourism. In fact, 'theory' is one of most misused terms today, for it is often construed as an expression of an unproven

About the authors

Prof Bob McKercher has been a tourism academic since 1990. Prior to that he worked in the Canadian tourism industry in a variety of advocacy and operational roles. He has broad ranging interest in tourism. He is a Fellow of the International Academic for the Study of Tourism, the Council for Australian University Tourism and Hospitality Education and the International Academy of Culture, Tourism and Hospitality Research.

Prof Bruce Prideaux commenced his academic career in 1992 at the University of Queensland where he later obtained a PhD. He began his professional life as a high school teacher and later moved to the public service where he worked as a transport analyst. His pre academia wanderings through Africa, Europe and Asia played a large part in developing his interest in tourism as both a traveller and later as a scholar. He has a wide range of research interests including sustainability issues, climate change, transition to carbon-neutrality, drive tourism, protected area tourism, rural tourism, remote area tourism and crisis management

To our families for putting up with us.

emergence of the dominant business studies/training orientation adopted increasingly in higher educational institutions. Stand-alone business tourism schools/departments or the emergence of stand-alone tourism specialisms within business schools provide an enhanced range of career opportunities for students. But these benefits come at the risk of students receiving a selective, cursory overview of the role social science research has played in our understanding of tourism, our beliefs about how tourism works and, importantly, our grounding in an understanding of how the social sciences explains much of the business of tourism.

The book is divided into five sections. Part 1 sets the stage. It begins with a review of what theory is and how models and concepts fit into this broad paradigm. A discussion of the challenge of defining tourism follows with a review of various models that describe the structure of tourism.

Part 2 examines the key features that explain how tourism works. Successful tourism is a function of attractions, access, politics and various supply and demand functions. This section begins with a detailed discussion of what attractions are and how they drive tourism. Because the term attraction is one of the most misused terms in tourism academia, a conceptual taxonomy of attractions is developed as a means of adding some structure to the topic. The issue of access is discussed next looking at the valuation of time and how both market access and physical access influence tourism flows. The following chapter focuses on the political economy of tourism and how that informs tourism as a form of international trade.

Part 3 looks at the evolution of tourism. A critical review of Butler's and Plog's lifecycle models is undertaken highlighting the conceptual strengths and limitations of each one. It also discusses unrecognized subtleties of the models. An alternate lifecycle model using complexity theory is then discussed. We argue it is a more reliable way to document destination evolution for its dynamic nature allows for multiple factors to play a role in influencing change.

Part 4 focusses on the tourist. It examines such issues as why tourists travel by exploring various motivational approaches and then examines the sociological and anthropological contributions to tourism. This section then examines the tourist as an individual and how tourists consume destinations.

Part 5 focusses on a range of pressing issues facing the sector, from an examination of social impacts through to the never ending challenge of sustainability and how tourism needs to be reconsidered in a dynamic, crisis-embedded world. The section concludes by reviewing planning models and calls for a rethink of how tourism is planned.

We hope you enjoy the book and that it makes you think about what tourism is and how it works.

Bob and Bruce.

Preface

This second edition of the book presents a conceptual overview of how different disciplines and fields of study have developed their own ideas about tourism, the tourist and the commercial sector that links the two. In doing so, it provides a unique insight into the complex idea of what tourism is and how we have come to think of it. The two authors have a combined 80 years or more of practical, industry and academic experience. We are still amazed at how little holistic knowledge of tourism there is, leading to the recognition of the need for a book like this.

What is tourism? Simple question. Difficult to answer. We all know it exists. We all know or think we know what is definitely tourism, who tourists are and what the tourism industry is. We also all know, or think we know, what is definitely not tourism, is not a tourist and what business sectors fall outside of the tourism industry. But tourism is neither absolute nor exclusive. Instead, it is classified by vague and fuzzy boundaries. That is what makes it such an exciting field of study, on the one hand, and such an exasperating phenomenon to study on the other.

As a result, much of what we think we know about tourism is well-known but not necessarily known well. Most topics are covered in tourism texts or cited frequently in the academic literature. But, the core ideas that drive how we think about tourism are often condensed into a short paragraph, a few sentences, or simply included as one of a series of references to illustrate a point. Knowledge for many comes from these brief summations and not from the foundational works themselves. We are hoping that through this text, people can reflect more on how we think about tourism and what epistemological and ontological processes we have followed to think the way we think.

The situation is complicated by the historic traditional disciplinary silos that have influenced tourism studies, which have produced discrete sets of definitions, rules, guidelines and regulations. This book cuts through that haze by presenting the first truly comprehensive text that incorporates different academic approaches to tourism to create a holistic understanding of what it is and how it works. It examines the core theories and concepts that lead to our understanding of each of the component elements of tourism in an attempt to develop a common, cross disciplinary understanding. It then seeks to show the linkages, forward and backward, between each of these elements and the other constituent components of the tourism system. In doing so, it also pays homage to the pioneers of the field who first started to think about tourism and were largely responsible for much of what we believe to be true now.

A book like this one becomes increasingly important as the nature of tourism studies shifts from a discipline focused approach that uses tourism as a means to explore different disciplines and/or uses core disciplines to explain tourism to the

Acknowledgements

This book has been 30 years in the making, since the two authors began their journeys as tourism academics. And so, there are almost too many people to thank by name for making this project possible. However, many need to be acknowledged directly and indirectly.

We started at a time when tourism scholarship was undergoing unprecedented growth. Most of the new programmes were situated in business schools or commerce departments, while most of the tourism scholars came from non-business, social science backgrounds. Our knowledge was based on the first generation of scholars who began to explore this area of study beginning in the 1970s. Their work provided the foundation for how we think about tourism the way we think about it today. The works of those pioneering scholars form the basis of this book. We, and indeed, all of tourism academics, owe a debt of gratitude to them.

In addition, we have to thank the many thousands of students we have taught over the years for challenging us, forcing us to rethink much of what was accepted to be true about tourism, yet had little real supporting evidence, and for being sounding boards for us to bounce ideas off them.

While many people can share rightfully in any success the book may have, the two authors take full responsibility for any of its deficiencies.

Contents

and often unprovable opinion. In reality, fact and theory go hand in hand for each is necessary for the other to be true (Cooper and Schindler 2014).

Theory

Theory in a social science context is an ill-defined notion, meaning different things to different disciplines and discourses (Stergiou & Airey, 2018). In fact, there are almost as many (mis)definitions of theory as there are users. Part of the problem is that theory means something very different in lay language than it does in science (Ghose, 2013). To most people, theory is just an idea, when in fact it is much more. As Feldman (2004:565) notes when talking about the management discipline, "*theory is often in the eye of the beholder. What is theory to one reader may not be theory to another*." Abend (2008) says sociologists use theory in seven different ways, while information systems researchers also struggle with the term (Lee, 2014). It is not surprising that such a diverse field of study as tourism faces even greater challenges, leading Smith and Lee (2010) to conclude there is an increasingly inconsistent use of the word.

Let's start at the beginning. What is theory? Cooper and Shindler (2014) define theory as the systematically interrelated concepts, definition and propositions that are advanced to define and predict phenomena. Yeoman and Currie (2019) say it is a body of logically interconnected propositions that provides an interpretive basis for understanding phenomena. Lee and Lu (2023) state theories provide well-reasoned and logical arguments concerning the nature of causal relationships between causes and outcomes. Within a tourism context, Dann, Nash and Pearce (1988: 4) defined theory as the body of logically interconnected propositions that provide an interpretive basis for understanding phenomena. Others add theory is an abstract description of the relationships between concepts that helps us to understand the world (Varpio et al., 2020). It has four basic purposes:

- To describe or answer questions, such as what is happening
- To explain or answer questions as to how and why something happens
- To predict or answer questions as to what will happen
- To control or answer questions as to how something should happen (Infante et al. 1997).

Theory then, should be a well-defined scientific principal that is supported by convincing experimental and observational evidence that has strong explanatory power, helps understand a phenomenon and makes predictions about future events (Sarokin, 2017). Sutton and Staw (1995) add it must answer the question why, while Gorelick (2011) feels theory involves the formation of hypotheses that can be tested empirically and rejected or not rejected. Perhaps Stewart et al. (2011)

summarise it best when they feel theory is an explanation that offers accounts of how and why things are as they are that intends to explain something.

Clearly, theory involves more than just unproven ideas or suppositions. Summarizing the literature, Bricker et al. (2015: 3) see theory as serving as a tool which guides inquiry and analysis. It guides inquiry in the form of shaping the development of hypotheses and it shapes analysis by aiming to explain observations through concepts. Theories must be refutable, testable or falsifiable. They should also be verifiable through observation. Wacker (1998) states theory must have four basic criteria. It must have:

- *Conceptual definitions* where existence and properties are asserted in order to account for what is observable that define who or what is included and specifically excluded. Moreover it should use common and clear language that is understandable by all;
- *Domain limitations* which specify the exact setting where the theory can be applied;
- *Relationship-building qualitie*s which logically assemble a causal relationship between items and state how variables are related or unrelated to others;
- *The ability to predict relationship*s and offer internally consistent predictions in the real world.

As Wacker (1998: 363) notes "*theories carefully outline the precise definitions in a specific domain to explain why and how relationships are logically tied so that the theory gives specific predictions.*" The the goal of 'good' theory is a clear explanation of how and why specific relationships lead to specific events. Sutton and Stack (1995: 378) add:

> "*theory is about the connections among phenomena, a story about why acts, events, structure, and thoughts occur. Theory emphasizes the nature of causal relationships, identifying what comes first as well as the timing of such events. Strong theory, in our view, delves into underlying processes so as to understand the systematic reasons for a particular occurrence or nonoccurrence.*"

Wacker (1998), Ritchie et al. (2008) and Smith et al. (2013) state 'good' scientific theory must have the following characteristics. It must demonstrate:

- *Uniqueness* where one theory must be differentiated from another and if two theories are identical, then they are really one theory;
- *Accuracy* within its domain;
- *Parsimony* whereby if two theories are similar in most respects, the one making fewer assumptions and needing fewer definitions probably is better;
- *Conservativeness* in that it cannot be replaced until a new, competing theory is shown to be superior;
- *Generalizability* where the more areas a theory can be applied to, the better;

- *Fecundity* where theory can be used to expand the area of enquiry into new directions; and it should have broad scope with explanations that go beyond those it was initially used to explain;
- *Internal consistency* and consistency with other accepted theories that are related to the same phenomena;
- *Abstraction*, and;
- *Independence* in time and space.

It is for this reason that Lee and Lu (2023) argue for the adoption of necessity theory in tourism and hospitality studies. Necessity theory asserts that a direct cause and effect relationship must exist that cannot be substituted for other causes. They illustrate that for an outcome to occur, the presence of a necessary cause is a fundamental requirement, for without it the outcome cannot come into existence. This issue is critical in social science research for often an unlimited number of variables can be considered to predict an outcome. Many of these variables may show a correlation, but correlation does not equate to causation. Instead, it is essential to look more deeply for the true cause of an outcome.

In other words, the principle of proximate cause, or in law *sine qua non* ("but not for") has to be considered. *Sine qua non* states an event would not have occurred without this action happening. It is defined as the nearest cause of an event (CILA, 2012) and can be considered as the active, direct, and efficient cause of something that sets in motion an unbroken chain of events (BD, 2016). Why is it important in tourism? *Sine qua non* distinguishes the difference between observed correlations and deeper causal factors.

In short, theory should be able to explain existing phenomena and also predict new phenomena. Theories, therefore, are sets of statements, some of which state laws, while others are singular factual or existential claims. Bramwell (2015: 205, 206) illustrates that theoretical activity seeks to find general accounts or explanations of the world. In doing so, it can increase understanding through a systemised structure to thinking that involves relatively abstract ideas, categories, groupings or interpretations. He adds that theoretical thinking is necessary because the world is complex and by simplifying complex processes we can identify things that are more important and look for underlying causes or relationships that affect how things work. Looking at the literature, Bramwell (2015: 206) argues empirical investigation is often bypassed so that attention can be focussed on mental processes, generalisation and explanation. In doing so, it can direct future empirical research.

Importantly, theory is not set in stone, for it represents the best knowledge at the time of its development (Brunner, 2019). That means that when or if new and better knowledge emerges, theories can be modified, rejected and/or new theories developed. This issue is especially critical in tourism studies where many of our theories have evolved from other disciplines that do not consider the unique nature of tourism. Indeed, it is surprising to us that when some academics find a gap between theory and practice, they conclude the practice must be wrong, instead of noting that, perhaps, the theory needs modification.

Theory plays a key in driving and/or shaping research. Bricker et al. (2015) argue theories prevent research errors, assist in making sense of observed patterns and shape and direct research efforts. Udo-Akang (2012) feels theory and research are interrelated. While research contributes to the explanatory power of theory, theory provides the framework to develop the research process. Wacker (1998) suggests three other relationships. First, theory provides the framework for analysis where differences of opinions exist and, importantly provides commonality of language. Second, it represents an efficient method for field development by reducing errors in problem solving by building on existing knowledge. Third, it establishes clear explanations for the pragmatic world, which give the basis for empirical support of rules, advances knowledge and guides future research.

It has proven difficult to define theory in a clear and concise manner, with the conflation of the terms paradigm, theory and model all to frequent in social science research (Khan, 2002). Instead, it is often easier to identify what is not theory. Sutton and Staw (1995) address many of the issues that seem to be confusing to many tourism academics. They identify five cases where things have been identified as 'theory' when they are not. They are summarised in Table 1.1.

Table 1.1: What Theory is NOT

NOT theory	Why
References	References provide the basis for grounding work conceptually. They are sometimes used like a smoke screen to hide the absence of theory where people use a string of references to hide the fact that we really do not understand the phenomenon in question. References are valuable to explicate which concepts and causal arguments are adopted from cited sources and how they are linked to the theory being developed or tested.
Data	Empirical evidence plays an important role in confirming, revising, or discrediting existing theory and in guiding the development of new theory. Theory and data each play a distinct role in behavioral science research. Data describe which empirical patterns were observed and theory explains why empirical patterns were observed or are expected to be observed.
Lists of variables or constructs	Especially relevant given the vast amount of structural equation modelling (SEM) work being produced. Theory must also explain why variables or constructs come about or why they are connected. Comparative tests of variables should not be confused with comparative tests of theory, because a predicted relationship must be explained to provide theory; simply listing a set of antecedent or casual ordering does not make a theoretical argument.

Diagrams	Diagrams show the nature of relationships but are not causal by design. Diagrams and figures should be considered as stage props rather than the performance itself. While boxes and arrows can add order to a conception by explicitly delineating patterns and causal connections, they rarely explain why the proposed connections will be observed.
Hypotheses or predictions	Hypotheses can be an important part of a well-crafted conceptual argument but are not theory in and of themselves. Hypotheses are concise statements about what is expected to occur, not why it is expected to occur.

Source: From Sutton and Staw (1995)

Use and misuse of theory in tourism

Renowned academic Steven Smith has explored the use and misuse of the term theory in tourism research. His first foray into this area (Smith & Lee, 2010) determined that there has been a substantial increase in the misuse of the term, perhaps as a result of tourism research being embedded in non-traditional disciplines, which means that academics may not have as deep a grounding in a core discipline as their predecessors.

Later he and a number of colleagues (Smith et al., 2013) identified seven ways the term theory is used or misused in tourism. Table 1.2 identifies them, defines their key features and their constituent elements. Essentially, there are two main themes to this study. The first is that the term 'theory' is used very loosely in much of our literature. Indeed, they suggest that apart from the first and second types, the term theory is misused as a synonym for different research methods, or as an untested set of opinions.

The second theme, and one that is more important, relates to theory of the first and second type. In the hard sciences, only one theory can exist. It must be testable and can be only rejected or not rejected. The scientific method is both cautious and skeptical in nature. If a theory can be rejected, that means another competing theory may be more appropriate. Theories can only be 'accepted' if they are tested repeatedly and not rejected. The more a theory is tested and not rejected, the more robust it becomes. But the key feature is that only one theory can exist at any one time and that when competing theories purport to explain the same phenomenon, only one can work.

But, they also identify theory of the second type in the social sciences where different disciplines may look at the same phenomenon and develop discipline-specific theories to explain that phenomenon. Here, competing theories may be valid. For example, in tourism, lifecycle evolution theories have been proposed from the social science disciplines of human geography, social psychology and complexity. Each one seeks to explain the evolution of destinations from its own narrow, disciplinary perspective and each is valid within its own domain. None of them necessarily explains fully the phenomenon of destination evolution.

The challenge with something like tourism is that it has been explored from dozens of different disciplinary perspectives, with each one driven by its own ideology, boundaries, methods and frameworks. As a result, multiple, complementary theories have emerged to try to explain tourism. Many, or most, of these theories fall into what Smith et al. (2013) would describe as theory of the third to seventh type. Yet, in reality, regardless of how one looks at tourism, four key issues arise: the person partaking in the activity, the setting where the activity occurs, the phenomenon of the activity and the providers of goods and services to the person traveling. The key challenge with tourism, then, is to understand the perspective or disciplinary focus used to describe the phenomenon and, by the same token, to ensure that one does mistakenly apply one discipline's theoretical constructs to explain another discipline's findings unless they are compatible.

Does tourism have theory? Good question. The short answer is it depends how one defines theory. Dann and Cohen (1991: 167) assert that no all-embracing theory of tourism exists because "tourism, like any other field of human endeavor, is a target field," simply providing fodder for the theoretical approaches of other disciplines.

Tourism, like many social science disciplines has an interesting relationship with theory. Sanchez et al (2019) argue there is little original tourism theory found in most research. Instead scholars tend to borrow concepts from other disciplines and apply them to tourism. They label this act as a second level of originality. The reasons are manifest and relate to the belief that tourism has not yet reached the status of being a discipline in its own right, coupled with its applied nature that allows it to be studied from other disciplines. This situation fosters the borrowing of extant theories and concepts from other disciplines, which offers both benefits and costs. The benefits of doing so enables tourism to shed new light on existing ideas, while the cost may be a constraining of originality.

These authors also identify a second approach, which they have labeled theory integration. Integration is less well regarded for it represents only incremental increases in knowledge. Here, the tendency is to conclude that studies support existing knowledge, but perhaps from a slightly different perspective, leading to many papers published offering little except confirmation. Others see great potential in theory integrations (Shaw et al., 2018), providing it involves genuine cross discipline integration of different ideas into a new holistic idea. These authors are essentially writing about the potential to integrate Type 2 theory, which has great applicability to tourism, given its pandisciplinary roots.

Table 1.2: Seven ways theory has been used in tourism research

Key feature	Element
1. Theory of the form used in natural sciences	
Falsifiable hypotheses that have been tested multiple times with positive results Single verifiable 'truth'	Based on the belief that there is a knowable, objective reality that transcends the researcher's own opinions or biases. Normally only one theory can exist to explain a given phenomenon. If two or more competing theories exist, one (or more) will eventually be proven wrong. Can be independently tested by other researchers. Understand their theories may ultimately be proven to be incomplete or incorrect. Limited to explanatory and predictive models of some phenomena supported by repeated tests, logically linked to other concepts and theories that provide an integrated and coherent understanding of some aspects of reality, and produces significant falsifiable predictions. Repeated tests with positive results cannot conclusively prove a belief is correct. One contrary result can prove an assumption, model, or theory is wrong.
2. Theory of the form often used in social sciences	
Two or more valid competing theories Falsifiable hypotheses that have been tested multiple times with positive results	Usually associated with complex social science phenomena. Results of any test of a hypothesis may show only equivocal support for the theory. Multiple theories may exist with equal validity within the scope of available evidence because such theories address complicated phenomena for which data and understanding may be incomplete. The failure of a Type 2 theory to support a hypothesis is, by itself not usually seen as a sufficient reason to reject the theory, at least not until after repeated failures.
3. Theory is equated with statistical analysis	
No a priori theory – instead test an ad hoc model. (If testing an 'a prior' theory, then it is likely an empirical test of Type 1 or 2 theory)	Statistical models that are formulated and presented as theory but without a tested a priori conceptual mode. Falsifiable in that an independent researcher can check the results or even replicate the study to determine whether the results are reliable using the data in question.
4. Theory is an untested/untestable verbal or graphic model	
Models with no empirical, testable hypotheses	Similar to Type 3 but not falsifiable by an independent observer. Verbal or graphic models that are expressed in such a form that independent, objective testing is not done or may not even be feasible.

5. Epistemology or a research design presented as theory	
Self-fulfilling prophecy See the world through the 'theory' and shape studies to confirm it	Epistemology - presents a worldview and identifies which questions and data are appropriate for scholarly enquiry and which are not. e.g. Postcolonial theory, conflict theory, feminist theory. Epistemologies cum theory specify what questions are legitimate to ask, what data may be collected, how data may be collected, and how data are to be analyzed and interpreted. Does not produce objectively testable hypotheses. Rather, it produces interpretations that are based on the assumptions of the epistemology. Self-fulfilling prophecy type of 'theory' tends to be self-perpetuating in that those who hold a particular epistemological perspective tend to view all the evidence they collect in terms of their chosen intellectual filter rather than allowing for the possibility of contrary findings. Tendency to interpret any results as confirming one's predetermined conclusions. Boundary between theories of the fourth type (untested models) and the fifth type (epistemology) can blur in casual reading. Theories of the fourth type are expressed in terms of a specific model or concept, whereas theories of the fifth type are formally articulated epistemologies described as theories.
6. Grounded theory	
Grounded theory - method	Inductive methodological approach of data collection, analysis, and interpretation utilizing a systematic process of iteration and constant comparison for coding transcripts and other data from which theories emerge.
7. Theory as a casual term or used as an analogy: 'Other'	
Analogy – idea cannot be tested Conclude ' theoretical' insights without offering theory	All other uses of the term, primarily the use of the word in a casual sense such as 'speculation'. In this group of theories, findings may be described as offering theoretical insights but these insights are not based on either a falsifiable hypothesis or an a priori model. Borrowing from one field as an analogy to tourism – i.e. chaos theory. Theory as analogy.

Source: Smith et al., 2013.

Smith and Lee (2010) feel we are beginning to see the nascent development of theory. Likewise, Ritchie et al. (2008) see the beginnings of the development of core tourism theory as distinct from theory developed in other disciplines and applied to tourism. To them, *"part of the problem that creates controversy over tourism as a field of inquiry or a discipline may be viewed as centering on defining what should be considered core tourism theory versus what is theory from other disciplines or foundational theory."* They feel these difference can be viewed as lying along a continuum with three key distinguishing characteristics used to separate core from foundational tourism theory. The first issue is generalizability, where theory that is core to tourism is specific to tourism and not readily generalizable to other disciplines, whereas foundational tourism theory aims to develop generalizations beyond tourism. The second issue relates to theory that is being built upon or critiqued. Here they argue theory that is core to tourism builds on or critiques existing core tourism theory, while foundational tourism theory builds on or critiques theory from its own discipline. Finally, the last issue is concerned whether or not the theory is contained within a single traditional discipline. Here core tourism theory is inherently multidisciplinary while foundational theory is confined to one discipline. This last point is critical for core tourism theory integrates different disciplinary perspectives to explain tourism and especially to fill in the gaps where one discipline's perspective is lacking.

Concepts

Concepts are abstractions of ideas (Sarokin, 2017). A concept can be broad or narrow and can be based on experience or the imaginary. Concepts are terms that refer to the characteristics of events, situations, groups and individuals that are studied. They may be impossible to observe directly, such as justice or love, or may be observable directly, such as a tree (Pizam, 1994: 95). Importantly, they are not theories, models or paradigms. Instead, they are mental boxes into which things are thrown that are thought to have things in common (Bosch, 2011). The level of abstraction can vary depending on the generalizability of the topic. Something like tourism is a very abstract concept. We know what it is, but as you will find out in the next chapter, it is virtually impossible to define in absolute terms. Other concepts, like paying for goods, are more concrete and observable, but payment is still an abstract concept at some level.

Bosch (2011) identifies four key features when defining concepts:

- Keep the definitions clear – which can be hard when we use words to describe words;
- Make the definition appropriate – use one that is consistent with what is generally meant by the idea;
- Avoid defining concepts with related concepts – try not to define tourism as a concept by describing tourists, another concept; and
- Avoid circular arguments.

There is a difference between a theoretical and conceptual framework (Varpio et al., 2020). While a theoretical framework is a logically developed and connected set of concepts and premises created to frame a study, a conceptual framework is the justification for why a given study should be conducted. Varpio et al. (2010) highlight that it describes the state of known knowledge, identifies gaps in our understanding of a phenomenon or problem; and outlines the methodological underpinnings of the project. A theory may be developed based on either conceptual or empirical evidence (Stergiou & Airey, 2018). A conceptual framework, on the other hand, is constructed to answer the question of how the research can add to or challenge what is already known.

Models

The term model is also problematic in that it too is used in different ways by different groups of scientists (Ghose, 2013). Essentially, in the social sciences, models are often visual representations that help build or explain theory (Gay & Weaver, 2014). Models, as a human constructs, help us better understand real world systems (anon, 2019) by presenting a simplified set of relationships that hopefully show some causal direction. They are felt to have some input, some process function and some output (anon, 2019). As Bricker et al. (2015) note, models are useful to simplify theory and demonstrate its relevance and applicability to real life. Cooper and Schindler (2014) explain that the difference between a model and a theory is that theory's role is explanation, while a model's role is representation.

In a tourism context, Getz (1986:22) identified two types of models based on their nature and purpose. Theoretical models describe aspects of the tourism system while process models describe sequences aimed at achieving particular outcomes.

Theoretical models can be further subdivided into:

- Descriptive models which define components of the tourism system,
- Explanatory models which show how a system or subsystem works with or without the specification of a causal relationship,
- Predictive models that enable forecasting.

Process models can also be subdivided into:

- Subjective types based on a particular dogma or idiosyncratic format,
- Problem solving that follows a sequence,
- Systems theory.

While Getz (1986) was concerned about the use of these models in a planning context, his classification has a broader application to the use of models in a tourism context.

Leiper (2004) discusses four benefits of models. A model represents a simplified version of reality that tries to capture the essence of something. In doing so, it can be used as a visual representation of a theoretical construct that can aid in analysis of a problem. Third, models should work at different scales, depending on fit for purpose. A good tourism flow model, for example, should be applicable at a scale ranging from travel between a hotel and an attraction as well as to understand global tourism flows. Finally, a model can serve the needs of different disciplines.

Conclusion

Theory is a difficult concept to understand fully, for there are almost as many theories about theory as there are definitions. But, one can envision theory, concept and model as existing in a nested manner (Figure 1.1), whereby models represent simplified operationalization of a concept and a concept represents an abstraction of a theory. Theory in turn represents a broad, logical set of interconnected propositions that provide a basis for understanding a phenomenon.

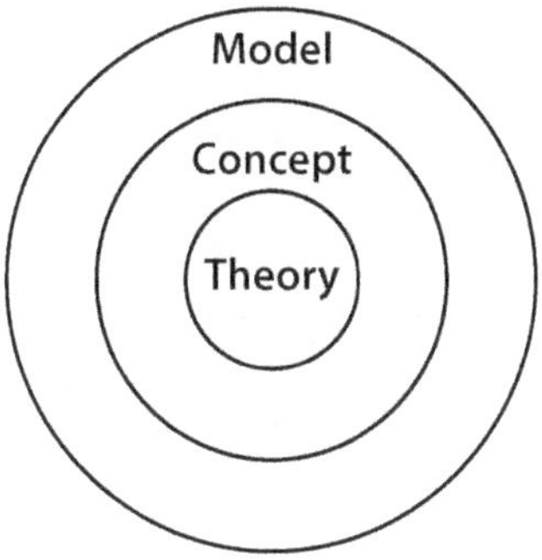

Figure 1.1: Nesting of Theory, Concept and Model

■ Discussion points / Exercises

1. Review five papers on a related theme and determine what type of Smith's theories have been used in these works, and indeed, if any theory has been used?
2. Discuss whether you feel tourism scholars understand the concept of Type 1 or 2 theory or are more likely to use it as Types 3 to 7.

Points to consider for revision

- Theories, concepts and models are interrelated
- A theory is a well-defined scientific principal that is supported by convincing evidence and has strong explanatory powers.
- Theory is often misused in tourism research, to such an extent that seven different, increasingly less powerful uses have been identified
- Concepts are abstractions of ideas
- Models are visual representation of issues, sometimes use to build theory and sometimes used to explain phenomena

Suggested additional readings

Ritchie, J. Sheehan, L & Timur, S. (2008) Tourism sciences or tourism studies? implications for the design and content of tourism programming. *Teoros* 27(1) http://teoros.revues.org/1621.

Smith, S., Xiao, H., Nunkoo, R., & Tukamushaba. E. (2013) Theory in hospitality, tourism and leisure studies. *Journal of Hospitality Marketing and Management* 22, 875-894.

Wacker, J. G. (1998). A definition of theory: research guidelines for different theory-building research methods in operations management. *Journal of Operations Management*, 16(4), 361-385.

References

Abend, G. (2008) The Meaning of Theory. *Sociological Theory* 26(2), 173-199.

Anon (2019) What is a Model. https://serc.carleton.edu/introgeo/models/WhatIsAModel.html <downloaded Oct 24, 2019>.

BD (2016) Business Dictionary.com http://www.businessdictionary.com/definition/proximate-cause.html

Bosch R. (2011) Concepts, variables and measurement. http://polisci.usca.edu/apls301/Text/Chapter%204.%20Concepts,%20Variables,%20and%2Measurement.htm <downloaded Dec 2, 2019>.

Bramwell, B. (2015) Theoretical activity in sustainable tourism research. *Annals of Tourism Research.* 54, 204-218.

Bricker, K., Donohoe, H., Becerra, L & Nickerson, N. (2015) Theoretical perspectives on tourism – An introduction. *In Bricker, K. & Donohoe, H. (eds) Demystifying Theories in Tourism Research.* CABI: Boston. Pp 1 – 6.

Brunner, B. R., (2019). What Is Theory?. In Brunner, B. (Ed) *Public Relations Theory: Application and Understanding*, Hoboken: Wiley. p.1-12.

Cooper, D. & Schindler, P (2014) *Business Research Methods.* New York: McGraw Hill

CILA (2012) *Introduction to the Insurance Industry*. Chartered Institute of Loss Adjusters. www.cila.co.uk/files/Certificate/Chapter%208.pdf <downloaded Mar 10, 2016.>

Dann, G., & Cohen, E. (1991) Sociology and tourism. *Annals of Tourism Research.* 18, 155-169.

Dann, G., Nash, D., & Pearce, P. (1988) Methodology in tourism research, *Annals of Tourism Research*, 15, 1-28.

Feldman, D., (2004) What are we talking about when we talk about theory? *Journal of Management.* 30(5) 565–567

Gay, B., & Weaver, S. (2011) Theory building and paradigms: A primer on the nuances of theory construction. *American International Journal of Contemporary Research.* 1(2) 24-32

Getz, D. (1986) Models in tourism planning. *Tourism Management* 7(1), 21-32.

Ghose, T. (2013) "Just a Theory": 7 Misused Science Words. *Scientific American,* April. www.scientificamerican.com/article/just-a-theory-7-misused-science-words/ <downloaded Oct 24, 2019>.

Gorelick, R (2011) What is theory? *Ideas in Ecology and Evolution* 4,1-10.

Infante, D., Rancer, A. & Womack, D., (1997). *Building Communication Theory*. 3rd ed. Prospect Heights: Waveland Press.

Khan, H. (2002) On paradigms, theories and models. http://www.cirje.e.u-tokyo.ac.jp/research/dp/2002/2002cf156.pdf <downloaded Oct. 25, 2019>.

Lee, A. (2014) Theory is king? But first, what is theory. *Journal of Information Technology.* 29(4), 350-352.

Lee, W., & Lu, L. (2023). Good theories predict: unveiling the untapped potential of "necessity" theorizing. *Journal of Hospitality & Tourism Research*, 0(0). https://doi.org/10.1177/10963480231188666

Leiper, N. (2004). *Tourism Management (3rd ed).* French's Forest: Pearson Education Australia.

Pizam, A. (1994). Planning a Tourism Research Investigation, in J. Ritchie & C. Goeldner (eds.). *Travel, Tourism and Hospitality Research, A Handbook for Managers and Researchers* (2nd ed), Toronto: John Wiley and Sons, pp. 91–10

Ritchie, J. Sheehan, L & Timur, S. (2008) Tourism sciences or tourism studies? Implications for the design and content of tourism programming. *Teoros* 27 (1) http://teoros.revues.org/1621 <downloaded Dec 2, 2019>.

Sarokin, D. (2017) Differences between concepts, theories and paradigms. *Sciencing.com.* https://sciencing.com/differences-between-concepts-theories-paradigms-8415723.html <downloaded Oct 24, 2019>.

Sánchez, I. R., Makkonen, T., & Williams, A. M. (2019), Peer review assessment of originality in tourism journals: critical perspective of key gatekeepers, *Annals of Tourism Research*, 77, 1-11. https://doi.org/10.1016/j.annals.2019.04.003.

Shaw, J.D., Tangirala, S., Vissa, B. & Rodell, J.B., (2018). New ways of seeing: Theory integration across disciplines. *Academy of Management Journal, 61*(1), 1-4.

Smith, S. and Lee, H. (2010) A typology of 'theory' in tourism. *In Pearce, D., and Butler, R (eds) Tourism Research: A 20-20 Vision.* Oxford: Goodfellow, Pp 28-40

Smith, S., Xiao, H., Nunkoo, R., & Tukamushaba, E. (2013) Theory in hospitality, tourism and leisure studies. *Journal of Hospitality Marketing and Management* 22, 875-894.

Stergiou, D., & Airey, D. (2018) Understanding tourism theory. *Tourism Review* 73(2): 156 – 168.

Stewart, J., Harte, V., & Sambrook, S. (2011) What is theory? *Journal of European Industrial Training* 35(3): 221 – 229. https://www.emerald.com/insight/content/doi/10.1108/03090591111120386/full/pdf <downloaded Oct 28, 2019>.

Sutton, R. I., and Staw, B. M. (1995). What theory is not. *Administrative Science Quarterly*, 371-384.

Udo-Akang, D. (2012) Theoretical Constructs, Concepts and Applications. *American International Journal of Contemporary Research.* 2(9), 89 – 97.

Varpio, L., Paradis, E., Uijtdehaage, S. & Young, M., (2020). The distinctions between theory, theoretical framework, and conceptual framework. *Academic Medicine, 95*(7), 989-994.

Wacker, J. G. (1998). A definition of theory: research guidelines for different theory-building research methods in operations management. *Journal of Operations Management*, 16(4), 361-385.

Yeoman, I. & Curry, A.,(2019). Bridging theory and practice. *World Futures Review*, 11(2), 103-107.

2 What is tourism?

By the end of this chapter, the reader will be able to:

- Describe why tourism is a multi-discipline
- Discuss the similarities and differences between the two types of definitions used to describe this phenomenon
- Describe why tourism is an extreme form of leisure
- Identify and analyse the force field of tourism knowledge
- Understand the development of myths about tourism and assess their validity.

Introduction

Tourism is not absolute. That is what makes it such an exciting field of study, on the one hand, and such an exasperating phenomenon to study on the other hand. Life is much easier when things are framed in black-and-white terms: good versus bad; healthy versus unhealthy; left versus right; right versus wrong. Increasingly, as well, people like to have the world presented to them in a simplistic manner where they are faced with apparently easy choices. Life would be easy, if only the world was simple. Yet, in reality the world is a rich tapestry of colours. What on the surface appears to be a simple yes/no choice, in reality becomes a very complex situation when one scratches below the surface.

Tourism is no different. We can all come up with absolute examples of what is and what is not tourism and who is and who is not a tourist. Visitors from China who spend five days on a package tour sightseeing and shopping in London are clearly tourists. Their activities and their resultant expenditures are also clearly tourism related. By contrast, if someone goes shopping for groceries near their home, they are clearly not a tourist. Other absolute categories such as, business travellers, people visiting friends and relatives and special interest tourists can also be defined. But after that, what constitutes tourism and the tourist enters a grey area. What if you have a second home and spend every weekend there? Are you a tourist? What if you are studying abroad for 11 months? Are you a tourist? The answer depends on how tourism and the tourist are defined.

By the same token, identifying all businesses that represent the tourism sector and the people who work in this sector is challenging (Leiper, 1999). People talk about theme parks as tourist attractions. Yet, a large portion of their visitors are local residents. Most so called tourist retail shops, restaurants and even travel providers also rely heavily on local traffic to keep them viable. Even the hotel sector is reliant to a certain extent on the local market for functions, weddings, meetings, special events, food and beverage and even accommodation.

Who works in the tourism industry? Again, we talk about tourism employees, but as Leiper (1999) mentions there is a huge grey area. You could probably say that hotel, airline, travel agent, theme park and tour operations staff work in the tourism industry, although certainly not all of their income comes from tourists. But, what about immigration officers, doctors, accountants, lawyers, insurance agents, the police and many others? Most people would not think they are part of the tourism sector, but they deal with tourists on a daily basis or serve the tourism sector indirectly.

This chapter will examine some of the inherent complexities in tourism and explain the resulting difficulties that have arisen through the academic inquiry of this activity.

Tourism is a complex topic of enquiry

Our understanding of tourism, and its resulting theories, concepts and models are a reflection of this complexity. How can we understand tourism as an absolute entity when we cannot even agree on whether it is a discipline, a field of study, a post disciplinary subject of enquiry, a pandiscipline (Coles, Hall & Duval, 2006; Hall et al., 2014; Tribe, 1997; Weaver, 2023) or if it is even an industry (Lew, 2012)?

Tourism has been the subject of academic inquiry for well over 80 years. A huge body of literature has developed over that time. More than 3000 universities worldwide teach tourism and more than 300 academic journals publish refereed, scholarly research. More than 10,000 academic papers are published each year (McKercher & Dolnicar, 2022), with the volume of published material doubling every five years (McKercher & Tung, 2015). Many things promoted as being core truths about tourism are accepted as fact, even though the basic premises behind them either have no sound evidence or have been disproven (McKercher & Prideaux, 2014).

Yet, in spite of the 80 years of research some key issues remain understudied. How much good work has been conducted on gender and tourism? Why is so little work conducted on drive tourism when it is the most common form of tourism? Today, when we have access to more data than ever before, more sophisticated research instruments, a much larger and better educated research community and many more outlets to share our knowledge, much of the research being produced is derivative and not very exciting.

The multi-disciplinary nature of tourism

The issue is complicated even further by the lack of single disciplinary focus. Instead, tourism can be classified as a multi-discipline, or as Weaver (2023) suggests, a pandiscipline which synthesizes concepts and theories from other disciplines to better describe and explain tourism-related phenomena. Jafari and Ritchie's (1981) foundation work is still relevant today, although the 'wheel' described in their paper has been expanded significantly by Goeldner and Ritchie (2006) and Cheng et al. (2011).

Figure 2.1: The evolving framework of Tourism Studies. Adapted from Cheng et al. 2011

The model adopted by Cheng et al. (2011), shown in Figure 2.1 illustrates how different disciplines have been added to the study of tourism. Unlike anthropology or psychology that have a single disciplinary base, the study of tourism has been influenced by a range of disciplines and fields of study. Each discipline brings with it its conceptual baggage and preferred methodological approaches that informs and guides research. This diversity explains the richness of tourism, but it also explains the great difficulty in coming up with a central, unified model, or even a single agreed upon definition.

The range of disciplines and fields of study that inform tourism is immense. Early research was based on five core disciplines of economics, sociology, psychology, geography and anthropology. Recently, Crouch and Perdue (2014) identified that our research now fits into 50 broad disciplines and fields of study, including the foundational areas, and such diverse fields as atmospheric sciences (climate change), biological sciences (biodiversity and sustainability), computational theory (enhanced and virtual reality), historical studies, transport, urban and protected area studies, as well as many more.

It is not surprising that Benckendorff and Zehrer (2013) and Laws and Scott (2015) say tourism is a fragmented area of study, with Gibson (1998) adding that it is typified by multiple communities of discourse with historically little cross fertilization between communities. Kouropalatis et al. (2023) and Koseoglu et al. (2023) note fragmentation is accelerating. Franklin and Crang (2001) noted 20 years ago that as we explore smaller and smaller units of analysis, each author tries to justify the importance of that research area.

Why has tourism become so fragmented? The answer can be traced to its origins, the background of academics studying and teaching in the area and the nature of tourism itself. In its simplest form, Dann (1997) asserts there is no single truth underlying tourism enquiry and as such, the field must be explored from various perspectives. Taillon (2014) adds no agreed-upon definition exists due primarily to its multidisciplinary nature. He, along with Ateljevic (2014), comment that the field has grown historically in a fragmented and disjointed manner driven by different disciplinary perspectives, institutional and political agendas and journals with different foci. Indeed, Kunwar (2018) asserts no single discipline can accommodate, treat or understand tourism. Instead it must be studied only if disciplinary boundaries are crossed and if multidisciplinary perspectives are sought and formed. Laws and Scott (2015) point out the tourism has to be considered as a system of interlinked parts with the body of knowledge described as a mosaic.

Historically, tourism studies has struggled to find a single home and instead began as a way that academics from different disciplines could gain deeper insights into their own disciplines. As a result, tourism was conceptualised in such different ways that each discipline developed its own set of definitions to describe and therefore examine this phenomenon (Tribe 1997). Xiao and Smith (2006) illustrate that academics from well-established disciplines such as sociology, geography and anthropology were the first to explore tourism in the late 1970s, while the field broadened to include management, economics and environmental issues in the 1980s. Research interest further broadened in the 1990s as leading journals ran special issues on arts, pilgrimage, heritage and a range of other topics. This situation continues today, for Butowski (2016) found that a slight majority of academics conduct their research on tourism within various core disciplines, while just under half based their research on other paradigms that inform tourism.

The backgrounds of academics add to the multi-disciplinarity of the subject area. Ateljevic (2014), for example, points out that 70% of tourism programs in the UK are located in business schools, while almost 90% of individual academics have a background in social sciences. Xiao, Li and Lin (2011) and Weiler, Moyle and McLennan (2012) analyzed doctoral dissertations and found that social sciences have informed the greatest number of dissertations, even now when programs are housed in business schools. An analysis of 52 doctoral dissertations in Turkey found they fell into six broad discipline areas (Şalvarcı & Aylan, 2019), while Xu and Zhang (2021) indicated that Chinese doctoral students could study in any one of 13 discipline areas.

Perhaps, though, the single greatest reason why tourism must be considered as a multi-discipline rests in the fact that the range of topics we explore is so broad. Some 30 years ago, Gunn (1991) speaking at a Canadian Association of Geographers conference, talked about the need for multidisciplinary tourism education, for he felt that the component parts of tourism could only be understood if several different disciplines were applied. He stated the element of travel had to be considered from the perspective of transportation, political science and geography, while visitor activities encompassed consumer behaviour, psychology, sociology, history and marketing. Economic impact could only be studied from a finance, economics, management, entrepreneurship and business sciences perspective, while the impacts of tourism needed both business and environmental fields of study. Liszewski (2010) concurs by arguing that tourism is a system of elements comprising the person, the process of tourism the destination entrance of places, activities and the impacts or consequences of tourism. Different disciplines are best suited to examine the processes, mechanisms, rules and regularities of each of the component parts.

The very multi-disciplinarity means that we must take a cautionary approach when considering what is meant by the word 'tourism', for each discipline or field of study defines it in its own terms, conflates ideas and adopts a self-serving approach to its study. The net result is that the same word can have quite different connotations depending on the author. What an economist considers as tourism or a tourist may be different from an anthropologist, sociologist or marketer, leading to much confusion and silofication. Pritchard (2012) comments that the field remains beset by many ontological, epistemological and methodological shortcomings.

Defining tourism - Why definitions are important

Agreeing upon a precise definition represents perhaps the greatest challenge that is holding back the development of a unified theory of tourism. The English language is very subtle, making it easy to use the same term to discuss completely different concepts unless it is clearly defined. A definition, according to the English Oxford dictionary is *"an exact statement or description of the nature, scope, or meaning of something."* Lisle (2017) indicates it is a statement or series of statements that explain a word's meaning that acts as a linguistic token that represents something. Gupta (2015) suggests definitions vary in function and can provide precision in meaning, accuracy, clarity or fecundity. Ultimately, they describe the essence of the thing, enabling us to use a common vocabulary when discussing or reading about an issue (Whitfield, 2012).

A good definition, therefore helps us assess a situation better and make better decisions, while imprecise or weak definitions make it difficult to even agree on what is being discussed and how to address or resolve a problem (Andrees, 2014; Veerasamy, 2013). Owens (2015) highlights how variations in definitions may actually change what we are talking about or who we are referring to and, therefore, limit the generalizability across different situations.

Veerasamy (2013) states definitions have to satisfy two criteria. First, they must be precise and not convoluted, by having as few elements as possible to capture the essence of the item. Second, they must be falsifiable, in that the definition not only captures what is embodied by the term but also what is not embodied or what is excluded.

Table 2.1: Benefits of definitions in tourism

Components of a good definition
Precise
Use clear terms
Falsifiable
Benefits to tourism / tourist of a good definition
Have a common understanding of the concept
Use common language so different groups are talking about the same thing
Set agreed upon boundaries to describe what is and is not included in tourism or who is/ is not a tourist
Describe and classify
Count
Analyse it according to the definition
Determine how to examine
Determine what you to examine

Ideally, then a good definition of tourism or the tourist will provide the types of benefits identified in Table 2.1. It will provide a common understanding of the subject and allow the use of common language to describe what we mean so that different groups using the term can discuss the same topic. We can then set boundaries that can define what is or is not included in tourism, and distinguish tourism or the tourist from things that are not considered tourism or the tourist. In doing so we can describe and classify the phenomenon, count it if needed, analyse it and examine it according to the definition, and also define or determine what it is that we wish to examine when we look at this broad concept called 'tourism'.

One definitional challenge is the attempt to distinguish between a 'tourist' and a 'traveller'. Are tourists really different from travellers? Tourism is an activity that occurs rarely in our daily life, is driven by personal motives and provides a range of personal benefits. We want to feel special when we travel and moreover, feel we are having a unique experience. Yet, we are often confronted by large crowds as shown here at the Trevi Fountain in Rome. How do many people react? They tend to make value judgements about others that somehow place themselves in a more superior position by saying "I am a traveller, while everyone else is a tourist." The use of such language carries with it a negative connotation of the other (Leiper, 2004). Being a 'traveller' infers greater standing and simultaneously distances oneself from 'tourist'. What is the real difference between a tourist and a traveller? None! We are all tourists. The only difference is semantic and perspective. From my perspective, I am a traveller, you are a tourist! But from your perspective, you are the traveller and I am the tourist!

The impossible task of developing a precise definition of tourism

On the surface, tourism appears to be a simple concept. The Oxford English Dictionary defines it using a circular argument as the *"theory and practice of touring, travelling for pleasure."* This simple concept, however, is exceedingly difficult to define in any meaningful manner. Hundreds of definitions have been proposed over the years and none has captured the essence of the subject clearly and precisely. Two big challenges emerge. The first is the boundaries of what constitutes tourism and non-tourism are imprecise and fuzzy, while the criteria to distinguish between tourists and non-tourists are even more blurred. The second is that tourism embodies both business considerations, where precise terms are required, and social science considerations, where a looser idea of the phenomenon enables better research (Tribe, 2006).

■ The core concept - a journey away and back

Lieper (1979) reminds us that the term *tour* traces its roots to the Greek term 'tornos' and the Latin term 'tornare' which describes a lathe or a circle. The essential feature of tourism, then, is a round trip that ultimately returns to the point of departure. The suffix *ism* refers to the action, process or typical behaviour while the suffix *ist* refers to the individual doing the action (Leiper, 2004). Combining *tour* and *ism* suggests tourism is the action or movement around a circle, while *tour* and *ist* is the person doing the action.

■ Technical vs heuristic definitions

Beyond that simple concept though, definitions begin to run into trouble, for no single one can capture the full essence of tourism and still satisfy all criteria of a good definition. The definitional challenge is further exacerbated by the use of circular arguments. As Morley (1990) discussed, tourism has often been defined through the use of such terms as tourist and the supply of tourist facilities. Ultimately, Leiper (2004) feels two main types of definitions exist: technical and heuristic. Each is quite different and yet, both are essential.

Technical definitions

Defining the size and scope of the sector is one of the great challenges facing governments, tourism organisations and tourism academics. To do so, requires absolute technical or statistical definitions that establish clearly defined boundaries. The first such definition was developed by the League of Nations in 1937 and described a tourist as a person who visits a country other than that in which he habitually lives for a period of at least 24 hours (Leiper, 1979).

In 1963, the United Nations Conference on Travel and Tourism developed definitions of international visitors and tourists that, with some modifications, were adopted by the World Tourism Organization. An 'international visitor' is any person who travelled to a country other than that in which they had their usual residence, the main purpose of whose visit is other than the exercise of an activity remunerated from within the country visited and who is staying for a period of one year or less (WTO as cited in Morley, 1990). This definition distinguished between *international tourists* and *international excursionists*. An international tourist was defined as a person staying at least one night, but not more than a year in a country, whereas an international excursionist was someone who did not stay overnight in a country.

Other technical definitions of both domestic and international tourism have applied arbitrary and artificial criteria of a minimum distance travelled (varying between 30 and 160 km), a minimum length of stay requirement and finally, trip purpose. Some definitions specifically exclude business travel, while others embrace it.

The oft cited UNWTO definition of a tourist for the satellite accounting purposes as "*any person travelling to a country other than that of his/her usual residence, who stays there at least 24 hours (or at least one night) and whose purpose of trip corresponds to leisure (pleasure, holidays, health, studies, religion and sports) or business, family, mission, meetings, and is other than the exercise of an activity remunerated from within the country visited*" is a classic example of a technical definition.

These definitions serve a useful purpose helping governments and tourism bodies come to grips with the size of the sector. As well, they serve to establish common criteria by which global tourism can be measured. From a geographical perspective, such definitions recognize the spatial aspects of tourism and the spatial interaction that arises from the tourists' movements (Pearce, 1987).

The use of an arbitrary definition serves statistical purposes, but it has limited utility for other uses. As Morley (1990) suggests, the real world is rarely statistically convenient. While it may be useful to try to distinguish between locals and visitors, in reality it may be almost impossible to make that distinction. The use of an arbitrary distance or minimum stay criterion may mean that two people doing exactly the same thing in the same space and time and exerting the same impacts may be defined differently. To overcome this problem some researchers propose rejecting such criteria and instead discuss tourism motivated trips regardless of the distance travelled or time spent away from home (Jansen-Verbeke & Dietvorst, 1987).

Heuristic approaches

The alternative approach is to adopt a heuristic definition. Heuristic definitions are broader and less precise in scope. They serve three main purposes: to help focus thinking; remove ambiguities; and aid communication. They do not have to conform to technical or statistical rules (Leiper, 2004). The first such definition was developed as early as 1942 by Hunziker and Kraph (as cited in Burkart and Medlick, 1974). They defined tourism as "*the sum of the phenomena and relationships arising from the travel and stay of non-residents in so far as they do not lead to permanent residence and are not connected with any earning activity.*"

Similarly, Jafari (1987) suggested that tourism was the study of man away from his usual habitat, of the industry which responds to his needs, and of the impacts that both he and the industry have on the host's socio-cultural, economic and physical environments. Pearce, recognized the link between tourism and recreation and suggested that "*tourism may be thought of as the relationships and phenomena arising out of the journeys and temporary stays of people travelling for leisure or recreational purposes*" (Pearce 1987:1).

These definitions appear to have greater legitimacy in the study of tourism as a social phenomenon for they attempt to recognize its component parts. But, as Leiper (1979) among others has argued, while being superficially attractive, they tend to be too broad, too vague and lack a clear spatial focus.

The need to consider both technical and heuristic definitions

In fact, both technical and heuristic definitions have their own relative strengths and weaknesses. As Table 2.2 indicates, the very strength of one represents the inherent weakness of the other. But you cannot consider defining tourism without addressing both its heuristic and technical dimensions.

Technical definitions tend to be simple, explicit and set clear boundaries to differentiate what is and is not a tourist or tourism. They tend to have set time limitations and a specific distance criterion. Many of these definitions also have fairly narrow prescriptions on who is a tourist. They are essential if one wishes to document the number of tourists, associated tourism businesses and the economic impacts of tourism. They are also falsifiable, in that it is easy to identify what is not tourism and who is not a tourist.

Their biggest weakness though is that the individual is reduced to nothing more than a number. Moreover, their arbitrary nature means some aspects of tourism and some tourists are either excluded or included and may not be counted. For example, someone who is away for only 23 hours may really be a tourist, but if the criterion is set at 24 hours, the person would not be counted. The same applies if an arbitrary distance threshold is not met, or if arbitrary time thresholds are exceeded. They do not consider the experiences people have, why they travel and what impacts their actions may have on host communities or the environment.

Heuristic definitions differ from technical definitions in a number of important ways. First, no physical distance needs to be specified. Second, there is no minimum or maximum time duration. Third, the boundaries of who are tourists can be much more flexible, while fourth, the boundary of what constitutes tourist attractions can also be flexible. These types of definitions are essential when considering the social and environmental impacts of tourism, motivations, and the experience generated. They are precise within their domain of trying to understand impacts, motives and the industry and also use clear terms, although they are often convoluted. But they are not falsifiable, for one cannot define who is not a tourist or what is not tourism unless some boundaries are embedded.

Table 2.2: Comparing technical and heuristic definitions of tourism

	Technical	**Heuristic**
Strength	Sets clear boundaries Can count tourists Can document tourist businesses Can calculate statistics and economic impacts Essential for census and statistical organisations Can calculate income, leakages, employment, etc. Allows business disciplines to study tourism	Inclusive Focuses on the tourist Considers motivation Considers experience Considers impact Allows non-business disciplines to study tourism
Weakness	Exclusive Does not consider motivation Does not consider the tourist Does not consider experience Does not consider impact	Fuzzy boundaries Cannot count tourists Cannot conduct any economic or statistical analysis

Fundamentals of tourism

The preceding review of some of the approaches used to define tourism reflects the challenges faced by people from multi-disciplinary backgrounds who share an interest in tourism. Perhaps, it can be argued that it is impossible to develop a precise definition of tourism. Perhaps as well, it is not necessary to do so and instead focus discussions of what tourism is on common themes rather than specific terminology. In fact, the title for this section is a misnomer, for we will not be defining tourism. To do so may be impossible.

Netto (2009) identifies fundamentals of tourism that summarise the key elements that encompass both technical and heuristic considerations (Table 2.3). Some of these elements are self-explanatory, while others require brief elaboration. The subject is the tourist, for without tourists there would be no tourism. This principle is fundamental in all of the concepts that will be discussed throughout the book. Moreover, the tourist has some clear characteristics that distinguish him or herself from non-tourists. The ideas of displacement and return relate to the core Latin concept where tourism involves a journey away and back. Motive is critical, for as Netto (2009:57) states *"there will always be a reason to travel be it explicit or not."* Hospitality represents the in-destination and in-transit zone sectors that serve the interests of the tourist. Tourism is comprised of a range of experiences that can be active or passive. Tourists will need to communicate with other tourists, workers in destination regions and others in order to have hospitality and experiences. Finally, any kind of tourism uses some kind of technology, with transportation and communications technology being the most important.

Table 2.3: Fundamentals of tourism. Source: Netto 2009

- Subject
- Displacement
- Principle of return
- Principle of motivation
- Principle of hospitality
- Principle of experience
- Principle of communication
- Principle of technology

Pleasure tourism is an extreme form of leisure

The pleasure aspect of tourism has attracted most of the research interest, although it must be appreciated that other forms of tourism exist, notably business tourism. Our thinking about pleasure tourism evolved from leisure and recreation and is intrinsically linked to the leisure paradigm, as outlined in Figure 2.2 (Murphy 1985). Tourism and recreation share many of the same facilities and compete for the same consumer dollar and level of government support (Jansen-Verbeke and Dietvorst 1987, Mathieson and Wall, 1982; Mill, 2008). Both also produce common social and psychological outcomes (Fedler, 1987), with similar motivations for participation (Iso-Ahola 1983). And, as Smith and Godbey (1991) suggest, both fields share the same heavy ideological baggage, be it a search for authenticity, a spiritual search for self-fulfilment, or an existential quest for meaning in an industrialized society.

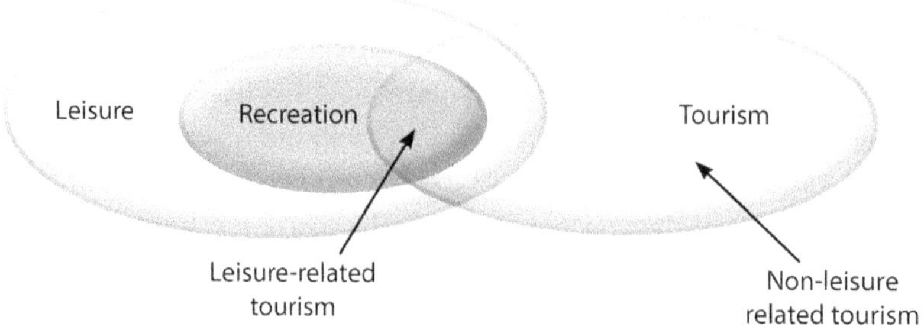

Figure 2.2: Tourism as an extreme form of leisure. Source: from Murphy, 1986

Pleasure tourism has been called an extreme form of leisure while business travel and some forms of unethical tourism, such as child sex tourism, fall outside what would normally be considered as part of the leisure paradigm. Leisure theory suggests that leisure is essentially free or unobligated time. Recreation is a voluntary activity that is done during leisure time, principally for enjoyment, intrinsic and extrinsic benefits. Because tourism generally occurs over large amounts of free time (days or longer) and usually involves some type of recreational activity, it therefore is seen to lie at one end of the leisure spectrum.

Tourism is an extreme form of leisure, and some aspects of tourism remain embedded in the leisure paradigm, for tourists and leisure participants share the same facilities and derive similar benefits from participation. Moreover, each activity vies for a share of the consumer dollar. Many so-called tourist attractions could not survive without a core local leisure market. Beechworth, a largely intact historic gold mining town in North-eastern Victoria, Australia, relies heavily on the local day trip market to support its many shops, restaurants and historic attractions. These visitors both supplement and complement the overnight tourism market.

Modern tourism could only develop in a society that had leisure time. Historically, leisure was an activity that was strictly the purview of the affluent. The Industrial Revolution changed all that, leading to the growth of leisure time and a recognition by 19th Century society that re-creation, or non-work time played a vital role in enhancing productivity of staff (Godbey, 1984; Kraus, 1984).

Jansen-Verbeke & Dietvorst (1987), state that leisure and tourism are strongly interwoven. Colton (1987), discusses the symbolic interactionist's view of the link between leisure recreation and tourism. Leiper (1979) suggests that tourism offers a special medium for recreation. Mathieson and Wall (1982) and Pearce (1987) discuss the linkages between leisure, recreation and tourism.

Therefore, the key difference between tourism and leisure is that tourism requires MORE than leisure: more time, money, effort, commitment, risk, etc. As a result, tourists' expectations are more and the product must also be more, as shown in Table 2.4. This issue is critical for many destination management organisations for they often think what they have to offer will satisfy tourists' needs because it will satisfy the needs of local residents. But, they forget the 'more' element. Few tourists will travel to see a movie, but they will travel to see a film festival, because it offers more, and can meet higher expectations created by greater involvement.

Tourism requires more!

What can satisfy the local leisure market may not have strong enough appeal to satisfy a tourist market, for tourism involves more time, effort, commitment, risk, possible cost and a range of other factors. As such, tourists expect something more: bigger, better, rarer, a better experience, etc. The product, therefore, must also offer more if it is to appeal to the tourist market. Many local destination areas make the mistake of assuming that something that satisfies local residents' needs will also appeal to tourists.

These two examples highlight the difference. Local residents may enjoy indoor skiing in a place like Hong Kong that lacks snow. But, this activity has little drawing power outside the immediate area. On the other hand, alpine regions in Germany offer an exceptional experience that act as a magnet to draw international tourists.

Table 2.4: Tourism requires MORE

Tourism requires MORE than leisure	Tourism expectations are MORE than leisure	Tourist experiences must be MORE than leisure
More time More effort More expense More commitment More pre-planning More skill More risk	More excitement More novelty More intense (either more active or more passive)	Bigger, better Different Unique Not available at home

Do we know the truth about tourism?

Do we really know the truth about tourism, or is much of what we believe a function of mythmaking and discipline influence. John Tribe asked this very question when he wrote (2006:360):

> *Is it possible to tell the truth about tourism? Surely an elaborate and expensive knowledge production system (universities, research departments, journals, conferences) exists just for that purpose. But could it be that the thousands of research articles that have been published have failed to uncover the truth?*

He feels the force field lens under which tourism operates distorts reality, caused by a gap between the fact of tourism and the phenomena of the collective 'truth' or knowledge held about the subject (Figure 2.3).

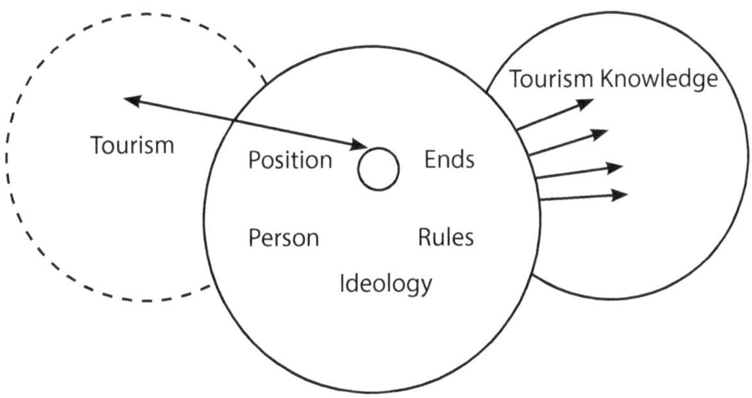

Figure 2.3: Force-field of tourism knowledge. Source: from Tribe, 2006

The following section summarises his work. The gap is caused by a five component 'force field' that knowingly or unknowingly distorts knowledge from fact:

- The person doing the research
- The rules guiding research
- The position of the individual

- The ends for which the research is conducted, and
- The ideology that drives research.

Person

All research is influenced directly or indirectly by the researcher. As much as we may try to be unbiased, it is impossible to separate the individual from the research process. At a minimum, any research must be filtered through the researcher's own cultural bias and influenced by personal tastes and interests. Geographers see the world one way, while psychologists see it another way. How we see it will influence how we, individually and collectively, investigate the phenomenon of tourism. The individual's own perception of the merits of tourism, and possibly his or her own positive or negative experiences, as either a tourist or a resident of a destination, will influence how they study it.

Sarniak (2015) identifies nine cases where the person, either the subject of the research or the researcher may bias the findings, as shown in Table 2.5. Respondents often want to satisfy researchers and give what they feel are the 'right' answers to questions, even if they do not necessarily feel that way. This issue is especially present in sensitive research topics where a high degree of social desirability bias will influence opinions. In addition, a degree of respondent fatigue may set in whereby the participant simply gets bored and starts to give a similar type of answer to subtley different questions.

Table 2.5: Potential for person bias

The subject	The researcher
Acquiescence bias – yes saying *Social desirability bias* – providing a socially 'correct' answer even if it does not reflect the respondent's views *Habituation* – providing the same answers to questions worded in the same ways, something that is especially risky with Likert scaled questions *Sponsor bias* where the sponsor is seeking a certain outcome	*Confirmation bias* or self-fulfilling prophecy research where you set out to 'prove' something and them proves it *Culture bias* where research is filtered through a cultural lens *Question order bias* where one question inadvertently directs the person to a certain response in following questions *Word bias and/*or use of leading questions *Halo effect* – seeing something positive based on a single attribute

Source: Sarniak 2015

Researchers are also not bias free. Most of the research published in English language media is produced by North Americans, Australians or English-speaking Europeans. This Eurocentrism has to influence perceptions of the world and how tourism functions. Self-knowledge and its counterpart global awareness also influence tourism. Surprisingly, some tourism scholars have travelled little.

Confirmation bias is certainly something that rears its head, especially when much special interest research is conducted by fans of the specialist interest, or when it seems much social impacts research is conducted by people who hold a dislike for tourism.

Rules

Research is driven by a series of the accepted rules, traditions, methods and protocols. Academic researchers stress that research must be grounded within some discipline based concept, which in turn drives how research is undertaken. A great deal of conservatism drives research, as few people are willing to challenge accepted methods or try alternative methods to examine the phenomenon. As a result, a level of inertia sets in where research is replicated using the same method which, not surprisingly, produces the similar results, which severely limits the ability to explore new ideas. The end result is a stagnation of ideas. Mazanec (2009: 313) is particularly scathing, noting how the rules used to conduct research are "*delaying or even hampering scientific progress and that is why I address them as myths - beliefs that lose substance when scrutinized thoroughly.*"

Indeed, the replication of the same method that produces similar results is an aspect of rule adherence that is problematic. That the same method produces the same results does not necessarily uncover the truth. Instead it may simply reinforce the fact that the one method will produce one set of results. Method bias may occur that obscures the quest for the truth. The veracity of claims can only be made when different methods are tested. If they unveil different results, then accepted dogma needs to be challenged. If they confirm previous findings, then perhaps the truth has been uncovered.

Ends

The purpose of the research is a third factor, for the pursuit of knowledge is never interest free. Questions relating to whom the research is being conducted for, whether it is commercial or conceptual, and whether it is deep or shallow, need to be considered. In addition, the ulterior motive behind the research needs to be considered, especially if it is done primarily to satisfy research assessment exercises or to reach arbitrary productivity thresholds to ensure tenure or promotion. It is especially challenging when research is driven by a small number of 'champions' who may control knowledge development and dissemination in a particular field of study. They may resist alternative explanations for topics they hold close to their hearts.

The issue of commodification of research to meet neo-liberal performance criteria is a critical matter these days. Fennell (2019:424) argues that this practice leads to gamesmanship whereby the pursuit of standards, audits, and administrative objectives becomes more important than the pursuit of knowledge. Indeed, various incentive structures around research impact and the pressure to produce research that will generate citations, regardless of the quality and innovation involved influence practice. As an example, the emergence of various research assessment exercises has had the unintended consequence of changing what is now considered to be good research (Brauer, 2023). As an example, early career academics are advised to work in highly populated, highly cited research areas, which hinders the development of research in niche and emerging areas (Pritchard, 2018). The result is the ends (productivity) count more than the means (good ideas and novel research).

Position

As Tribe (2006) mentions, position relates to both geographic and intellectual positions. He notes that the seemingly innocuous concept of geographical location influences the thinking of tourism in a number of subtle ways, for it impacts on ethnocentricity, departmentalism and academic tribalism. Departmentalism describes the organisational effects of universities on research, where tourism programs are embedded in different schools or departments. They, in turn, influence which type of staff are hired, the perspective of research undertaken and the interpretation of tourism. Tribe (2010) writes about tribalism and recognizes that researchers are located within academic communities, each with its own the behaviour patterns, rules, norms and hierarchies. Leaders within a community may exert undue influence on that community.

Ideology

Finally, ideology refers to the common sense set of beliefs (often implicit) permeating society which guides thought and action. Like it or not, most of the concepts in tourism have been developed either from an American and European ideological perspective, where the individual rules supreme, or from a European social responsibility perspective. Little research has been developed from a collectivist Asian perspective. Likewise, Western tourism theories and concepts are grounded strongly in the Judeo-Christian philosophy and not in other philosophies.

Myth creation

The difficulty in identifying the truth about tourism has led to the creation of many widely held myths. Myths are demonstrably false beliefs that are long-standing and never subjected to deep inspection (McGee, 1985). Frantz (2006) identifies them as wrong ideas widely held by authority figures. Kompier (2006) is much more succinct, classifying them as believed truths. Myths are created when simple truths are taken out of context, repeated over time and narrowed to such a simplistic extent as to become nothing more than jargon. As Heydenryck (1993) states the original truth becomes so diluted that only a kernel of truth remains. Barthes (1984) adds myths distort facts, so much so that the symbolic or metaphorical value of the myth often exceeds its absolute truth (Alexander & Archer, 2000). They are common in all areas of academic study, for more than any other reason they serve to legitimize a field of study (Hansen & Williams 1999).

A number of causes of myths have been identified (Kompier, 2006; Lance, 2011; Mazanec, 2009; Mercer, 2006; Vandenberg, 2006), including:

- The origins of a discipline, where the work by many early scholars remains dominant, even though their ideas have been confronted by contradictory evidence
- Inertia
- Entrenchment by the peer review process and doctoral student training
- Results that are too good not to be true
- Poor research
- Failure to read and place research in the proper context
- Self-fulfilling prophecy research, where research sets out to 'prove' something and does
- Fallacy of misplaced rationalism (Schaeffer, 2008), where a great deal of time and effort is devoted to pursuing elaborate, rational explanations for phenomena, without first demonstrating their existence
- False attribution of cause and effect without considering other causal agents
- Stakeholders.

A study of senior academics identified over 40 myths that fall into six broad categories, as shown in Table 2.6 (McKercher & Prideaux, 2014). The self-interest myths about the size of tourism, its existence as an industry and arguments it is a discipline serve the valuable role of legitimizing it as a field of study, especially in its early days when tourism research was seen as a frivolous or inconsequential issue (Nash, 1979). Promulgation of these myths helped create a shared sense of belonging among academics (Dann, Nash & Pearce, 1988; Tribe, 1997). Negative foundation myths formed the basis of the social science inquiry into tourism as a response to pro-development oriented attitudes in the 1970s that had little regard for the adverse impacts tourism could have on the environment and host communities. Doxey's (1975) Irridex is probably the best known, though least read of these types of myths. The positive reactive stakeholder myths, especially about the new type of tourist engaging in a new type of ethical behaviour, was a response to the negative origin myths, while convergent myths sought balance between the two ideologies. The too good not to be true myths emerged at a time when tourism studies exploded globally in the 1990s and a whole series of new academics discovered sustainable development and special interest tourism. Lastly inherited myths are those that were imported from other disciplines and applied to tourism, often poorly.

Table 2.6: Myths about tourism

Type of Myth	Examples
Self Interest	Tourism is the world's largest industry Tourism is an industry Tourism is a discipline
Negative Foundation	Tourism destroys tourism Tourism perpetuates colonialism / tourism is neo-colonialism Tourism generates some positive economic benefits, but they are offset by massive social, cultural and environmental impacts Tourism leads to conflict Tourism commodifies culture
Reactive Positive Stakeholder	Tourism is a path to economic independence Tourism revitalises culture Tourism protects the environment Tourism provides a range of positive social benefits People damage the environment, not tourism
Convergent	Size, importance and differences of ecotourism/ecotourists and other forms of alternative tourism/tourists Employment Small is preferable A traveller is better than a tourist
Too Good Not To Be True	Special interest tourism as the new mainstream Tourism and peace Economic and employment benefits of tourism Tourists seek authentic experiences Ecotourism and ecotourism operators are more ethical than others Tourism is a clean sector Sustainability and the triple bottom line can be operationalized Quality of tourism jobs Multiplier effect of tourism
Inherited	Relationship between satisfaction, loyalty, intent to return and actual repeat visitation rates Carrying capacity works in tourism Willingness to pay Tourism lacks theory

Source: McKercher and Prideaux 2014

Conclusion

What is tourism and why do we think about it the way we do? This chapter explored a range of issues but explicitly did not try to define the term other than to identify broad parameters that involve a round trip taken for some reason that creates some type of impact. Who is a tourist? Again, we have taken the decision not to try to define the term other than to appreciate that a tourist is the person who takes the trip.

Is this a satisfying outcome? No. As stated at the outset of this chapter, it would be nice to come up with absolute terms. But that is impossible given the number of disciplines that explore tourism, the force field lens under which tourism is investigated and the impossibility of rectifying the inherent dissonance between heuristic and technical definitions.

None the less, this chapter does set the stage for examining a range of issues explored throughout the rest of the book, for each of the different theories, concepts and models developed adopt unique disciplinary foci. Flexibility in defining the term is thus required to make sense of why we think about tourism the way we think about it.

■ Discussion points / Exercises

1. Tribe (2006) discusses the force field of tourism and how it influences research. This force field is evident even in the journals that publish tourism research. Identify any five journals and go online to review their mission statements, aims, scope, and objectives. Develop a chart showing how each uses the force field of tourism to create its own version of tourism knowledge.
2. Many myths have been created about tourism. Identify as many as you can and explain the type of myth from Table 2.6 and why the myth has emerged.
3. Develop your own definition of tourism. Analyse whether it reflects a technical or heuristic viewpoint and try to identify the limitations of the definition from the perspective of the viewpoint not considered.
4. Tourism is MORE. As a result, tourists' expectations are GREATER, and the type of service or facilities they demand must be BETTER than they can find at home. Analyse your own destination's offerings by identifying attractions and then analysing them by identifying how tourists' expectations are greater than local residents, and how they must demand better offerings than found at home.

Points to consider for revision

- Tourism is a multi-discipline informed by 50 or more disciplines or fields of study. They have each made their own contributions, but also may have defined tourism in different ways to suit their needs.
- Tourism is extremely difficult to define for it has to consider both technical and heuristic elements, which in many ways are mutually exclusive.
- Pleasure tourism traces its roots to leisure, but it requires more, meaning the product on offer must provide a better, bigger or different experience than a local leisure or recreational product in order to appeal to tourists.
- How the knowledge force field gap influences what we think about tourism and why we think about it the way we do.
- As a result, many myths of tourism have emerged that do not stand up to rigorous scrutiny.

Suggested Additional Readings

Tribe, J. (2006) The truth about tourism. *Annals of Tourism Research,* 33(2): 360-381

McKercher, B. and Prideaux, B. (2014) Academic myths of tourism. *Annals of Tourism Research.* 46: 16-28.

References

Alexander, D. & Archer, S. (2000) On the myth of 'Anglo-Saxon' financial accounting. *The International Journal of Accounting* 35(4): 539-557

Andrees, B. (2014) Why definitions matter. *International Labour Organisation.* www.ilo.org/global/about-the-ilo/newsroom/news/WCMS_234854/lang--en.

Ateljević, I (2014) Mapping a history and development of Tourism Studies field. *Turizam* 62(1), 75-101

Barthes, R. (1984) *Mythologies,* translated by Annette Lavers, Hill and Wang, New York Excerpt from http://xroads.virginia.edu/~DRBR/myth.html.

Benckendorff, P. & Zehrer, A. (2013) A network analysis of tourism research. *Annals of Tourism Research.* 43,121-143.

Brauer, R. (2023) Tourism studies and the changing research ecosystem. *Tourism Recreation Research,* DOI: 10.1080/02508281.2023.2188707

Burkart A. & Medlick, S. (1974) *Tourism: Past Present and Future.* Heinemann

Butowski, L., (2016). The issue of disciplinarity and non-disciplinarity of tourism studies. *Téoros. Revue de Recherche en Tourisme,* 35(1). journals.openedition.org/teoros/2899

Cheng, C., Z. Li, J. Petrick, & O'Leary, J. (2011) An examination of tourism journal development. *Tourism Management* 32, 53-61.

Coles, T. Hall, C. M. & Duval, D. (2006): Tourism and POST-DISCIPLINARY ENQuiry, *Current Issues in Tourism*, 9(4-5), 293-319

Colton, C. (1987) Leisure, recreation and tourism: a symbolic interactionism view, *Annals of Tourism Research*, 14, 345-360.

Crouch, G. & Perdue, R. (2015), The disciplinary foundations of tourism research: 1980 – 2010. *Journal of Travel Research*, 54(5), 563-577.

Dann, G. (1997) Paradigms in tourism research. *Annals of Tourism Research*, 24: 472–474

Dann, G., Nash, D., & Pearce, P. (1988) Methodology in tourism research, *Annals of Tourism Research*, 15, 1-28.

Doxey, G. V. (1975) A causation theory of visitor-resident irritants: methodology and research inferences. In *Travel and Tourism Research Associations Sixth Annual Conference Proceedings* (195-198), San Diego, September.

Fedler, A. J. (1987) Introduction: Are leisure, recreation and tourism interrelated, *Annals of Tourism Research*, 14(3), 311-314

Fennell, D. (2019) The ethics of excellence in tourism research. *Journal of Travel Research* 52(4), 417-425.

Franklin, A., & Crang, M. (2001). The trouble with tourism and travel theory? *Tourist Studies*, 1(1), 5–22.

Frantz, J (2006) *Academic Myths*. http://frantzmd.info/Social%20Commentary/Academic_Myths.htm.

Gibson, H. (1998) Sport Tourism: A critical analysis of research. *Sport Management Review* 1, 45-76.

Goeldner, C. & Ritchie, B (2012) *Tourism Principles, Practices and Philosophies*. Hoboken: John Wiley and Sons.

Gunn, C. (1991) The need for a multidisciplinary tourism education. In R.D. Bratton, F. Go, & J. R. Brent Ritchie (eds) *New Horizons in Tourism and Hospitality Education, Training and Research*, Calgary: University of Calgary, pp. 27-33.

Gupta, A. (2015) Definitions. *Stanford Encyclopedia of Philosophy*. https://plato.stanford.edu/entries/definitions.

Hall, C. M., Williams, A. M. & Lew, A. A. (2014) Tourism: Conceptualisations, Disciplinarity, Institutions and Issues. In Lew, A., Hall, C. M., & Williams, A. M. (eds) *The Wiley Blackwell Companion to Tourism*, Chichester: John Wiley.

Hansen, L. & M. Williams (1999) The myths of europe: legitimacy, community and the 'crisis' of the EU. *Journal of Common Market Studies* 37(2), 233-249

Heydenryck, K. (1993). A systems model for political communication: A case study of academic mythmaking. *Critical Arts: South-North Cultural and Media Studies* 6(2), 27-48.

Iso-Ahola, S. (1983) Towards a social psychology of recreational travel, *Leisure Studies*, 2(1), 45-56, DOI: 10.1080/02614368300390041

Jafari, J. (1987) Tourism models: the sociocultural aspects. *Tourism Management* 8(2), 151-159

Jafari, J. & Ritchie, B. (1981) Toward a framework for tourism education: Problems and prospects, *Annals of Tourism Research* 8(1), 13-34

Jansen-Verbeke, M. & Dietvorst, A. (1987), Leisure, recreation, tourism: a geographic view on integration, *Annals of Tourism Research,* 14(3), 361-37.

Kompier, M. (2006) The "Hawthorne effect" is a myth, but what keeps the story going? *Scandinavian Journal of Work, Environment and Health.* 32(5), 402-412.

Koseoglu, M., Mehraliyev, F. Xiao, H .& King, B. (2023) Tourism knowledge: its creation and dissemination by region, *Anatolia,* 34(2), 190-209, DOI: 10.1080/13032917.2021.2005647

Kouropalatis, Y., Zenker, S., Morgan, R. E., & Lindgreen, A. (2023). Diffusion and future influence of tourism research: addressing the how and when questions of tourism knowledge structures. *Journal of Travel Research,*. https://doi.org/10.1177/00472875231164986

Krause, R. (1984) *Leisure and Recreation in Modern Society, 3rd Ed.* London: Scott Forseman and Co.

Kunwar, R. (2018) Tourism education, curriculum spaces, knowledge production, and disciplinary pluralism. *The Gaze Journal of Tourism and Hospitality.* 9, 83-155

Lance, C. (2011) More statistical and methodological myths and urban legends. *Organizational Research Methods* 14(2), 279-286

Laws, E, & Scott, N. (2015) Tourism research: Building from other disciplines. *Tourism Recreation Research.* 49(1), 48 – 58.

Leiper, N. (1979) The framework of tourism: towards a definition of tourism, tourist, and the tourist industry. *Annals of Tourism Research* 6(4), 390-407.

Leiper, N. (1999). A conceptual analysis of tourism-supported employment which reduces the incidence of exaggerated, misleading statistics about jobs. *Tourism Management,* 20(5), 605-613

Leiper, N. (2004). *Tourism Management (3rd ed).* French's Forest: Pearson Education Australia.

Lew, A. (2012) Tourism is NOT the world's largest industry – so stop saying it is. *TGJournal'sTourism Place.* http://tourismplace.blogspot.hk/2008/04/tourism-is-not-worlds-largest-industry.html.

Lisle, D. (2017) *Logic: The Importance of Definitions.* https://biblicalscienceinstitute.com/logic/logic-the-importance-of-definitions.

Liszewski, S. (2010) Tourism Studies: Situated within multiple disciplines or a single independent discipline? Tourism 20(2), 37-44.

Mathieson A., & Wall, G (1982) *Tourism: Economic, Physical and Social Impacts,* Longman, London.

Mazanec, J. A. (2009) Unravelling myths in tourism research, *Tourism Recreation Research* 34(3), 319-323.

McGee, R (1985) Lies we live by: Some academic myths and their functions. *Teaching Sociology* 12(4), 477-490.

McKercher, B. & Prideaux, B. (2014) Academic myths of tourism. *Annals of Tourism Research.* 46, 16-28.

McKercher, B. & Tung, V. (2015) Publishing in tourism and hospitality journals: Is the past a prelude to the future? *Tourism Management.* 50, 306-315.

McKercher, B. & Dolnicar, S. (2022) Are 10,752 journal articles per year too many? *Annals of Tourism Research.* https://doi.org/10.31235/osf.io/afp28.

Mercer, D. (2006) The trouble with paradigms. A historiographical study on the development of ideas in the discipline of castle studies. *Archaeological Dialogues* 13(1), 93-109

Mill, R. (2008) The inter-relationships between leisure, recreation, tourism, and hospitality in Brotherton, B. & Wood, R. (eds) *The SAGE Handbook of Hospitality Management* London: Sage.

Morley C (1990) What is Tourism?. *The Journal of Tourism Studies* 1(1), 3-8

Murphy P (1985) *Tourism: A Community Approach*, Routledge, London.

Nash, D (1979) Tourism as an Anthropological Subject. *Current Anthropology* 22(5), 461-481.

Netto, A. (2009) What is tourism? Definitions, theoretical phases and principles. In Tribe, J (Ed) *Philosophical Issues in Tourism.* Bristol: Channel View Publications

Owens, J. (2015) Why definitions matter: Stalking victimization in the United States. *Journal of Interpersonal Violence.* 31(12), 2196-2226.

Pearce, D. (1987) *Tourism Today: A Geographic Analysis*, Harlow: Longman Scientific and Technical.

Pritchard, A. (2012) Critical tourism research, *In (Eds)* Ateljevic, I., Morgan, N., & Pritchard, A. *The Critical Turn in Tourism Studies; Creating an Academy of Hope.* Oxford: Routledge, pp 11-13.

Pritchard, A. (2018). Predicting the next decade of tourism gender research. *Tourism Management Perspectives, 25,* 144-146.

Şalvarcı, S. & Aylan, F. K. (2019). Analysis of PhD dissertations in tourism (2014-2018) . *Journal of Gastronomy Hospitality and Travel,* 2(1), 29-37. https://dergipark.org.tr/en/pub/joghat/issue/47297/596275

Sarniak, R. (2015) 9 types of research bias and how to avoid them. www.quirks.com/articles/9-types-of-research-bias-and-how-to-avoid-them.

Smith, S., & Godbey, G. (1991) Leisure, recreation and tourism. *Annals of Tourism Research.* 18(1), 85-100.

Taillon, J. (2014) Understanding tourism as an academic community, study or discipline. *Journal of Tourism and Hospitality.* 3(3), DOI: 10.4172/2167-0269.1000131

Tribe, J. (1997) The indiscipline of tourism. *Annals of Tourism Research* 24(3),638-657.

Tribe, J. (2006) The truth about tourism. *Annals of Tourism Research,* 33(2), 360-381

Tribe, J. (2010) Tribes, territories and networks in the tourism academy. *Annals of Tourism Research* 37,7-33.

Vandenberg R. (2006) Introduction: Statistical and methodological myths and urban legends: where, pray tell, did they get this idea. *Organizational Research Methods* 9(2), 194-201

Veerasamy, V. (2013) *The Importance of Good Definitions (Or: How To Think Clearly).* www.referralcandy.com/blog/importance-of-good-definitions.

Weaver, D., (2023). Tourisation theory and the pandiscipline of tourism. *Journal of Travel Research, 62*(1), 259-265.

Weiler, B., Moyle, B., & McLennan, C. (2012) Disciplines that influence tourism doctoral research: The United States, Canada, Australia and New Zealand. *Annals of Tourism Research* 39(3), 1425-1445.

Whitfield, G. (2012) *The Importance of Proper Definition* https://piadvice.wordpress.com/2012/06/13/the-importance-of-proper-definition.

Xiao, H. & Smith, S. (2006). The making of tourism research. *Annals of Tourism Research,* 33, 490–507.

Xiao, H., Li, M., & Lin, E. (2011) Diffusion patterns and knowledge networks: an inductive analysis of intellectual connections in multidisciplinary Tourism Studies, *Journal of Travel & Tourism Marketing*, 28(4), 405-422, DOI: 10.1080/10548408.2011.571574

Xu, H. & Zhang, Q., 2021. Tourism PhD programs in China. *Hospitality and Tourism Education in China,* pp.129-153.

Part 2 - How Tourism Works

The first part of the book set the stage, by describing what tourism is and the issues related to theory. This next section of the book focusses on understanding how tourism works as both a social and commercial activity. It is important to appreciate the key factors that drive tourism and influence the intersection between supply and demand.

Chapter 3 discusses the role of attractions or demand generators, without which the pull of the destination may not exist. Traditionally, attractions have been thought of in fairly narrow supply side terms. Recently, though, the UNWTO and others have started to think of them from a demand perspective, meaning that almost anything a tourist does in a destination could be considered as an attraction. Yet, the epistemological discussion of attractions, activities and products has been narrow traditionally. This chapter introduces the idea that attractions can be considered as existing on one of a number of tiers in a taxonomic structure, from the very broad to the very specific.

Chapter 4 considers the other key element of access, for without good access, the ability of a destination to compete in the global marketplace is limited. The chapter begins with a discussion of time and how varying perspectives of the use of time influence tourist behaviour. It then proceeds to examine distance decay, the first law of geography, and its related concept of market access. The chapter concludes with a review of various itinerary models.

Chapter 5 examines some of the political economy issues that affect international tourism. It must be appreciated that nowhere in international law does anyone have the right to enter a country. Instead, it is a privilege granted by both the departing and receiving nation or territory. The chapter explores how political economy and economic organisation affect both people's desire to travel and the ability of the economy to supply suitable experiences. It concludes by recognising that access to international travel is a fundamentally political act that grants unequal access to different people.

3 Attractions and demand generators

By the end of this chapter the reader will be able to:

- Conceptualise an attraction
- Critique the tourism attraction system and demonstrate how to apply markers
- Explain how the same attraction can play different roles for different market segments
- Analyse a taxonomy of attractions
- Apply the tourism attractions taxonomy in real world situations

Introduction

Tourism is driven by attractions and access. Attractions explain why people come to a destination, while access influences the volume of visitors. The two are interrelated for strong attractions create demand to enhance access, while weak access effectively limits demand and thus the size and scale of attractions. The opposite is also true. Weak attractions create little necessity to improve access, while strong access creates the opportunity to enhance the breadth of attraction offerings.

Tourism attractions play a fundamental role in the successful development of any destination. They serve as demand generators that give the customer a reason to visit (Cooper, 2021). Moreover, consumption of attractions is often considered to be the central theme of the tourism experience. Goeldner and Ritchie (2012) argue that attractions drive people to travel and represent the main motivations for travel. More importantly, they note that without attractions there would be little need for all other tourism goods and services. Mill and Morrison (2012) illustrate that attractions at a primary destination have to have sufficient breadth and depth of appeal to encourage visitors to stay for many days. Holloway and Humphries (2016: 16) also note destinations must have adequate attractions to appeal to a large number of tourists and add further, the more attractions a destination has, the easier it becomes to market.

However, like most things in tourism, attractions are quite misunderstood, with Leask (2010) noting that even arriving at a universally accepted definition is a challenge. Trying to define demand generators is a reflection of the complexity of tourism systems. Issues such as classification, inclusive or exclusive definitions, size, aesthetic appeal, management regime and the target market all come into play. It is for this reason that some prefer the use of the term 'visitor attraction' instead.

A bigger issue, though, is that the term has been misused, with the concept of what constitutes varying levels of attractions and activities conflated and therefore confused. Holloway and Humphries (2016) observe even the term 'destination' is used synonymously with attraction. This chapter sifts through many of the issues relating to the conceptualisation of attractions and how they work. It begins by trying to define what an attraction is and how the tourism attractions' system works. Then it examines how the term has been abused, so that its use encompasses far more than the original meaning. This discussion leads to the need for some type of structured, taxonomic system to classify the various levels of an attraction. Finally, we look at the role attractions play in drawing visitors to a destination.

Conceptualising attractions

The idea of what constitutes an attraction was discussed in our literature some 40 years ago but has received relatively little attention since. Scholars agree in general that an attraction consists of a single entity bounded by time or space that is visited by tourists. There have been many attempts to categorise attractions over the years. Some have focussed on defining groups and sub groups according to the traditional triad of natural, cultural and built, with built attractions being divided into purpose-built and non-purpose-built (Holloway, 2002). Goeldner and Ritchie (2012) identified five broad classes of attractions, including cultural, natural, events, recreation and entertainment, while Walker and Walker (2011) note a continuum of attractions from natural to built, entertainment oriented. Douglas Pearce (1989) identified the source of attractions as including purpose-built facilities, features that are part of the regional fabric, places that have been transformed from their original function to serve tourism use or features of existing facilities that supplement or complement their original function. However, as Cooper (2021) notes, defining attractions is challenging because they come in a variety of forms, sizes, ownership structures and locations.

Three features are worth considering. First, if no tourists visit, then a place cannot be called a tourist attraction (Leiper, 1990). This point seems moot, but lies at the heart of the issue of phantom demand affecting many so-called attractions, where they were built for tourists but tourists show a disdain for them. Second, the narrow geographic or temporal focus means that a region or an extended environment is not considered as a single, holistic attraction (Pearce, 1991), even though attractions may have a linear component (Denstadli & Steen Jacobsen, 2011; Hardy, 2003). Instead, they have been identified as attraction nodes or spatial agglomerations of attractions. By extension a shopping district would not be considered an attraction per se, but instead, each shop might be considered as an attraction in its own right. Third, there is the issue of how the activities pursued fit into the equation. Technically, activities represent the consumption of a set of like attractions, such as shopping and sightseeing and, therefore, cannot be considered an 'attraction' in themselves. But Destination Management Organizations promote activities as attractions, suggesting that attractions may be conceptualised as existing at different levels. More on this later.

If no tourists visit, then the place cannot be called a tourist attraction. Phantom demand affects many places, such as the case of the Paddle Steamer Cumberoona in Albury, Australia. It was built as an Australian bicentennial project, with high demand expected because other communities along the Murray River had a thriving paddle steamer trade. However, high operating costs, fluctuating water levels and the closure of other local attractions led to its demise. It has since been sold and moved downstream to a more favourable location.

The conceptualisation of attractions began with a supply side perspective, but has only recently shifted to considering them from the demand side. Phillip Pearce (1991: 46), for example, offered a supply based operational definition of "*a named site with a specific human or natural feature which is the focus of visitor and management attention.*" By his own definition, good scenery is not a tourist attraction, but a scenic lookout which is named, managed and well used is.

Leask (2008) more or less adopts this approach in her substantial volume of work on visitor attractions. Her model, shown in Figure 3.1 identifies the main attractor of visits, such as the resource in question, interpretation, events and augmented services provided as the core of the model. From here, she looks at some of the structural issues, whether it is natural or built and whether admission is free or has a charge attached to it. The ownership structure follows, with two broad types of public or private ownership. Page (2019) notes four different ownership styles, including private, public, charity and voluntary ownership. Lastly, she considers the primary market or market mix. Some visitor attractions cater mostly to the local market, while others cater to a regional, national or international market.

Within the cultural tourism space, for example, Timothy (1997) identifies four different communities of cultural tourist based on the level of connectivity to the place in question. World heritage attractions may draw large numbers of tourists that invoke feelings of awe, but they probably do not invoke feelings of deep personal attachment. National cultural assets symbolize a society's shared recollections that represent much about how it expresses its identity through valorization of durable national icons that reflect national pride. Domestic tourists may feel a sense of pilgrimage by visiting these places and paying homage to national symbols that represent their shared identity. Local and personal sites engender progressively stronger feelings of connectivity and likely facilitate different depths of experience by the visitor.

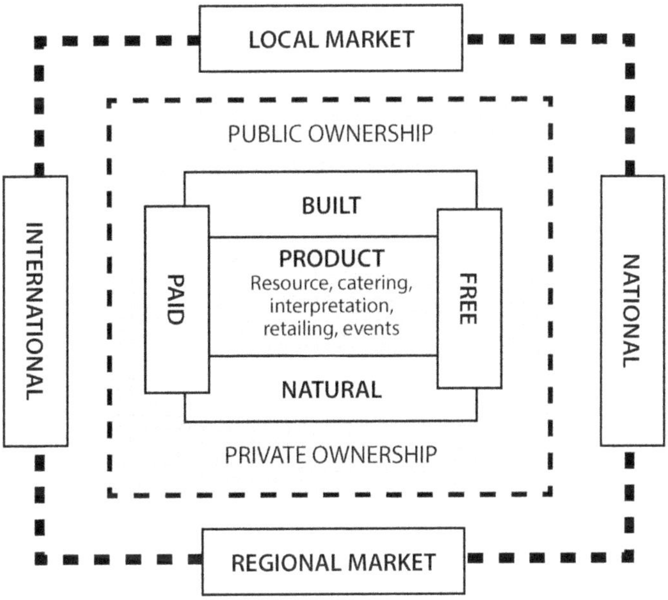

Figure 3.1: Model of visitor attractions. Source: from Leask (2008)

Lew (1987), also supported the supply side argument, but takes a more liberal view when he suggests attractions consist of all those elements of a 'non-home' place that draw discretionary travellers away from their homes. Importantly, he disagreed with Pearce that they needed to be signed, signalled and managed for tourists and instead suggests that the suite of potential attractions is much broader.

Recently, though, the focus has shifted to considering attractions from the perspective of the consumer. Koutoulas (2015) reminds us tourism products exist to satisfy the needs of tourists and for no other reason. He, therefore, suggests we need to look at them from the demand side. The UNWTO (2008a, b) also adopts a demand side approach when it states attractions are defined as much by the activities pursued by visitors as by the supplier, meaning that almost anything could be considered as an 'attraction' if it is used by tourists.

Traditionally, tourist attractions have been defined using a supply side perspective. They are either named or unnamed places, events or activities that draw tourists to a destination. Recently, the UNWTO has adopted more of a demand-side perspective, noting attractions are defined as much by the activities of tourists as by suppliers. This reconceptualization has fundamentally broadened the scope of what can be considered as an attraction to include, for example, people for the visiting friends and relatives' market, street performers (as here), places of business for the business tourism market, schools and universities for the educational tourist and many more.

This change in perspective is important for it clarifies a number of ontological issues. First, it expands the idea of what constitutes an attraction beyond the traditional triad of location-specific built, natural and cultural features. Instead, it recognises that any feature that attracts tourists may be considered as an attraction. Thus friends could be an attraction for the visiting friends and relatives market or a business could be an attraction for a business tourist. Second, at a more generic level, it expands the idea of an attraction to include non-location-specific experiences or activities pursued by tourists. In this manner, sightseeing, exploring, shopping, dining, relaxing on a beach and many other activities are identified as attractions at some level. Even though consumption still occurs at specific site (a scenic spot, a specific shop, beach or restaurant, etc.), the collective set of activities that constitute the experience suggests that attractions may be categorised in a hierarchical or taxonomic manner from the specific (a shop) to the more general (shopping). Third, the definition acknowledges that attractions do not have to be commercial or managed, in recognition that much of what tourists consume is part of the public domain, available free of charge (McKercher, 1993). Finally, the definition eliminates scale and spatial considerations as defining variables. Attractions can be as small as an historic artefact or as large as a ski resort.

A hierarchy of attractions exists at most destinations. The most important attractions play a critical role in shaping the image of a destination and in influencing visitation (Mill & Morrison, 2012). An obligation to visit icon attractions may define the entire trip. But, not all attractions drive the visit decision. In fact, most play little or no role in destination choice and instead act as complementary activities. The purchase decision becomes increasingly discretionary for lower order attractions, until visits to the lowest order ones are typified by low involvement purchases or by little effort required by the visitor. In some instances, they simply represent a productive use of free time and nothing more (Kantanen & Tikkanen 2006).

Traditionally, they have been placed in a simple hierarchy of first or primary order, second order and third order attractions (Cooper, 2021; Leiper, 1990). Primary attractions are what draw people to the region. One of the goals of tourism planning, therefore, is to try to create primary attractions that will provide the destination with that unique draw card. Secondary attractions generally have regional appeal. People will travel shorter distances to see them, or if they are already in the region will make a specific trip to visit the attraction. Tertiary attractions, on the other hand, have limited appeal. While primary attractions may drive development of a region, secondary and tertiary attractions are important players in completing the attraction mix and in encouraging people to stay in the region longer.

Of course, like most things in tourism, this type of classification is overly simplistic, for it assumes the market is homogeneous and will therefore rate all attractions the same way. It ignores the fact that tourists consume attractions to satisfy their needs and that needs vary depending on the segment or even individual involved. Mill and Morrison (2012) and Cooper (2021) further recognize that attractions appeal to different groups of tourists in a different manner. Certain attractions can be regarded as primary attractions for one segment of the travel market, but as lower order attractions for another group of travellers.

Moreover, needs vary. In some cases, needs are singular and specific, meaning that the likely range of attractions that can satisfy them is small. On the other hand, people often travel to have many broad and unspecified needs met, such as sightseeing, spending time with family or escape. In these situations, it may be difficult to ascribe visitations to any specific attraction. Instead, the perceived existence of more generic attraction classes or even attraction families that can satisfy needs may be the level of specificity that people want.

Tourism attractions systems

The preceding discussion suggests that attractions function as stand-alone entities. However, renowned tourism scholar Neil Leiper (1990) challenged this idea and instead argued that a tourist attraction does not exist in isolation, and instead functions as a component of a tourist attractions' system. The system includes three elements: the tourist; the attraction or nucleus, and; some type of marker or catalytic element that connects the tourist to the attraction or the attraction to the tourist. His model is shown in Figure 3.2. This contribution has not garnered the attention it deserves, for it is an important conceptual contribution to the field. Each of the components is discussed briefly.

Figure 3.2: The tourist attraction system. Source: From Leiper 1990

■ The tourist

Like everything else in tourism, the tourist lies at the heart of the tourism attraction system, for without tourists, places would not be thought of as tourist attractions. Leiper suggests the essence of tourist behavior involves a search for satisfying leisure experiences that can only be met by searching for suitable attractions or experiences. This quest involves looking for various signs, symbols and other markers that signify the place as something worth visiting (Leiper 1990:371).

■ The nucleus

The nucleus or specific attraction is *"any feature or characteristic of a place that a traveler contemplates visiting or actually visits"* (Leiper 1990: 372). Because tourists' needs and wants are individualistic, what constitutes a nucleus is also equally individualistic. Nuclei can have varying geographic scales, consist of a temporal dimension and exist along a hierarchy. The geographic scale can range from a single entity to touristic precincts where a range of attractions and activities are bundled in close proximity. The temporal dimension relates to short duration events such as festivals, trade shows and the like. The hierarchical consideration relates to the importance tourists place on any given nucleus. A primary nucleus is the key feature that draws tourists to an area, while a secondary nucleus is a known pre-visit but does not influence the visit decision, and a tertiary nucleus is an unknown pre-visit but is visited none the less.

Markers

It is the marker, though, that is the most important feature of the tourist attractions' system for it represents the catalytic link between the tourist and the attraction that triggers motivation (Clark, 2009). Markers are defined as *"items of information about a phenomenon that is a potential nuclear element in a tourist attraction"* (Leiper, 1990: 377). The contribution of earlier work by MacCannell (1976) is acknowledged. Culler (1988: 5) defines markers as *"any kind of information or representation that constitutes a sight as a sight: by giving information about it, representing it, making it recognizable."* MacCannell (2001) adds that markers signal the sight as something worth seeing.

What constitutes a marker is quite broad as Wong and McKercher (2013) summarise. Most of us can identify commercial markers as those developed by the attraction itself, the travel trade and travel media. These include advertising, promotional materials, press releases, web and social media presence, and others. But, markers can also include other forms of information, including common knowledge that is not attributable to any single source, movies, novels and magazines (Mercille, 2005), intangible heritage associated with destinations (Bessiere, 1998), information provided by key informants, including staff at tourist information centers (Wong & McKercher, 2013), souvenirs that remind the individual and others of the site (Culler, 1988), word-of-mouth from other tourists, lifestyle television programs and increasingly comments on social media sites (Tussyadiah & Fesenmaier, 2009). Interestingly, more seems to have been written about non-commercial markers than about commercial ones.

Markers are arguably the most important component of the tourist attraction system, for they signal a place as being worthy of visiting. Signage signals something is worthy of note can can act as a catalyst to visit. Crowds, though, are often the most powerful on-site marker, as shown here at the Hagia Sophia in Istanbul, for they signal a place as something worth seeing.

Markers may be more important than the attraction itself (MacCannell 2001) for without them, the tourist would not become aware of the attraction. Markers, therefore, stimulate desire to look for experiences that match their needs and wants. Clark (2009) reminds us that markers often contain information or present an image about what might be experienced at the sites targeted. As a result, a key goal of markers is to act as image enablers (Jenkins, 1999). Effective attractions' systems work when markers stimulate visitation, while the system may become defective when no markers are present, when they send conflicting messages or do not encourage visitation.

Leiper suggests markers can be found in three locations, in generating markets, in transit and in the destination (or as he labels them, *contiguous* markers). It is simpler, though, to consider them as existing at two levels, those found prior to arrival and those available once in the destination. Generating markers seem to be most important for large scale attractions that can afford to have a presence in source markets or reflect core destination images (Litvin & Mouri, 2009). They also play a key role in behaviour for tourists decide what they want to do prior to departure, and if their time is limited are unlikely to modify their behaviour much (Shoval & Raveh, 2004). Richards (2002), for example, found almost half of all cultural tourists surveyed indicated their decision to visit cultural sites was made prior to departure.

Lower order attractions have to rely on in-destination markers for their owners/managers do not possess the resources to engage in awareness building through generating markers (McKercher & Lau, 2007). Their inability to reach source markets prior to departure means that they are disadvantaged, as people who are unaware of the activity are unlikely to participate unless they happen to stumble across it on their way to other attractions.

The role of markers is variable. Tourists visit a destination's most popular attractions, irrespective of whether they are receptive to generating or in-destination markers. Those who are unaware of major attractions prior to arrival, become aware of them and, more importantly, visit. By contrast, the conversion rate for planned visits to middle or lower order attractions declines, irrespective of pre-departure awareness levels. Inertia appears to play a critical role in determining the behaviour of tourists. The most popular attractions will be visited because of their popularity, while lesser attractions will generate fewer visits by the simple virtue of their lower popularity. Thus, at a crude level, pre-departure awareness levels may provide the best indicator of likely visitation rates among tourists.

A taxonomy of attractions

The concept of an attraction has moved a long way from its humble origins as a specific entity bounded by space or time into a brand category that includes a vast array of like entities categorised under broad umbrella concepts. In fact, the term 'attraction' is used so loosely now that its meaning can include anything that tourists may consume, from a specific product to an entire category of products or activities that extends well beyond the specific. As an example, an historic site is classified as an attraction in the traditional sense of the word. But, by the same token, 'heritage', consisting of many historic sites, archeological sites and historic landmarks scattered around a destination is promoted as an 'attraction.' Even the more generic term 'culture', which includes not only historic sites, but other forms of tangible and intangible heritage, has also gained legitimacy as a form of 'attraction'. Yet, historic sites are constituent elements of heritage and heritage is a constituent element of culture. As one moves from the specific the general, then, the set of attractions, activities and entities that can be included grows exponentially but also becomes less focused. Conversely, as one moves from the general to the specific, what constitutes an attraction becomes more and more clearly defined.

Two challenges emerge. First and foremost, consumption still occurs at the specific attraction, event or entity level. Thus, while a destination may sell 'culture' or 'heritage' as part of its attraction mix, in reality, the tourist will not consume 'heritage' but instead will consume specific sites under the subset of heritage. But having said this, the second consideration is do we know what originally motivated the tourist to visit the destination? In some cases, it may be the specific attraction, while in many other cases, it will be the more generic attraction sets of, say, 'heritage' or 'culture.' In the latter case, the specific attraction may be relatively unimportant in the visit decision, providing enough opportunities exist to consume different heritage or cultural places.

So, the long and short of it is that attractions may be seen to exist in a hierarchical form from the specific site at the bottom to sets of like attractions, to sets of similar attractions, to sets of thematically related attractions, and so on. The set of thematically similar attractions and attractions sets increases significantly as one moves up the classification system, and likewise decreases significantly until reaching a single item as one moves down the system. The classification of each tier of the hierarchy may be conceptually valid.

It is worth repeating that tourism remains a fragmented and disjointed activity (Benckendorff & Zehrer, 2013; Echtner & Jamal, 1997; Laws & Scott, 2015) typified by multiple communities of discourse with historically little cross fertilization between them (Gibson, 1998). Nowhere is this issue more true than with attractions, where the need to develop a classification system has been long recognised

(Mehmetoglu, 2004), but not acted upon. Instead, a series of independent studies examining, different product types has been undertaken, such as Young et al.'s (2007) classification of VFR tourism with overlap, gaps and no internal cohesion (McKercher, 2016a)

In response, a taxonomy of tourism products was developed in order to add some structure to the understanding and classification of attractions (McKercher, 2016a). Taxonomic systems are empirically based and classify items using observable and measureable characteristics. They are formed based on the idea of a natural classification, where members of each tier of a taxonomy are, on average, more similar to each other than they are to members of other taxa at corresponding levels (Sokal, 1986). Reasonably homogenous groups are formed with reasonably sharp boundaries. It is for this reasons that the use of a common vocabulary is important (Hansman & Hunt, 2004; Hedden, 2010). Essentially, then a taxonomy is a hierarchical classification system to indicate natural relationships between items at different levels (Fenneman, 2013). Taxa are arranged vertically to reflect increasing specificity, with the top tier encompassing a broad comprehensive category that captures the essence of all groups and sub groups that fall under it. Lower level tiers are disaggregated into one or more narrower cohorts, until one reaches the bottom where individual entities are found (anon, 2016).

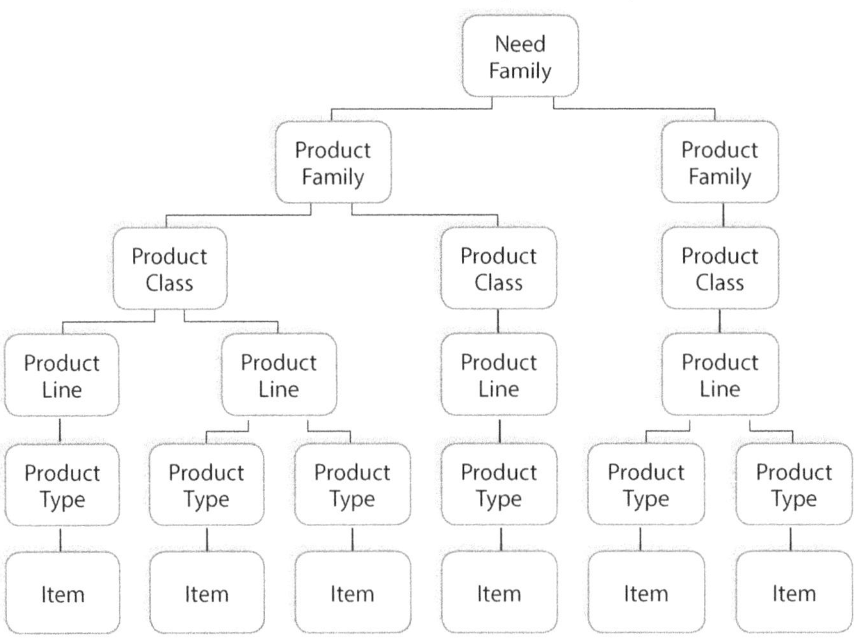

Figure 3.3: Product taxonomy. Source from Kotler and Keller, 2012

The tourism taxonomy represents a modified version of Kotler and Keller's (2012) marketing-oriented product taxonomy shown in Figure 3.3. It is explained simply using the example of drinks. At the top is a broad Need Family (products that can satisfy the need to quench one's thirst). The degree of specificity increases moving down through the taxonomy. For example, the Product family may include, dairy, sodas or juices. All subsequent products then fall under these families. Using sodas as an example, the Product Class might be flavoured sodas (colas, non-colas, etc.), while the Product Line may be sugared and sugar free. The Product Type could be the various brands, while the Item may be different sizes available. Thus, one can go from the broad family to the specific item by reading down the taxonomy or see the relationship between specific items and the need family by reading up.

A similar structure can be applied to tourism attractions that links broad Need Families at the top with specific items or attractions at the bottom, or conversely link specific Items at the attraction at the bottom to more generic needs at top. The idea is that the more specific a need is, the fewer attractions exist that can satisfy that need. For example, if your need was to have a historic theme park experience, then your options would be limited in most destinations. But, if you need was more general, then a vast array of built attraction Product Families could exist to satisfy that need.

Five broad attraction Need Families are identified (McKercher, 2016a). They include Pleasure, Personal Quest, Human Endeavour, Nature and Business. The first three layers of the taxonomy, encompassing the Need Family, Product Family and Product Class are shown in Figure 3.4. Pleasure captures most of what we could consider as leisure tourism. Personal quest reflects more meaningful touristic experiences, while human endeavour captures the full range of cultural and heritage activities, the creative arts as well as dark tourism. Nature looks at passive and active outdoor recreational opportunities, while business is self-explanatory.

Figure 3.5 gives a more detailed description of the Pleasure Need Family by drilling down to more specific layers. The categories shown here begin with the Product Family, and then show examples of the Product Class, Product Line and Product Type. Individual Items are too numerous to mention here but represent the specific temporal or physical space consumed.

Why is such a system important? To begin, it adds some clarity to what we normally call attractions by recognising they can be conceptualised at different levels of specificity. More importantly, it recognises that consumers' needs vary as do their motives for visiting places. A 'foodie' may be interested in sampling the culinary delights of a region, without reference to any specific place. As such his or her needs sit at the top of the taxonomy, and further mean that a wide array of attraction Classes, Lines and Types may satisfy their needs. By contrast, someone who seeks only Michelin starred restaurants has much more specific needs that can be satisfied by fewer choices. At an extreme someone may travel to a destination because of a specific attraction, be it a restaurant or a culinary festival. In this case, the need is precise and can only be satisfied at the item level.

Figure 3.4: Partial taxonomy of tourism products. Source: McKercher, 2016a

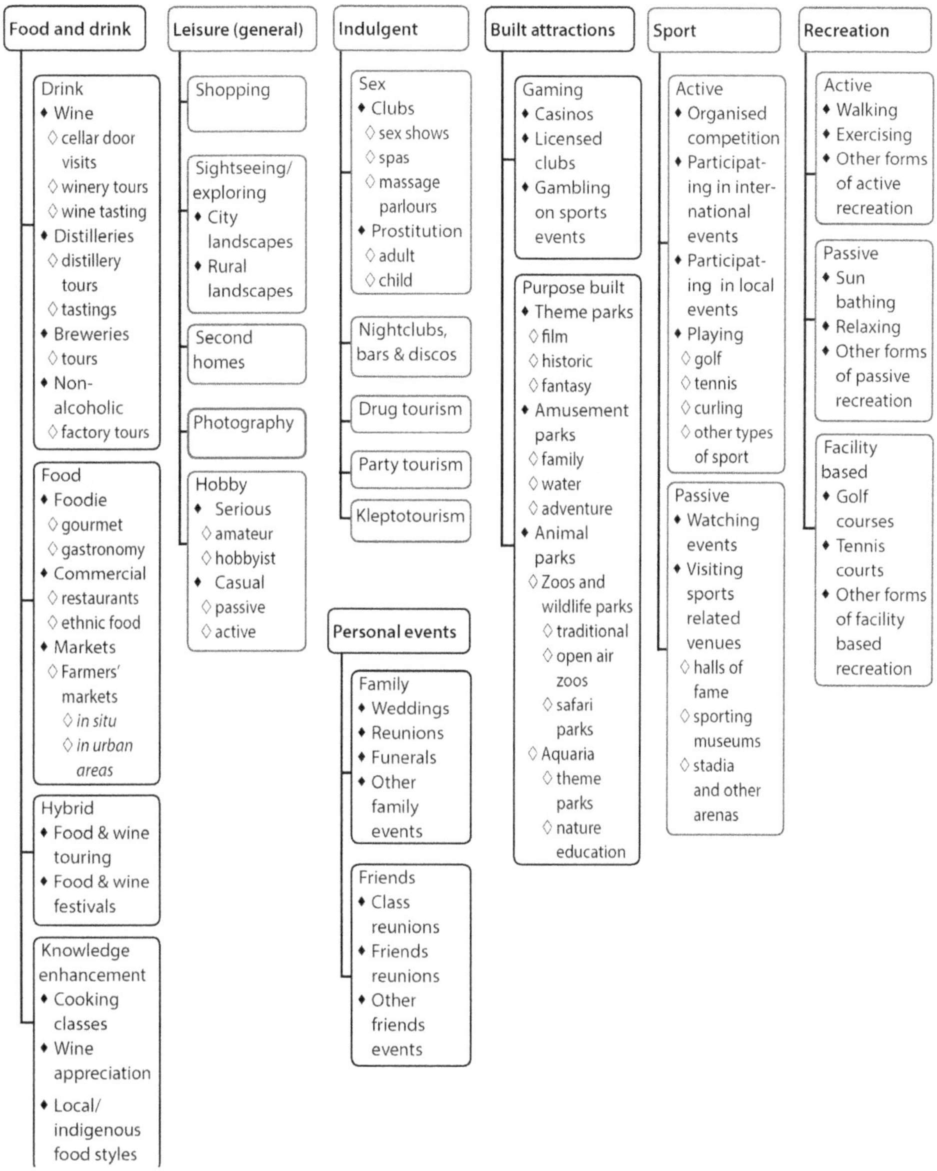

Figure 3.5: Detailed 'Pleasure' Need Family Taxonomy. Source: McKercher, 2016a

Where does a geyser in New Zealand fit into the attractions taxonomy? The term 'attraction' is one of the most misused terms in tourism, for it has been used to define anything from a single entity to a broad family of products. A taxonomy of tourism products has been developed to attempt to add some structure to our categorisation of attractions. It is based on Kotler's (2012) product taxonomy and tries to order attractions and/or products into family trees. Thus, at the bottom or most specific level of the family tree, this geyser could be considered as an Item. But it also can fit into a Product Type (geothermal), Product Line (nature based), Product Class (natural area), Product Family (natural) or Need Family (pleasure). The nature of the product becomes more clearly defined as one moves from the Need Family down through the taxonomy to the Item and increasingly inclusive as one moves from the Item up to the Need Family.

Do attractions attract tourists?

The taxonomy also raises the fairly simple but profoundly important question of do attractions actually attract tourists? Do people come to a destination because of a specific attraction or because of a collection of like attractions (i.e. shopping or dining), where the individual attraction plays little or no role in the visit decision? This question is important for destinations strive to create attractions under the assumption of the "if you build it they will come" principle. Historically, tourism marketers pushed products onto the consumer. This has changed in recent years as a more marketing-oriented management approach has been adopted in recognition that products exist to satisfy consumer needs. Instead, now experiences are pushed to satisfy needs ranging from the specific to the general. In reality, tourists' needs are quite diverse. And so, what constitutes an attraction may also be quite diverse. Two features are critical when considering the role of attractions in the visitation decision: proximate cause and the relationship between needs and level of the attractions' hierarchy.

Chapter 1 referred to proximate cause or *sine qua non* ('but not for') as the nearest cause of an event (CILA, 2012). It is the active, direct, and efficient cause of something that sets in motion an unbroken chain of events (BD, 2016). In insurance terms, *sine qua non* means that if you had not driven through a red light, the accident would not have occurred regardless of how inattentive the other driver was. Thus, if not for your actions, the accident would not have occurred, making you liable.

The same situation applies in tourism, for it is important to distinguish the difference between what people do in a destination and why they come. The two may be quite different, for we do a lot of things that may have little to do with the reasons why we visited. *Sine qua non* is tested rarely in visitor surveys, other than asking generic questions about trip purpose (usually limited to business, pleasure, visiting friends and perhaps a special event) and maybe some generic motivational questions. But it should be, for it is important to know why people come. Visiting popular attractions represents what we do, but does not necessarily act as a proxy for why we come. Proximate cause, therefore, distinguishes the difference between what features of the destination draw people from what they did. The 'attraction' or 'attractions' that cause people to come frame everything else they do. This concept aligns closely with the UNWTO's conceptualisation of an attraction as anything that tourists do in, or draw them to, a destination.

This observation also raises the importance of matching people's needs to the specific tier of the attraction's taxonomy. Tourists' needs vary from the generic to the specific and usually include a variety of both. Indeed, the possible range of needs is as equally vast at one level and as it is highly individualistic at another. Stebbins and Hartel (2006), writing about serious leisure suggest the trip may be motivated to have a single, specialist need met. The same applies to most forms of business travel. Others though suggest most travel decisions are influenced by a basket of needs (Herbert, 1996; Peters & Weiermair, 2000) that may comprise of a mix of specific and generalized needs (Mannell & Iso-Ahola, 1987; Dunn Ross & Iso-Ahola, 1991) or simply generalized needs (Crompton, 1979).

It is likely that specific needs can only be satisfied by a small number of products, or one specific product, found at the base of the taxonomy, whereas more generalist needs can be satisfied by attraction sets found at the higher order of the taxonomy. Where a combination exists, the balance between specific and generalist attractions comes into play. Alternatively, when needs are general or unspecified, almost any destination and any activity within a destination can satisfy them potentially (Caccomo & Solonandrasana, 2002; Botti, Peypoch & Solonandrasana, 2008). Indeed, Crompton (1979) determined that particular attributes of destinations are relatively unimportant if people travel to have broadly based social psychological needs of regression, escape and relationship development met.

An example will help explain this concept. Say you 'need' a holiday and just want to relax by escaping your home and exploring a destination, but have no deeper need. Your needs are very generic and therefore, almost any destination that fits your financial and time budgets as well as interests can satisfy them. You may be very active while travelling, but it does not matter what you do as long as the desire for an escape is met. Hence, broad Need or Product Family sets of attractions can satisfy your need. But, if you are a film buff and want to visit a certain film festival, your needs are specific and can only be satisfied by either a product Type (film festivals) or maybe even a single product Item (Cannes).

A model linking the specificity of need with the tier of the tourism attractions taxonomy to better understand the role that individual attractions play in the overall travel decision has been developed (McKercher, 2016b). Ultimately, all touristic consumption occurs at the Item level of the product taxonomy. You cannot go 'shopping' unless you visit specific shops, or pursue heritage without visiting heritage sites. But, thinking of proximate cause, was it the opportunity to shop at any number of stores that attracted the person or was it the chance to shop at a specific store that drew her or him to the destination? The key point becomes one of determining what role, if any, the individual attraction played in driving visitation, In some cases, the individual attraction is key, for were it not for that attraction, the person would not be visiting. *Sine qua non*!

Equally, the importance of individual attractions diminishes as specificity of need diminishes. If you travelled for escape or to spend quality time with your family members, then specific attractions may have played little or no role in the destination decision. Instead, the decision would have been made based on whether a bundle of like attractions or even a group of attraction sets (pretty scenery, lots to see and do) would have influenced the decision.

Figure 3.6 shows this relationship. A specific need (example *A*) can only be met by a specific attraction. Case *C* shows how a mix of specific and generalist needs can be met by a both attractions at a low order of the taxon as well as a more generalist high order attraction sets. 'Bleisure' is a common example, where the person comes for specific business reasons and then extends the trip for less well defined pleasure reasons. Here, the belief that the destination can offer sufficient range of highly substitutable shopping and dining opportunity sets, in aggregate, will satisfy needs. Lastly, example *B* shows the common situation where one travels for more generalized needs of escape, exploration, spending time with family and friends and the like. Here, the specific attributes of the destination are less important than the mix of activities it offers.

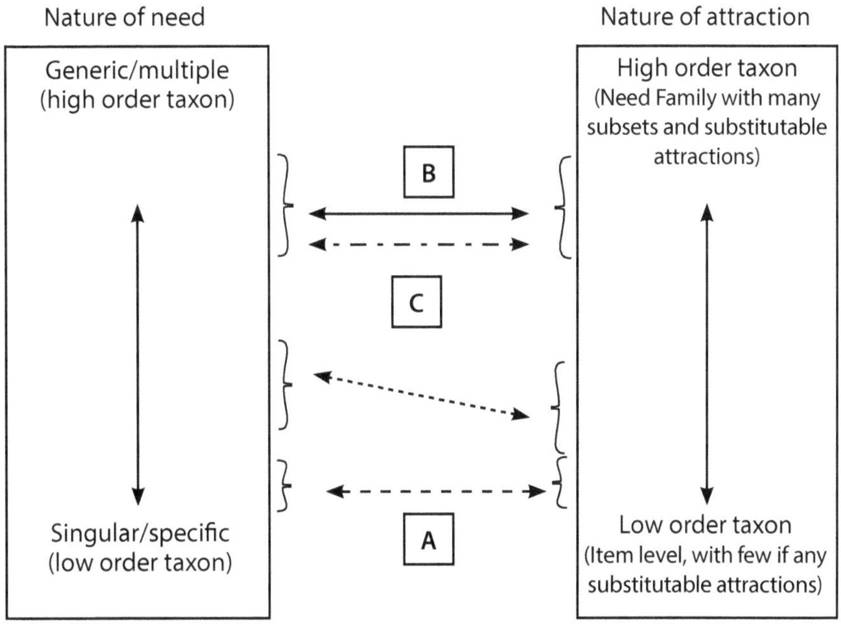

Figure 3.6: Matching a hierarchy of needs with the product taxonomy. *Source:* McKercher, 2016b

This model shifts the conceptualisation of what constitutes a primary, secondary or tertiary attractions from the traditional focus on the specific attraction itself to the tier of attraction that draws tourists. If needs are specific, then the Item level becomes the primary attraction. But, if needs are more generic, then the nature of the primary attractions shifts higher up the taxonomy.

Conclusion

Surprisingly little research has been conducted on attractions even though they are the focal point of tourism. This chapter reviewed the conceptualisation of attractions and an attractions' hierarchy before presenting a novel system that identifies a continuum of attractions organised in a taxonomic structure. This taxonomy also helps us understand better the role that single attractions play in tourism and explains better the link between motive and attraction.

One of the ironic features of tourism is that the vast tourism infrastructure exists to service people's desire to consume attractions, most of which are available free of charge or at a very low price. It costs nothing to walk down a street and experience a city's vibrancy, sights, sounds and smells. It costs nothing to enjoy views, sit on a beach, wander into a shop or spend time with friends and family. Entry into many publicly funded attractions is highly subsidised so one only pays a fraction of the actual cost of visitation. Indeed, the vast majority of what the average tourist spends is for amenities and services that facilitate the consumption of attractions and not for the attraction itself. Few people in the tourism industry actually work in tourism attractions. Most of them work for businesses that facilitate, or make possible, the opportunity to visit attractions. Hotels, airlines, tour operators, and restaurants are not attractions. Yet, without these products tourists could not visit attractions.

Discussion points / Exercises

1. Select a major attraction in your area and document all types of markers that signify the attraction as being worthy of a visit. Make sure you include both commercial and non-commercial markers. Be as creative as possible.
2. Select a tourism Product Family of your choice and fill in the taxonomic structure by identifying the different Classes, Lines and Types as well as examples of specific products under each Type. Analyse the table to identify logical linkages upward and downward through the taxonomy
3. What is the proximate cause for pleasure tourists to visit your region? In doing so, identify where attractions would fit in the taxonomy to match the needs of tourists.

Points to consider for revision

- Like most things in tourism, it is extremely difficult to define an attraction precisely. Historically, efforts have focussed on a supply side approach by categorising or grouping like attractions and developing suitable models. More recently, a demand side approach has determined that almost anything a tourist does in a destination can be considered as an attraction, which broadens the concept substantially.
- Attractions do not operate in isolation. Instead, they are part of an attractions' system that includes the tourist, the nucleus and markers. In many ways, markers are the most important element of the system for they act as the catalyst that links the consumer to the product.
- The term 'attraction' has been used very loosely in tourism. In reality, attractions/products can be organised in a taxonomic structure that groups them according to their inter-relationships. At the top of the taxonomy is the broad Need Family, and as one progresses downward the taxonomy becomes more specific until one reaches individual Items. Conversely, working upward, one can go from the specific to the general.

- Proximate cause is a key element for any destination management organisation. Understanding the reason why people come (*sine qua non* – if not for) is far more important than documenting what they did.
- In many cases individual attractions play little role in the decision to visit multi-product destinations. Instead, attraction sets comprised of like attractions play a more important role.

Additional recommended reading

Leask A. (2014) Visitor attraction management: A critical review of research 2009 – 2014. *Tourism Management* 57, 334-361.

Leiper, N. (1990) Tourist attractions systems, *Annals of Tourism Research,* 17(3), 367-384

McKercher, B. (2016) Towards a taxonomy of tourism attractions. *Tourism Management* 54, 196-208.

McKercher. B. (2016) Do attractions attract tourists? A framework to assess the importance of attractions in driving demand. *International Journal of Tourism Research.* DOI: 10.1002/jtr.2091

References

Anon (2016) Definition for three types of theme parks. www.familythemeparkvacations.com/three-types-of-theme-parks.html.

BD (2016) Business Dictionary. www.businessdictionary.com/definition/proximate-cause.html.

Benckendorff, P. & Zehrer (2013) A network analysis of tourism research. *Annals of Tourism Research.* 43, 121-143.

Bessiere, J. (1998). Local development and heritage: Traditional food and cuisine as tourist attractions in rural areas. *Sociologia Ruralis,* 38(1), 21-34.

Botti, L., Peypoch, N., & Solonandrasana, B. (2008). Time and tourism attraction. *Tourism Management,* 29(3), 594-596.

Caccomo, J. & Solonandrasana, B. (2002) Réflexions autour du concept d'attraction touristique: analyse et taxonomie *Revue de Recherche en Tourisme UQAM* (Canada), 3, 68–71

CILA (2012) *Introduction to the Insurance Industry*. Chartered Institute of Loss Adjusters. www.cila.co.uk/files/Certificate/Chapter%208.pdf .

Clark, I. (2009). Naming sites: Names as management tools in indigenous tourism sites – An Australian case study. *Tourism Management* 30(1), 109-111.

Cooper, C. (2021) *Essentials of Tourism. 3rd Edition.* London: Sage.

Crompton, J. L. (1979). Motivations for pleasure vacations. *Annals of Tourism Research,* 6(4), 164-182.

Culler, J. (1988). The semiotics of tourism. In J. Culler, *Framing The Sign: Criticism and Its Institutions* (pp. 1-10). Norman: University of Oklahoma Press.

Denstadli, J., & Steen Jacobsen, J. (2011) The long and winding roads: Perceived quality of scenic tourism routes. *Tourism Management.* 32(4): 780-789.

Dunn-Ross, E., & S. Iso-Ahola (1991) Sightseeing tourists' motivation and satisfaction. *Annals of Tourism Research,* 18(2), 226-237

Echtner, C. & Jamal, T. (1997) The disciplinary dilemma of tourism studies. *Annals of Tourism Research* 24(4), 868-883.

Fenneman J. (2013) What is taxonomy and where did it originate. *E-Flora BC* http://www.geog.ubc.ca/biodiversity/eflora/IntroductiontoPlantTaxonomy.html.

Gibson, H. (1998) Sport tourism: A critical analysis of research. *Sport Management Review* 1, 45-76.

Goeldner, C. & Ritchie, B (2012) *Tourism Principles, Practices and Philosophies.* Hoboken: John Wiley and Sons.

Hansman, S., & R. Hunt (2004) A taxonomy of network and computer attacks. *Computers and Society* 24(1), 31-43.

Hardy, A. (2003) An investigation into the key factors necessary for the development of iconic touring routes. *Journal of Vacation Marketing,* 9(4), 314-330

Hedden, H. (2010) *The Accidental Taxonomist,* Medford: Information Today.

Herbert D. T. (1996) Artistic and literary places in France as tourist attractions. *Tourism Management* 17(2), 77-85

Holloway, J (2002). *The Business of Tourism,* Prentice Hall, Upper Saddle River.

Holloway, J., & Humphries, C. (2016) *The Business of Tourism, 10th Ed.* Harlow: Pearson.

Jenkins, O. (1999). Understanding and measuring tourist destination images. *International Journal of Tourism Research,* 1, 1-15.

Kantanen, T., & Tikkanen, I. (2006). Advertising in low and high involvement cultural tourism attractions: Four cases. *Tourism and Hospitality Research,* 6(2), 99-110.

Kotler, P., & Keller, P (2012) *Marketing Management 14th Ed.* Toronto: Prentice Hall.

Koutoulas, D. (2015). *Understanding the Tourism Product.* Interim symposium of the Research Committee on International Tourism of the International Sociological Association 14-16 May, University of the Aegean, Mytilini, Greece. 10.13140/RG.2.1.2250.4806.

Laws, E, & Scott, N. (2015) Tourism research: Building from other disciplines. *Tourism Recreation Research.* 49(1): 48 – 58.

Leask, A, (2008). The nature and role of visitor attractions. www.researchgate.net/publication/284256972_The_Nature_and_Role_of_Visitor_Attractions. 10.1016/B978-0-7506-8545-0.50007-1.

Leask, A. (2010). Progress in visitor attraction research: Towards more effective management. *Tourism Management,* 31(2), 155-166

Leiper, N. (1990) Tourist attractions systems *Annals of Tourism Research,* 17(3), 367-384

Lew, A. (1987) A framework of tourist attractions research *Annals of Tourism Research,* 14, 553-575.

Litvin, S., & Mouri, N. (2009). A comparative study of the use of 'iconic' versus 'generic' advertising images for destination marketing. *Journal of Travel Research,* 48(2), 152-161.

MacCannell, D. (1976) *The Tourist: A New Theory of the Leisure Class.* New York: Schoken Books.

MacCannell, D. (2001). Tourist Agency. *Tourist Studies,* 1(1), 22-37.

Mannell, R. & Iso-Ahola, S. (1987) Psychological nature of leisure and tourism experience. *Annals of Tourism Research,* 14(3), 314-331

McKercher, B. & Lau, G. (2007). Understanding the movements of tourists in a destination: Testing the importance of markers in the tourist attraction system. *Asian Journal of Tourism and Hospitality Research,* 1(1), 39-53.

McKercher, B. (1993) Some Fundamental truths about tourism: understanding tourism's social and environmental impacts *Journal of Sustainable Tourism,* 1(1), 6-16.

McKercher, B. (2016a) Towards a taxonomy of tourism attractions. *Tourism Management* 54, 196-208. www.sciencedirect.com/science/article/pii/S0261517715300431

McKercher. B. (2016b) Do attractions attract tourists? a framework to assess the importance of attractions in driving demand. *International Journal of Tourism Research.* DOI: 10.1002/jtr.2091

Mehmetoglu, M. (2004) Tourist or Traveller? A typological approach. *Tourism Review,* 59(3), 33-39

Mercille, J. (2005). Media effects on image: The case of Tibet. *Annals of Tourism Research* 32(4), 1039-1055.

Mill, R. & Morrison, A. (2012) *The Tourism System 7th Ed.* Dubuque: Kendall Hunt.

Page, S. (2019) *Tourism Management 6th Ed* Oxon: Routledge.

Pearce, D. (1989) *Tourist Development 2nd Ed.* Harlow: Longman.

Pearce, P. (1991) Analysing tourist attractions *The Journal of Tourism Studies,* 2(1), 46-55.

Peters, M., & Weiermair, K. (2000). Tourist attractions and attracted tourists: how to satisfy today's' fickle'tourist clientele?. *Journal of Tourism Studies,* 11(1), 22-29.

Richards, G. (2002). Tourist attraction systems: Exploring cultural behaviour. *Annals of Tourism Research,* 29(4), 1048-1064.

Shoval, N. & Raveh, A. (2004) Categorization of tourist attractions and the modeling of tourist cities: based on the co-plot method of multivariate analysis *Tourism Management* 25(6), 741-750

Sokal, R. (1986) Phenetic taxonomy: Theory and methods. *Annual Review of Ecology and Systematics* 17, 423-442.

Stebbins, R & Hartel, J. (2006) *The Serious Leisure Perspective.* www.seriousleisure.net.

Timothy D. (1997) Tourism and the personal heritage experience *Annals of Tourism Research* 24(3), 751-754.

Tussyadiah, I., & Fesenmaier, D. (2009). Mediating tourist experiences: Access to places via shared videos. *Annals of Tourism Research,* 36(1), 24-40.

UNWTO (2008a) *International Recommendations for Tourism Statistics 2008 Draft Compilation Guide. Madrid: UNWTO.* http://unstats.un.org/unsd/tradeserv/egts/CG/IRTS%20compilation%20guide%207%20march%202011%20-%20final.pdf.

UNWTO (2008b) *Internationalb Recommendations for Tourism Statistics 2008 Madrid: UNWTO.* http://unstats.un.org/unsd/tradeserv/IRTS%202008%20edited%20 whitecover.pdf.

Walker, J. & Walker, J. (2011) *Tourism Concepts and Practices*. New York: Prentice Hall.

Wong, C., & B. McKercher (2013) Web markers for various tour products: the case of Hong Kong. *Tourism Management Perspectives.* 8, 126-130.

Young, C., Corsun, D. & S. Baloglu (2007) A taxonomy of hosts visiting friends and relatives. *Annals of Tourism Research*. 34(2), 497-516.

4 Access and the Spatial Interactions of Tourists

By the end of this chapter, the reader will be able to:

- Distinguish between the resource and commodity values of time and analyse how they affect travel patterns
- Describe distance decay and analyse how it influences both absolute demand and behaviour of tourists
- Define the Effective Tourism Exclusion Zone and analyse its impact on tourism flows
- Evaluate segment transformation and the impact of distance on behaviour
- Define market access and assess how it influences tourist behaviour
- Explain different itinerary models

Introduction

While attractions drive tourism, access, or the ability of visitors to get to and travel within a destination, plays a key role in its success. If tourists cannot get to a destination, then they cannot experience the attractions therein. If they cannot move around the destination easily, their actions will be constrained. Moreover, if access is limited, the ability of the destination to provide viable products is also limited. This chapter examines the role that access plays in tourism.

Time

Before we discuss access, it is important to introduce the concept of time, for tourist flows can be seen as a function of the relationship between time and space (Leiper, 1990), with time, arguably, being the most critical factor affecting movements. As Pearce (1998) notes, time availability can directly constrain or expand the number and range of potential activities available and the depth at which individual activities can be experienced. Indeed, it is about the only absolute in tourism and is also the scarcest resource (McKercher & Lew, 2004). Usually the amount of time one has is fixed either by the necessity to return to work or by flight reservations and other itinerary constraints. It is rare for people to extend their holidays, although tourists can modify plans if, for example, they encounter inclement weather (Becken & Wilson, 2013). Time is even more constrained among business travelers.

Time cannot be saved or stored for future use, but can only be transferred from one type of activity to another (Truong & Henscher, 1985). How people choose to spend their time and the decisions they make around its use shape behaviour (Lew & McKercher, 2006), and influence movement patterns, participation in activities, and satisfaction (McKean et al., 1995). Tourists, though, have a a great deal of discretion in how they choose to spend their time. Some may engage in fast tourism, collecting as many sites and destinations as possible in a short period of time, while others may engage in slow tourism (Oh et al., 2016), choosing to linger at a small number of places and enjoy a deeper experience. Some still will choose a combination of both slow and fast tourism.

Interestingly, the expenditure of time was discussed in detail 30 years ago but seems to have been forgotten more recently in this era of empirically driven approach to tourism research. Authors such as Chavas, Stoll and Sellar (1989), McKean, Johnson and Walsh (1995), and Walsh, Sanders and McKean (1990) were among the first to examine it, yet, as Hall (2005) discusses, it remains an under-explored, though vital, area of tourism. As a result, time use in general and decisions about spending transit time in particular, remain understudied phenomena. GIS (Geographic Information Systems) analysis effectively identifies the 'what' of time allocation but does not explain the 'why' of such allocation decisions (Hardy, 2020; Xu et al., 2021).

■ Transit Time

How tourists spend or allocate transit time between the home and destination area, or between destinations, is one of the most critical issues affecting time use decisions. Essentially, travel time can be perceived in one of two ways. On the one hand, it can have a commodity or utility value, where the act of travelling has value in itself and where the journey to the destination is as or more important than being at the destination. People are willing to trade off time spent at the destination for travel time. Here, it is argued that undirected travel can result in greater satisfaction and improved well-being (Hook et al. 2021). On the other hand, it can be seen as a cost or a resource where its expenditure is calculated based on an opportunity/cost manner (Chavas et al. 1989). In such cases, it is a necessary task that has no value in its own right. Instead, it may have a negative effect on the overall trip experience. If the perceived cost outweighs the opportunity, travel time will be forsaken for more time spent at the goal. Dickinson et al. (2011) labelled some people who adopt such an approach as cash rich and time-poor travellers. Table 4.1 compares the two concepts.

Table 4.1: Different valuations of time

	Commodity	**Resource**
Value of time	Opportunity	Cost
Transit time	Has value in and of itself	Has little value and needs to be minimized
Type of tourist	Process oriented	Goal oriented
Perceived amount of time availably	Much	Little
Trade-off	Time spent getting there means less time at the destination	Time spent at the destination at the cost of time spent travelling
Typical trip	Long haul Multiple destination Touring	Short haul Single destination

In a sense, therefore, one could argue that trips where time is seen as being a commodity are process-driven where the journey is as important as the destination. People who are more likely to take this type of trip tend to have more time, engage in multi-destination trips often with no main destination and are motivated as much for exploration and discovery as they are for being at the destination (McKercher & Zoltan, 2014). Touring, sightseeing and exploring are key features of these trips. Most are independent, rather than part of a package tour. Most long haul travel and much long duration short haul travel falls into this category, as does much adventure tourism, such a bicycle tours, hiking and canoeing trips. Here, the actual distance travelled each day is small, while the journey to get to the night's rest stop is the key goal. The same situation applies in destinations, where sightseeing and exploring, perhaps with no deeper goal in mind, are seen as valuable and rewarding ways to spend one's time.

Conversely, the type of trips where time is seen as a scare resource that must be spent efficiently can be described as goal-driven, where maximizing time in the destination or at a specific attraction is the primary objective. The person who values time accordingly tends to have a smaller time budget and/or a full itinerary and as such does not have the luxury of engaging in much touring. Instead, the choice is made to minimise travel time in order to maximize time elsewhere. Many families with young children engage in this type of trip, as do most short break, short haul tourists.

This binary distinction is overly simplistic (Mokhtarian & Salomon, 2001), and instead the valuation of travel time exists on a continuum, with most tourists showing a tendency to value time in part as a resource and in part as a commodity (McKercher & Tkaczynski, 2023). Some may lean towards one end of the spectrum, while others may value both dimensions equally. It must also be appreciated that transit time is both context and trip specific, meaning the same person may display a process or goal orientation on different trips.

An examination of how tourists value transit time, and the trade-offs they may need to make, resulted in a 10 phase continuum as shown in Figure 4.1 (McKercher & Tkaczynski, 2023). Those who travel to tour or explore tend to display a predominant process orientation. Similarly, those who find the journey to the destination relaxing, feel the holiday starts when the leave home and also have to make pragmatic decisions to break up a trip are likely to adopt more of a mixture of goal and process orientation, with a goal orientation beginning to dominate once pragmatic concerns come to the forefront. Practical considerations of travelling with children and pets may mean sacrificing travel time to get to the destination relatively quickly. Likewise, the desire to visit friends and relatives at the destination or to attend a special event pushes the consumer more toward a goal orientation. Those who dislike driving, are on a short trip or who feel the pull features of the destination are very powerful are likely to be goal-oriented tourists who seek to minimise travel time.

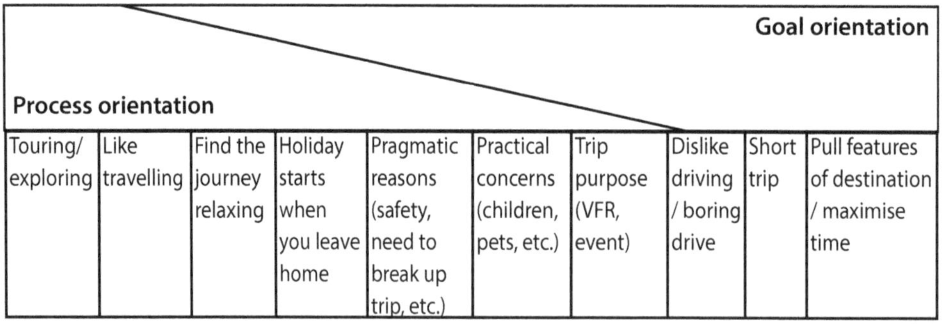

Figure 4.1: Valuation of travel time. Source McKercher & Tkaczynski (2023)

The time space relationship is not absolute, for time or space can effectively be 'bought' depending on the mode of transit taken and the willingness of the individual to pay a price premium for greater distance. Time can be bought by either taking a direct route to the destination and thus avoiding sightseeing along the way or by purchasing more efficient transport links to connect the source to the destination area. Distance can also be bought for the same amount of time depending on the mode of transport taken.

The same situation applies within a destination where effective trip planning, concentrating activities around primary attractions and the use of different transport modes influences time budget allocation (Cvelbar et al., 2018, Sirirak & Pitakaso, 2018). Indeed, Botti, Peypoch and Solonandrasana (2008) suggest tourists make rational choices about how much time they want to spend at each place, which attractions or activities represent primary and secondary objectives, and whether they are interested in exploring other lesser known places.

The trend toward more short break trips and fewer long duration trips has significant implications for the overall sustainability of tourism and especially climate change related issues. As Gössling, Scott and Hall (2018) note, limited time budgets are making more and more people take short duration trips where they go to must-see attractions and destinations, leading to a greater concentration of visitors in time and space. The advent of low cost carriers, though, means that tourists can travel further in a much more carbon-intense manner than previously available.

Distance decay

Tobler's (1970:236) First Law of Geography states that "*everything is related to everything else, but near things are more related than distant things*." It forms the basis for the distance decay concept. Distance decay states demand declines exponentially as distance increases, whether distance is measured by Euclidean distance, time, effort or cost. It has been found to apply in such a wide array of social and spatial situations that Waters (2017) says it complies with the three basic requirements of a 'law': universality, necessity and synthesis. Universality means it applies to all members of the class. Necessity indicates that the relationship is guided by some underlying principle, while the synthetic component joins at least two concepts, in this case consumer behavior with physical and human geography.

The decaying effect of distance on travel demand was first noticed by Greer and Wall (1979). Since then, a number of studies has applied it in a variety of tourism (Hooper, 2015; Lee et al., 2011; McKercher & Lew, 2003; McKercher et al., 2008; McKercher, 2021; Nyaupane & Graefe, 2008) and recreation contexts (Eagles et al., 2015; Hanink & White, 1999).

Distance decay theory (Figure 4.2) shows that demand for tourism should peak at some relatively close distance from the origin and then decline exponentially. This pattern recognises that people must travel a certain minimum distance to achieve a sense of escape (Greer & Wall, 1979). The shape of the curve reflects the intersection of different supply and demand functions influenced by time availability.

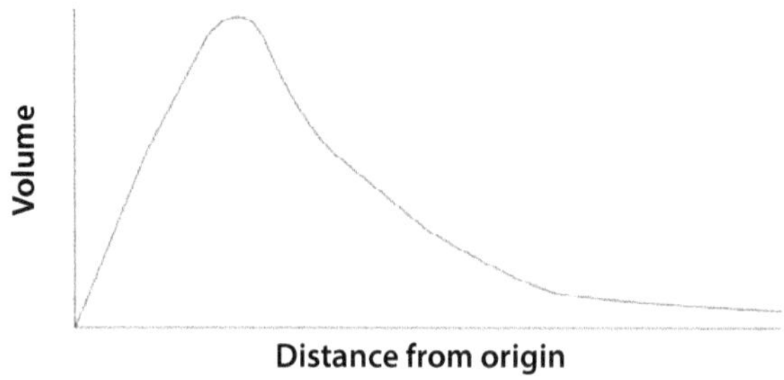

Figure 4.2: Classic distance decay curve

It is ubiquitous and enduring over time, as Lee, Denizci-Guillet, Law & Leung (2012) discovered when they examined 10 years of outbound travel from Hong Kong. They found that apart from minor variations, the overall spatial dispersion of tourists was broadly stable. Minor variations can be explained by changes in the general economic situation and/or the business cycle (Wong et al., 2016; Croes et al., 2018), for when the economy is robust, people will travel further, while when it is fragile, distance travelled shrinks. Indeed, Butler (2008) says the bulk of tourists go to the same regions and do the same things they have always done, while Lorde, Li, and Airey (2016) suggest the stability of the overall pattern is attributed to habit persistence.

Two features of distance decay must be appreciated. First and foremost, distance is not a deterministic variable. Nobody wakes up in the morning and says "*I only want to travel x km while x -1 km is not far enough and x+1 km is too far.*" Instead, it represents a valid proxy variable that reflects the combined impact of a bundle of tangible and intangible attributes (Sirakaya & Woodside, 2005), contextual influences (Decrop & Snelders, 2005), as well as physical geographic considerations. Factors include cost considerations, time availability, willingness to enter culturally distant environments, travel party composition, and a range of other factors.

The second feature of distance decay is that it can only be measured from the perspective of the source market and not the destination. Distance decay measures changes in demand from a source. It can either be presented in absolute terms (actual number of tourists from the source who travel a certain distance threshold) or in relative terms (share of tourists who travel a certain distance). The denominator remains constant, enabling valid like for like measures to be calculated.

Some researchers have tried to apply distance decay by looking only at the share of arrivals to a destination, found by divided the number of arrivals from the source market by all arrivals to that destination. There is a logical fallacy in this approach, for such a method cannot control for different population sizes of source markets. Because the denominators are different, meaningful comparisons cannot be made. Let's give a simple example of why this approach is so flawed. China generated about 154,000,000 outbound trips in 2019, before COVID-19 struck. Of these, about 1.4 million people visited Australia, representing about 0.9% of total outbound travel. New Zealand, with a much smaller population, generated some 3.2 million outbound trips, of which slightly over 1.4 million were to Australia. Australia's share of New Zealand's outbound travel was, therefore, about 44%, meaning that proportionately, 45 times more New Zealanders went to Australia than did Chinese. Yet, if you look at arrival figures only, both generate about 15% to 16% of arrivals. So, from the perspective of New Zealand outbound, Australia is a critical destination, while from the perspective of Chinese outbound, Australia is a relatively minor destination. In fact, 20 other countries and regions attracted a greater share of Chinese outbound than Australia in 2019.

Three types of distance decay curve have been identified, as shown in Figure 4.3 (McKercher & Lew, 2003). The classic curve shows a peak near the source and a geometric decline. A second curve shows a 'plateauing effect' where demand is stable over some distance before declining. The cause of this type of curve is due to the limited accommodation or destination options which effectively force people to travel further than they normally would prefer to go. In doing so, peak demand is extended until a threshold is met where demand drops exponentially. A third style has a secondary peak some great distance from the source, reflecting the unique drawing appeal of a distant attraction. This third style is caused by two factors. On the one hand the distant destination may have such great drawing power that its appeal overcomes the frictional effect of distance. Studies have shown, for example, that this pattern has been observed in the case of Australian's travel to the UK, European's travel to Spain and travel from South America to the USA. On the other hand, it may be caused by a gap in opportunity or an Effective Tourism Exclusion Zone.

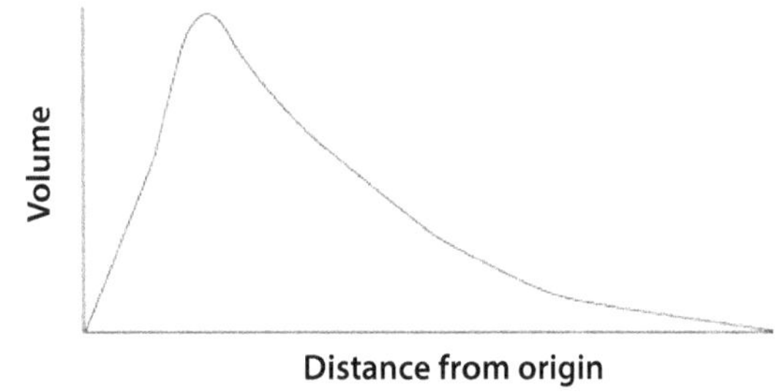

Figure 4.3a: Theoretical distance decay curve (Source: Bull, 1991)

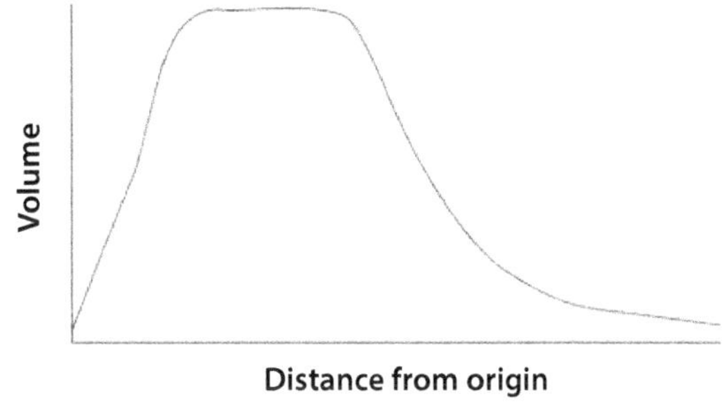

Figure 4.3b: Plateauing distance decay curve. *Source*: McKercher, 1998.

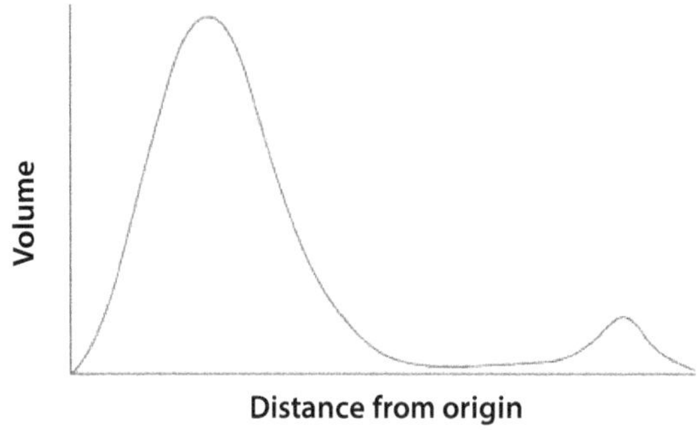

Figure 4.3c: Distance decay curve with secondary peak.

Effective Tourism Exclusion Zones

The presence of an Effective Tourism Exclusion Zone (ETEZ) exists where little tourism activity occurs (McKercher & Lew, 2003). Figure 4.4 depicts three types of ETEZs. The causes of this effective tourism exclusion zone may include regions that are lacking in tourism development, various geographic features such as impassable mountain ranges, deserts or oceans, places at war where insecurity is rife, destination areas that prohibit residents of certain markets from visiting or simply destinations that offer products that are unattractive to the source market. Fotheringham (1981) and Eldridge and Jones (1991) label this concept 'space warping' where equivalent distances have spatially uneven effects on interaction are common when more of fewer destination opportunities are present.

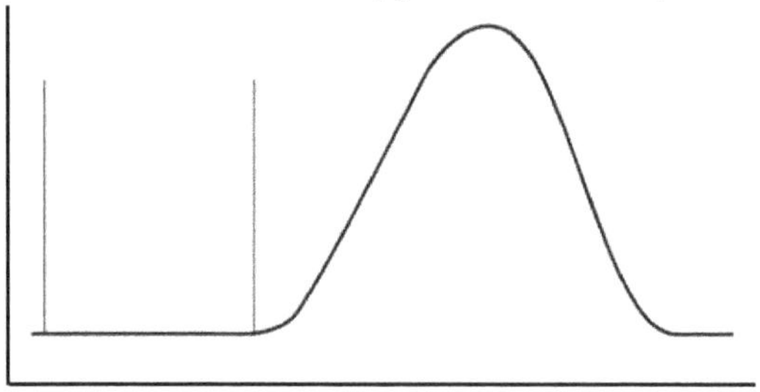

Figure 4.4a: Large ETEZ close to source market

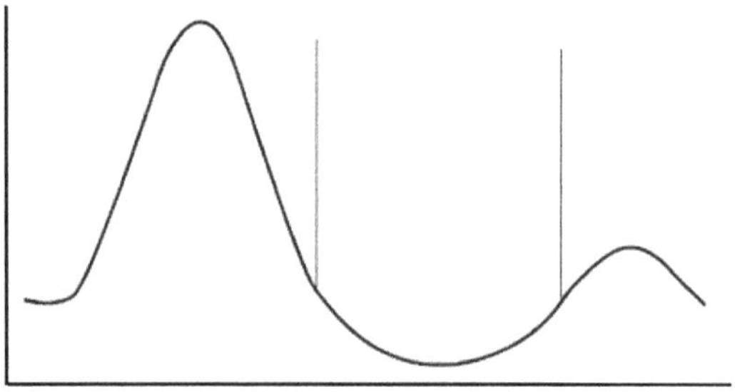

Figure 4.4b: ETEZ some distance from source market

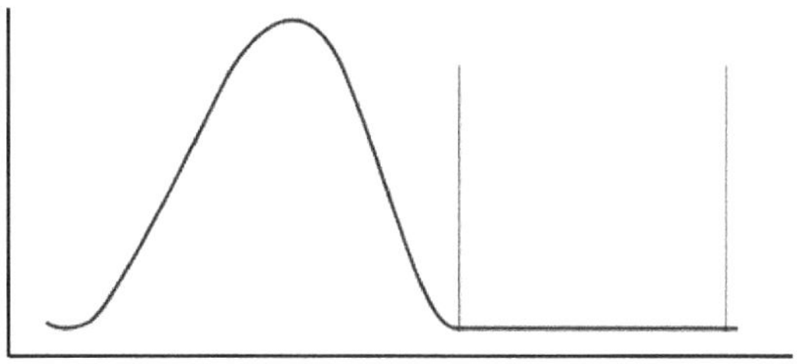

Figure 4.4c: Distance ETEZ has little impact on demand

The location of the ETEZ can have a dramatic effect on tourism movements. A small zone close to the source market will have little or no real impact on movements given that people have to travel a minimum distance anyways to have that sense of escape. A large ETEZ located close to the source market is typically found in isolated communities or, in the case of international tourism amongst island source markets (such as Japan, New Zealand and Australia), source markets with poor political relations amongst immediate neighbours or those whose neighbours are in a state of conflict. The effect is to elongate the area of low demand and to push the demand peak to more distant areas. A moderate ETEZ located some distance from the source market will tend to accentuate demand closer to the market and likely lead to a secondary peak while an ETEZ a great distance from the source market will likely have no impact on demand.

The inner edge of the standard ETEZ represents the limit of short haul travel where trips are typified by short break, short duration, repeat pleasure trips. The outer edge signifies the onset of long-haul travel which is typified more by longer duration, independent travel, multi-destination travel and more aspirational types of tourism.

■ The effect of distance decay on tourist flows

Figure 4.5 shows the results of the impact of distance on tourism flows (McKercher & Mak, 2019). The study analysed outbound travel from 92 source markets to over 4700 discrete origin-destination pairs. Travel to land neighbors dominates outbound travel, accounting for 53% of all departures. Almost 80% of all arrivals are to destinations with gateways cities less than 2000 km from a source market's gateway. Share of departures for travel greater than 5000 km was typically three percent or lower. In other words, more than half of all global travel is to immediate land neighbors and very little is to distant destinations.

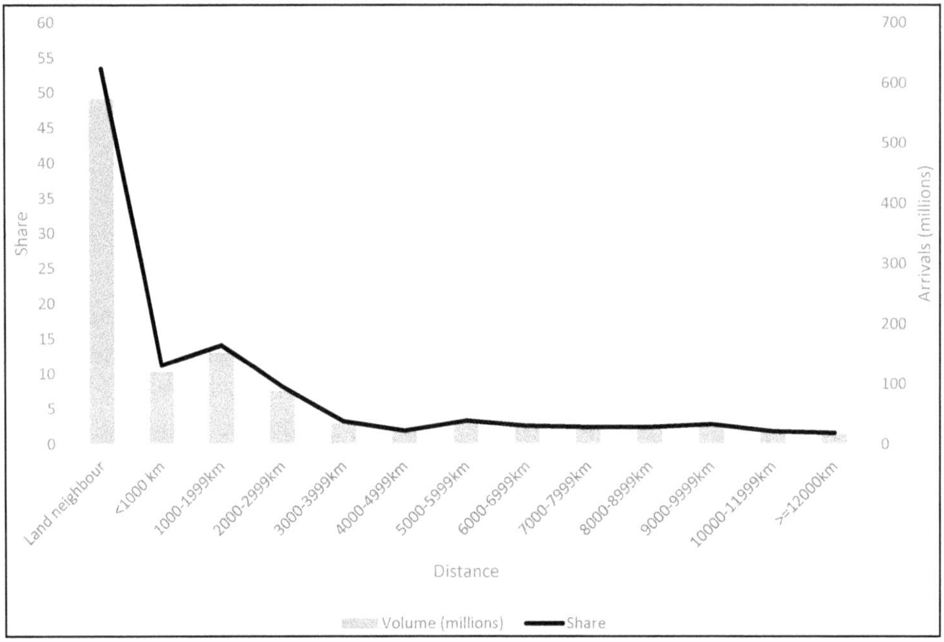

Figure 4.5: Arrival volume and departure share by distance 2016. Source: McKercher & Mak (2019).

■ Distance influences segment profile

Distance acts as a filtering agent. Some segments are more likely to travel short distances, while others may be predisposed to travelling longer distances. Anyone who can travel can travel short haul, but not everyone can or is willing to travel long haul. At its extreme, some segments may effectively disappear once a certain distance threshold is reached, while others may not emerge until a threshold is crossed. As shown in Figure 4.6, the solid line shows overall demand, while the dotted lines show the hypothetical demand curve for different segments. Segments A and B are particularly affected by distance and peak close to home before demand drops rapidly. Segments C and D are more decay resistant and peak progressively further from home before declining. Segment E is particularly distance resistant and emerges later than the others, peaking a great distance from home and only declining slowly. This decaying effect explains changes in market mix at different destinations, which in turn should explain different behavior.

Which segments are more sensitive and which are more robust? Market segments that are sensitive to distance include the young professional female or male segments (due to a combination of finances and limited vacation time), the least affluent members of a community (due to costs), families with young children (due to cost considerations, especially for transport, and an unwillingness to place children in a culturally unfamiliar environment), and older less well educated individuals (again due to cost considerations and unwillingness to enter culturally strange environments). Those that are more robust and are, therefore, more willing to travel longer distances include more affluent, well-educated empty nesters who have the financial ability and accrued leave time to travel long distances and who also are more willing to enter strange environments. In addition, adventuresome backpackers also travel long haul, even if they do not have large financial reserves. In essence, then, the segments that were most likely to be affected by the impact of distance on demand were akin to individuals who could be found in the psychocentric end of Plog's (1974) spectrum while those who were more likely to do so share attributes common the Allocentrics or Venturers.

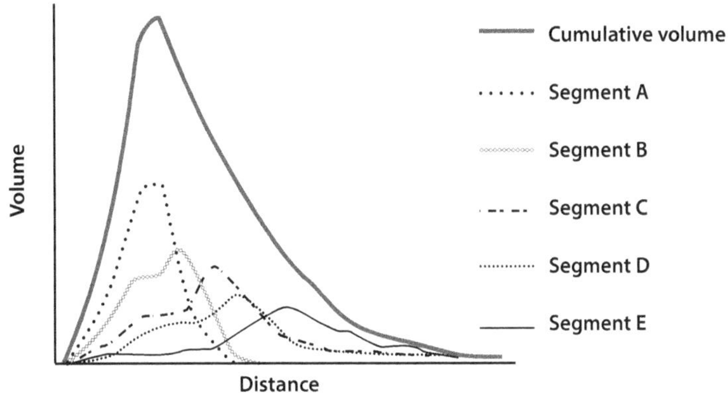

Figure 4.6: The filtering effect of distance on market segments. Source: McKercher (2008)

■ Impact of distance on behaviour

While distance decay is an interesting geographic concept, it exerts a much more profound impact of tourism than most people appreciate, for the filtering effect of distance influences who is likely to visit a destination, their motives and their resultant behaviour. This relationship is shown in Figure 4.7. Anyone who can travel can travel short distance, making this market quite homogeneous. The long haul market is much more exclusive, though as some groups are filtered out.

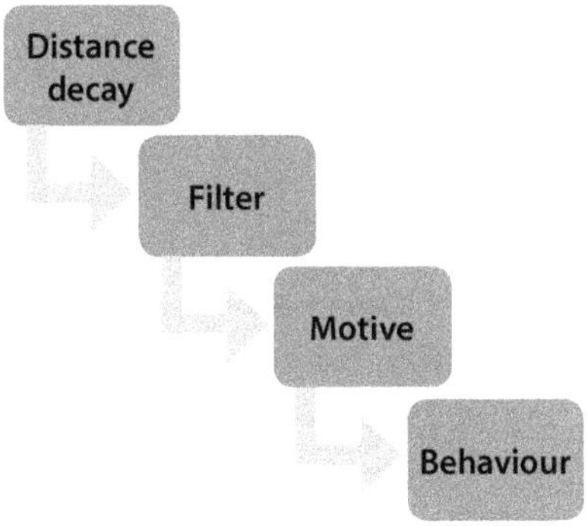

Figure 4.7: The implicit effect of distance on behaviour

Motives of different segments vary as well, and so if some segments are over or under represented by distance, then the average motive profile of the market will also vary by distance. Short haul travel is typical of the type of modal travel identified by Graburn (1983, 2018). It is the regular type of annual holiday people take where the primary motive is hedonism, escape, rest, relaxation and spending time with family. By contrast, Yeoman and Lederer (2005) write about the aspirational nature of much long haul travel and how it is viewed as a rare, often once-in-a-lifetime occurrence. In addition, Tideswell and Faulkner (1999) and Hwang and Fesenmaier (2003) observed that long haul tourists sought to have multiple trip purposes satisfied, while short haul tourists tended to take single destination, single purpose trips.

These differences in profile and motive translate into quite different behaviour patterns. Relatively speaking, long haul tourists also more likely to explore the cultural features of a destination, and in particular to visit historic sites, museums, outdoor markets and participate in culturally-oriented activities. By contrast tourists from nearby markets engage in activities that reflect the underlying motives of rest, relaxation and escape, including shopping, dining and visiting theme parks.

The other features of the filtering effect of distance influences the segments targeted by Destination Management Organizations. Table 4.2 identifies market segments targeted by the Hong Kong Tourism Board.

Table 4.2: Hong Kong's Target Markets

	China	Indonesia	Malaysia	Philippines	Singapore	Thailand	Japan	Korea	Australia	USA	Canada	UK	France	Germany
Young (16 – 30) - backpackers	X												X	X
Young professionals (20s to late 30s)		X	X	X	X	X	X	X	X					
Family (any age travelling with children)	X	X	X	X	X	X								
Mid-career (31 – 45)	X							X	X	X	X	X	X	X
Achiever (46+)		X			X				X	X	X	X		X

Source: HKTB Market Summaries, 2018.

Hong Kong is used here as an example, but most other destinations target similar segments, although the names vary somewhat. What is apparent is the change in target market by distance. While Destination Marketing Organizations target specific market segments, equally certain segments are predisposed to visit destinations depending on how far they are willing or able to travel. The challenge for the DMO is to identify those segments that are most likely to visit, even if they are not as large or attractive as others that are less likely to visit. It targets the younger and family segments among the short haul market because they are most likely to respond positively to calls to action. But, notice that it targets a much older segment among long haul tourists. It is not that long haul markets do not have attractive younger segments. Instead, these segments are far more likely to travel short haul. By contrast, achievers, affluent seniors and backpackers are able to travel long haul, and so they are targeted. So, while the young office lady segment from Asia might be interested in a short haul trip to Hong Kong, this destination holds little realistic appeal to the same segment that resides in Europe. Instead, they are likely to visit Paris, London or Rome.

Market access

Market access is a more difficult concept to grasp. It has received almost no attention from tourism academics since Pearce (1989) first introduced the idea, yet is of vital importance. It is related to distance decay, but instead of looking at absolute distance, it looks at the number of intervening opportunities one must pass before reaching a desired destination.

It is predicated on the assumption that tourists are rational consumers who, given the choice of similar opportunities, will choose the most proximate one. Essentially, market access argues that more proximate destinations have a competitive advantage over destinations offering similar products that are less proximate (McKercher, 1998), especially among those who value time as a resource. This supposition is based on the argument that destination choice is often influenced by convenience and that, given a choice between similar destinations, the tourist will tend to choose the more convenient one. Market access can be measured by the relative difference in the time, cost, distance, or effort required to access different destinations (Pearce, 1989).

There are three main differences between market access and distance decay. The first is that it is a relative term, where absolute distance is less important than relative distance between competing destinations and the number of destinations offering similar experiences that must be bypassed. As such, second, it is product or tourism type specific. The third issue is that distance decay has a consumer orientation by asking "how far do I want to travel?", while market access has a destination orientation by considering "how many destinations must the prospective consumer pass before reaching the desired one?"

An example of the relative nature of market access will help explain the concept better. Consider a person who lives in Sydney, Australia. That person has a choice of close to ten beaches in the greater Sydney area including world-famous beaches like Bondi and Manly. A beach located 50 km away would be seen to have very weak market access because the consumer would have to pass literally dozens of opportunities offering similar experiences before reaching this beach. Now, consider the same person who wants to go skiing. The nearest ski fields to Sydney are located in Kosciusko National Park, some 500 km away with the next nearest ski field located in the State of Victoria some 750 km from Sydney. Thus, Kosciusko ski fields enjoy strong market access because they represent the most proximate skiing opportunity. Relatively speaking, therefore, a beach some 50 km away may be competitively disadvantaged, while a ski field 500 km may have a strong competitive advantage due to the lack of intervening opportunities.

Main difference between market access and distance decay

Distance decay and market access represent opposite sides of the same situation, as shown in Figure 4.8. Distance decay examines aggregate outbound travel patterns while market access is activity specific. In effect, distance decay represents the sum of individual demand and supply curves as defined by market access. Market access is a relative term whereby a distant destination may have strong market access if few if any intervening opportunities exist. Yet in aggregate for all types of travel, cumulative demand will likely be low.

The main difference between market access and distance decay can be shown by considering the *x* axis of a graph. Distance represents the *x* axis in a distance decay curve, while intervening opportunities represent the *x* axis in a market access curve.

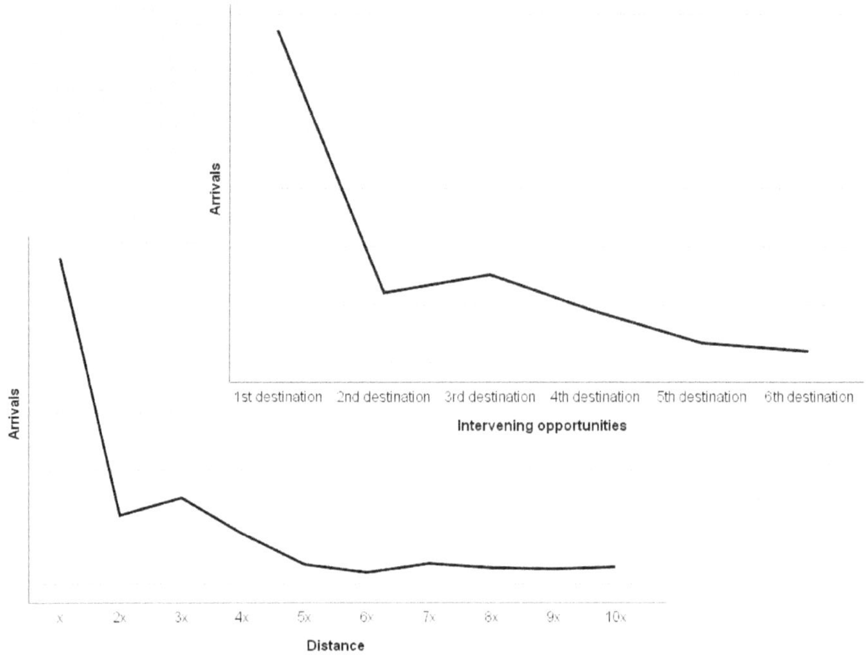

Figure 4.8: Difference between distance decay and market access

One of the few studies of this type found destinations with both strong and weak market access had their own competitive advantages, for the impact of market access is influenced by time availability and, to a lesser extent, travel party type (McKercher, 1998). Short break tourists with the least amount of time were far more likely to select destinations with strong market access. By contrast, those with larger time budgets were more likely to bypass nearby destinations and head further afield, where they could escape the crowds. Likewise, families, especially those with young children chose destinations with strong market access, while couples tended to travel to places with weaker market access.

Market access has a number of development implications. Destinations with stronger market access are more likely to experience weekly cycles of weekend crowding and midweek slumps but are less likely to be affected by huge seasonal variations in visitation. Those with weaker market access are more likely to be affected by seasonality, but because of the distance involved are less likely to be affected by the weekly cyclical variations during their high seasons.

While market access may be a powerful source of competitive advantage it is also highly vulnerable, for the development of a new tourism product or destination area close to the source market can fundamentally change market access from being strong to weak. If the competitive advantage is based solely on lack of intervening opportunities, the creation of an intervening opportunities can eliminate that one source of competitive advantage.

Can market access be modified?

While the impact of distance is more or less universal, the impact of market access is variable. Market access can be modified by a number of factors. These include:

- *Removing artificial or political barriers to entry*. For example, Assaf and Josiassen's (2012) global studies revealed visa restrictions have a negative impact on the flow of travelers, while Balli et al. (2013) and Lawson and Roychoudry (2013) suggest the elimination of visas can result in the tripling of arrivals.
- *Changing travel time.* The advent of low cost carriers has fundamentally transformed Spain's tourism sector, with Rebello and Baidal (2009) noting both quantitative and qualitative changes in tourism activity. The same situation can be found when high speed rail services are introduced (Wang et al., 2012, 2014). The volume of tourists has increased, and importantly, more changes have been seen in the rise of independent travel, changes in accommodation choices and changes in visitor profile.
- *Changing costs.* Changes in Spain's tourism sector have been affected by the low price charged for air travel. It is now cheaper to fly to Spain than it is to take a train to Paris, effectively enhancing Spain's competitive position.

- *Changing effort*. All one has to do is look at ski resorts. The development of ski lifts has transformed skiing from an activity that only the young and fit could participate in to a mass activity available to all, young and old. The lift overcomes the challenge of having to climb a mountain. Instead, one can let a machine do the work that you would otherwise be expected to do, and you can enjoy the effect of gravity to take you down the mountain.

Itinerary models

Modelling inter-destination itinerary movements was a popular topic in the late 1980s and 1990s (Flognfeldt, 1999; Lue et al., 1993; Mings & McHugh, 1992; Oppermann, 1995), but has once again seems to have fallen out of fashion. Mings and McHugh (1992) identified four types of touring routes taken by domestic American tourists. These models include a "direct route" itinerary taking the most direct path to and from the destination; the "partial orbit" which consisted of a long transit leg followed by a short circular tour; the "full orbit" itinerary which involved a circular tour and not taking the same route twice; and the "fly-drive" itinerary which is a modification of the partial orbit except the transit leg was taken by air. Lue, Crompton and Fessenmaier (1993) identified four types of itineraries: the "en-route" model, where tourists could make a number of stops on their way to a main destination; the "base camp" model where they based in one place and did a number of trips from there; the "regional tour", involving a number of stops on a tour and the "trip-changing" pattern that involved a multi-foci, touring trip, visiting a number of destinations without overlapping any leg of the trip. Oppermann (1995), focusing on international travel, identified seven possible itinerary types, including the "open jaw loop" model which applies to tourists who enter a country through one gateway and leave through another and the "multiple destination areas loop" which basically consists of all other destinations mashed together.

Indeed, more than two dozen different inter-destination itinerary styles were identified. Yet, on closer inspection, they fall into four main types (McKercher & Zoltan, 2014). They are shown in Figure 4.9 and are:

- Single destination, there-and-back trip that may or may not include side trips.
- Transit leg to the destination area, followed by a circle tour, stopping overnight at different places.
- Circle tour with or without multiple access and egress points.
- Hub and spoke itineraries where tourists base themselves in a destination area and take side trips to other destinations

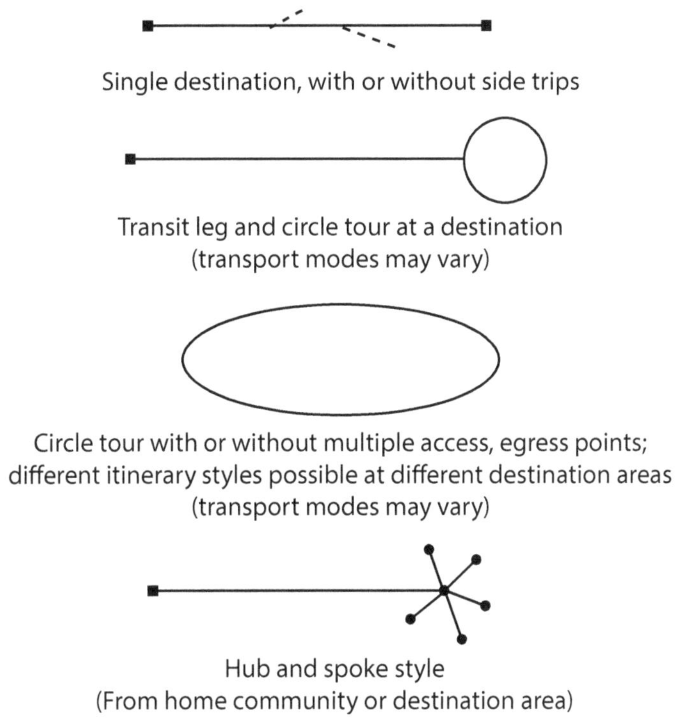

Figure 4.9: Inter-destination itinerary types, Source: McKercher & Zoltan, 2014

Conclusion

The impact of distance on tourism is often undervalued, yet it exerts a profound impact on all aspects of travel from defining the size of the market, to influencing motives, behaviour and defining tourists' experiences. This chapter has explored a range of issues related to tourism flows and how the valuation of time, distance decay and market access influences tourism.

■ Discussion points / Exercises

1. Discuss which groups of tourists are more likely to see travel time as a commodity and a resource, and evaluate what impact that may have on their travel patterns.
2. Given that distance decay is a universal construct and that distance acts as a filter for some segments, consider the Chinese and Indian outbound markets, analyse which destinations will see a large influx of tourists as these markets grow and which are likely to attract the high end more mature market.
3. Identify an ETEZ near your location and examine the effect it has on tourism.
4. Conduct a market access analysis of three special interest segments of your choice.

Points to consider for revision

- How much time people have and how they value travel time plays a key role in tourist movements. People who see travel time as a commodity, see value in the journey itself as much as being at the destination. Those who see it as a resource, see travel time as a cost that should be minimised
- Distance decay is a universal construct in geography and when tested empirically is found to apply to tourism. More than half of all international travel is to land neighbours, while once distance passes a threshold of 5,000 km or more, source markets generally contribute little to overall demand.
- The shape of the distance decay curve is influenced by the positioning of the Effective Tourism Exclusion Zone, a zone where little or no tourism exists. If located near the source market, it will likely push demand further than if no ETEZ existed. A moderate distance from the source market, will likely accentuate the peak and then create a further secondary peak some distance away.
- Distance acts as a filter, advantaging some segments and disadvantaging others. As a result of this filtering effect, differences are noted in the profile, motives and in-destination behaviour as distance changes.
- Market access is like distance decay, but it is a relative term that is consumer specific. It evaluates the number of intervening opportunities that must be passed before reaching the destination. It too shows a typical decaying curve. However, market access can be modified.

Suggested additional readings

McKercher, B. & Mak, D. (2019) The impact of distance on international tourism demand *Tourism Management Perspectives* 31, 340-347.

McKercher, B. & Tkaczynski, A. (2023) Valuation of travel time. *Annals of Tourism Research* https://doi.org/10.1016/j.annals.2023.103573

References

Assaf, A.G. & Josiassen, A. (2011). Identifying and ranking the determinants of tourism performance: a global investigation. *Journal of Travel Research*, 51(4), 388-399.

Balli, F., H. O., Balli & K. Cebeci (2013). Impacts of exported Turkish soap operas and visa-free policy on inbound tourism to Turkey. *Tourism Management*, 37, 186-192.

Becken, S., & Wilson, J. (2013) The impacts of weather on tourist travel. *Tourism Geographies.* 15(4), 620-639

Botti, L., Peypoch, N., & Solonandrasana, B. (2008). Time and tourism attraction. *Tourism Management,* 29(3), 594-596.

Bull, A. (1991) *The Economics of Travel and Tourism,* Pitman, Melbourne.

Butler, R. (2008) Modern tourism and its development in a post modern age. In Babu, S., Mishra, S., & Parida, B (Eds), *Tourism Development Revisited: Concepts, Issues and Paradigms.* Response: London, pp 54-63.

Chavas J. P., Stoll, J. & Sellar, C. (1989) On the commodity value of travel time in recreational activities *Applied Economics* 21, 711-722.

Croes, R., Ridderstaat, J., & Rivera, M. (2018) Asymmetric business cycle effects and tourism demand. *Journal of Travel Research.* 57(4), 419 – 436.

Cvelbar, L., Mayr, M., & Vavpotic, D. (2018) Geographical mapping of visitor flow in tourism: A user-generated content approach. *Tourism Economics* 24(6), 701-719.

Decrop A. & Snelders, D. (2005) A grounded typology of vacation decision-making. *Tourism Management* 26(2),121-132.

Dickinson, J. E., Lumsdon, L. M., & Robbins, D. (2011). Slow travel: issues for tourism and climate change. *Journal of Sustainable Tourism,* 19(3), 281-300. https://doi.org/10.1080/09669582.2010.524704

Eagles, P., Johnson, P., Potwarka, L. & Parent, C. (2015) Travel distance classes for tourism destinations: a proposal from Ontario Provincial Park camping, *Journal of Ecotourism,* 14(1), 64-84, DOI: 10.1080/14724049.2015.1071829

Eldridge D. & Jones J. P. (1991) Warped space: a geography of distance decay. *Professional Geographer* 43(4), 500-511.

Flognfeldt, T. (1999) Traveler geographic origin and market segmentation: The multi trips destination case. *Journal of Travel and Tourism Marketing,* 8(1), 111-124.

Fotheringham A. S. (1981) Spatial structure and distance decay parameters. *Annals of the American Association of Geographers* 71(3), 425-436.

Gössling, S., Scott, D & Hall. M. (2018) Global trends in length of stay: implications for destination management and climate change, *Journal of Sustainable Tourism,* 26(12), 2087-2101, DOI: 10.1080/09669582.2018.1529771

Graburn, N. (1983) The anthropology of tourism. *Annals of Tourism Research* 10, 9-33.

Graburn, N. (2018) Secular pilgrimage: A general theory of tourism. In Gmelch, S. & Kaul (eds) *Tourists and Tourism: A general reader 3rd Ed.* Long Grove: Waveland, pp. 17-28.

Greer, T. & Wall, G. (1979) Recreational hinterlands: A theoretical and empirical analysis in G. Wall (ed.) *Recreational Land Use on Southern Ontario,* Waterloo University, Waterloo, Canada, pp 227-246.

Hall, C. (2005) Time, space, tourism and social physics, *Tourism Recreation Research,* 30(1), 93-98, DOI: 10.1080/02508281.2005.11081236

Hanink D. M., & White, K. (1999) Distance effects in the demand for wildland recreational services: the case of national parks in the United States. *Environmental and Planning Annals* 31, 477-492.

Hardy, A. (2020). *Tracking Tourists: Movement and mobility*. Oxford: Goodfellow.

Hook, H., De Vos, J., Van Acker, V., & Wilcox, F. (2021). On undirected trips, satisfaction, and well-being: Evidence from Flanders (Belgium). *Transportation Research Part D, 99*, 103108. https://doi.org/10.1016/j.trd.2021.103018

Hooper, J (2015) A destination too far? Modelling destination accessibility and distance decay in tourism. *GeoJournal* 80, 33-46.

Hwang Y. & D. Fesenmaier (2003) Multidestination pleasure travel patterns: empirical evidence from the American Travel Survey, *Journal of Travel Research* 32, 166-171.

Lawson, R., & Roychoudhury, S. (2013) Do travel visa requirements impede tourist travel? *Journal of Economics and Finance*. 40(4), 817-828.

Lee, H. A., Deniczi-Guillet, B. D., Law, R., & Leung, R. (2011). Robustness of distance decay for international pleasure travelers: a longitudinal approach. *International Journal of Tourism Research*. 14(5), 409-420.

Leiper, N, (1990) *Tourism Systems: An Interdisciplinary perspective*. Department of Management Systems. Palmerston North: Massey University

Lew A. & B. McKercher (2006) Modeling tourist movement: a local destination analysis *Annals of Tourism Research* 33(2), 403-423.

Lorde, T., Li, G., & Airey, D. (2016) Modeling Caribbean tourism demand: an augmented gravity approach. *Journal of Travel Research* 55(7), 946-956.

Lue C., Crompton, J., & Fessenmaier, D. (1993) Conceptualization of multidestination pleasure trips. *Annals of Tourism Research*, 20(2), 289-301

McKean J., Johnson, D. & Walsh, R. (1995) Valuing time in travel cost demand analysis: An empirical investigation *Land Economics* 71(1), 96-105.

McKercher, B. (1998) The effect of market access on destination choice, *Journal of Travel Research* 37, 39-47

McKercher, B. (2008) Segment transformation in urban tourism. *Tourism Management*. 29, 1215-1225.

McKercher, B. (2021) Who kept travelling and where did they go? Domestic tourism in Australia. *Tourism Geographies*. https://doi.org/10.1080/14616688.2021.2016937.

McKercher, B. & Lew, A. (2003) Distance decay and the impact of effective tourism exclusion zones on in international travel flows *Journal of Travel Research* 42(2), 159-165.

McKercher, B. & Mak, D. (2019) The impact of distance on international tourism demand *Tourism Management Perspectives* 31, 340-347.

McKercher, B. & Tkaczynski, A. (2023) Valuation of travel time. *Annals of Tourism Research* https://doi.org/10.1016/j.annals.2023.103573

McKercher, B. & Zoltan, J. (2014) Tourist flows and spatial behaviour, in Hall, C.M., Williams, A. & Lew, A. (eds) *A Companion to Tourism*, Blackwell, pp 34-44.

McKercher, B. Chan, A. & Lam, C. (2008) The impact of distance on international tourist movements, *Journal of Travel Research* 47(2), 208-224.

Mings, R. C., & McHugh, K (1992) The spatial configuration of travel to Yellowstone National Park. *Journal of Travel Research,* 30, 38-46.

Mokhtarian, P. L. & Salomon, I. (2001). How derived is the demand for travel? Some conceptual and measurement considerations. *Transportation Research Part A,* 35(8), 695-719. https://doi.org/10.1016/S0965-8564(00)00013-6

Nyaupane G., & Graefe, A. (2008) Travel distance: a tool for nature-based tourism market segmentation, *Journal of Travel & Tourism Marketing,* 25(3-4), 355-366, DOI: 10.1080/10548400802508457

Oh, H., Assaf, A. G., & Baloglu, S. (2016). Motivations and goals of slow tourism. *Journal of Travel Research,* 55(2), 205-219.

Oppermann, M. (1995) A model of travel itineraries. *Journal of Travel Research,* 33(4), 57-61.

Pearce, D (1998) Tourism time budgets. *Annals of Tourism Research* 15, 106-121.

Pearce, D. (1989) *Tourist Development 2nd Ed.* Harlow: Longman.

Plog, S. (1974) Why destination areas rise and fall in popularity, *Cornell HRA Quarterly* 14(4), 55-58.

Rebello, J., & Baidal, J. (2009). Spread of low-cost carriers: tourism and regional policy effects in Spain. *Regional Studies,* 43(4), 559-570.

Sirakaya, E. & Woodside, A. (2005), Building and testing theories of decision making by travelers. *Tourism Management,* 26(6), 815-832.

Sirirak, W., & Pitakaso, R. (2018) Marketplace location decision making and tourism route planning. *Administrative Sciences* 8(4), 72.

Tideswell, C. & Faulkner, B. (1999) Multidestination travel patterns of international visitors to Queensland *Journal of Travel Research* 37, 364-374.

Tobler, W. R. (1970) A computer movie simulating urban growth in the Detroit region. *Economic Geography,* 46, 234-240.

Truong, T. & Henscher D (1985) Measurement of travel time values and opportunity cost model from a discrete-choice model, *The Economic Journal,* 95 (June), 438-451.

Walsh, R., Sanders, L. & McKean, J. (1990) The consumptive value of travel time *Journal of Travel Research,* Summer 17-24.

Wang, Z., Huang, S., Zou, T., & Yan, H. (2012) Effects of the high speed rail network on China's regional tourism development. *Tourism Management Perspectives* 1, 34-38

Wang, D., Qian, J., Chen, T., Zhao, M., & Zhang, Y (2014) Influence of the high-speed rail on the spatial pattern of regional tourism – Taken Beijing–Shanghai High-Speed Rail of China as Example, *Asia Pacific Journal of Tourism Research,* 19(8): 890-912, DOI: 10.1080/10941665.2013.818049

Waters, N. (2017) Tobler's First Law of Geography. In Richardson, D., Castree, N., Goodchild, M., Kobayashi, A., Liu, W., & Martson, R. (Eds) *The International Encyclopedia of Geography*. Chichester: John Wiley and Sons.

4

Wong, I., Fong, L., & Law, R (2016) A longitudinal multilevel model of outbound travel behavior and the dual cycle model. *Journal of Travel Research*. 55(7): 957 – 970.

Xu, Y., Li, J., Park, S. & Li, Q. (2021). Tourism geography through the lens of time use- a computational framework using fine-grained mobile phone. *Annals of the American Association of Geographers, 111*(5), 1424-1440. https://doi.org/10.1080/24694452.2020.1812372.

Yeoman, I. & Lederer, P. (2005) Scottish tourism: Scenarios and vision, *Journal of Vacation Marketing* 11(1), 71-87

5 The Political Economy of Tourism and International Tourism

By the end of this chapter, the reader will be able to:

- Identify and describe the various approaches to the political economy of tourism
- Describe factors that influence bilateral tourism flows
- Define tourism as a form of international trade in services
- Explain the politics of visa requirements.

Introduction

International tourism is influenced by both small and big 'P' politics. Political factors influence who can visit a country and who is not welcome. Tourism is not devoid of political interference, nor is the political economic situation under which it evolves value free. The political economy of tourism influences the path governments take to develop the sector, including the level of government intervention as owner, operator and policy decision-maker. Artal-Tur et al. (2015) note the impact of diplomatic relations on tourism flows is immense. Political relationships influence international air service agreements and through the General Agreement of Trade in Services, controlled by the World Trade Organization (WTO), influence the internationalization of tourism .

Political economy of tourism

Tourism scholars have generally paid little attention to issues relating to the political economy of tourism. Understanding how we arrived at our system of economic production is important for it provides insights into the stages of economic evolution.

■ Capitalism

Fletcher, Fyall, Gilbert and (2018, p. 615) define capitalism as "*a social system based on individual rights where goods and services are produced and exchanged with minimal government interference*". This simple definition encapsulates three important elements: a social system, exchange, and minimal government interference. Obviously, this definition is not universal, for the understanding of the term 'capitalism' is contested and continues to evolve. Views on the role that government should play in marketplace operations constitute the key points of difference in the standpoints adopted by various schools of economists. Li, Liu and Song (2019) identify three standpoints, the first being those who believe that the market is able to operate efficiently with little or no government intervention. This viewpoint is typified by the laissez-faire approach advocated by Adam Smith in the 18th Century and more recently by the neoclassical school that rose to prominence in the 1980s. The second group is broadly identified as the Keynesian school that believe that government intervention is essential for stabilising economic activity. The third and most recent standpoint is characterised by new Keynesian economics which models macroeconomic operations with micro foundations. In their analysis, Li et al. (2019) ignored the neo-Marxist school which takes a contrasting view of economic organisation based on the role of labour and capital.

Leaving aside issues related to the structure of 'social systems' outlined in Fletcher et al.'s (2018) definition of capitalism, many economies operate with a mix of State Owned Enterprises (SOE) and Privately Owned Enterprises (POE) and with varying levels of government intervention in the economy. The Chinese version of capitalism for example features both SOEs and POEs, both of which are subject to high levels of state oversight. Other countries such as the USA and the UK practice a version of capitalism that lies much closer to the principles of classical laissez-faire capitalism where labour markets are largely deregulated, and the state has largely withdrawn from operating state-owned enterprises. From a tourism perspective, an understanding of the form of capitalism adopted by specific countries is important. In China for example, POEs generally have close connections with state agencies and actors and are sometimes assigned roles in carrying out government policy (Milhaupt & Zheng, 2015). This is not the normal practice in western economies.

Laissez-faire economics

Adam Smith, an 18th Century scholar often referred to as the father of economics, postulated that competition and rational self-interest can lead to prosperity. His theories advocated minimal intervention by governments in the market in the belief that removal of government controls would lead to ever-increasing prosperity. His theories underpinned the emergence of the capitalist economic system in the late 18th Century and economic liberalism, described as an economic and political philosophy that advocated support for private property and the market economy. The contributions by Smith and other scholars, including David Ricardo, John Stuart Mills and Thomas Malthus, led to the emergence of classical economics where the so-called natural law of production and exchange lead to a largely self-regulating economic system, such that government intervention is not required. By extension, classical economics is the antithesis of non-capitalist economic systems of production, such as planned economies and socialism.

The problem with this form of economic organisation is that crisis events (wars, major natural disaster, political upheaval, monopolies and pandemics) can severely disrupt self-regulating economic systems forcing government intervention. The COVID-19 pandemic provides an example of how a self-regulating economy is unable to deal effectively with crisis events where the solutions involved public health considerations rather than economic considerations (Prideaux et al., 2020). During the COVID-19 pandemic many governments including Japan, USA, UK and Australia funded large stimulus programs to prop up national economies. In Japan, the size of fiscal measures introduced to combat the COVID-19 was equal to 21.1% of national income while in the US the size of fiscal measures were equal to 14.7% of its national income (Emmerson & Stockton, 2020). Interventions of this type have parallels with the economic policies introduced by governments during the latter years of the Great Depression (1929 to late 1930s) and the 2007/08 Global Financial Crisis (GFC).

Keynesian economics

Keynesian economics is based on the view that aggregate demand is unstable and volatile in market economies and to avoid economic downturns governments need to intervene in the market with both monetary and fiscal policy to stabilise it. In general, Keynes favoured a managed market economy based on government intervention when required to smooth out the business cycle. This approach contrasts to earlier aggregate supply side focus that was a characteristic of classical economics. During the period from the latter part of the Great Depression to the mid-1970s Keynesian economics was viewed as the standard approach to economic policy development. During this period, states managed economic downturns through macroeconomic interventions and welfare policies (Williams, 2004).

One of the enduring legacies of Keynesian economics was the formalisation of the concept of the multiplier effect (Rusu, 2011). Within a tourism context the multiplier effect refers to the direct, indirect and induced effects of tourism spending and is often used to measure of success in terms of job creation and income generation. Direct and secondary (indirect and induced) spending in other areas of the economy that are generated by tourism expenditure include income, investment, exports and government spending (Archer, 1989).

Another important concept introduced by Keynes (1935) was the idea of uncertainty. In periods of uncertainty, consumers may postpone or cancel basic spending resulting in a fall in national economic activity (Gozgor & Ongan, 2017) and a change in consumption patterns. Tourism is one activity that is affected by economic uncertainly (Demir & Gozgor, 2018). In fact, tourism is often a leading indicator of economic uncertainty, for people will postpone travel or alter their travel plans (shorter distance and less time) in the lead up to recessions and then start to travel again prior to the official end of such periods of constraint.

Keynesian and later post-Keynesian economic approaches were replaced by neoliberalism economics from the early 1980s. This development has been attributed to the failure of Keynesian macroeconomics to manage stagflation during the 1970s, the collapse of socialist planning systems in the 1980s and the Latin American debt crisis of the 1970s (Lin, 2012). One of the major responses to these perceived failures of Keynesian economics was the growth of globalism and the subsequent weakening of controls able to be exercised by states over their domestic economies. More recently, the failures of neoliberalism typified by the need for significant government interventions in the economy during the GFC have led to renewed interest in Keynesian economic and the use of public spending to stimulate growth and employment creation (Bramwell & Lane, 2009).

Neoliberalism

Neoliberalism is the most recent incarnation of capitalism and occupies a dominant position in the global political economy. However, its meaning is contested, widely debated and subject to constant evolution as the circumstances of national economies and the global economy change. For example, in a discussion on the evolution of the term over a number of decades Boas and Gans-Morse (2009: 137) observed that *"neoliberalism has undergone a striking transformation, from a positive label coined by the German Freiberg School to denote a moderate renovation of classic liberalism, to a normatively negative term associated with radical economic reforms in Pinochet's Chile."*

Peck, Brenner, and Theodore (2018: 6) describe neoliberalism as *"an idealised vision of market rule and liberal freedoms, combining a utilitarian conception of market rationality and competitive individualism with deep antipathies to social redistribution and solidarity."* From a similar perspective, Wearing, McDonald, Taylor and Ronen

(2019) describe neoliberalism as a process that aims to privilege free-market forces by reducing the role of government decision making in all areas of life.

Neoliberalism is strongly supported by major international organisations including the World Trade Organization, the International Monetary Fund (IMF) and the World Bank (Wearing et al. 2019). Contemporary neoliberalism is associated with the views of influential economist Milton Friedman of the Chicago School of Economics. Friedman, a critic of government intervention, placed considerable emphasis on the efficiency of the market versus the distortions that he argued are commonly associated with government intervention and market regulation. Liberalism in this context has a direct linage to laissez-faire economics which emphasises the role of the market (Palley, 2004).

In its current form, neoliberalism is broadly characterised by laissez-faire style economic liberalism and free market capitalism and its defining features include deregulation, privatisation, globalisation, free trade, austerity and an increasing role for the private sector in the economy and society. As Jones, Parker and Ten Bos (2005) note, Neoliberalism as a set of economic ideals rose to prominence from the mid to late 1970s and is associated with economic policies introduced President Ronald Regan in the United States and Prime Minister Margaret Thatcher in the UK. Jones et al. (2005, p. 100) describe neoliberalism as an *"updated version of older ideas about 'liberal economics' which has long argued that markets should be free from intervention by the state. In its simplest version, it reads: markets good, government bad"*.

Neoliberal economic policies were enshrined in the so called 'Washington Consensus' described as a set of ten policy settings advocated by the US Treasury, the World Bank and the IMF that were prescribed for countries facing economic crises. As Wearing et al. (2019) observed, the Washington Consensus represents a common consensus between North American and European politicians, multinational corporations, corporate executives, regulatory agencies and media owners about the promotion of free-market ideologies as a superior mechanism for managing economies and equitable allocation of resources. One of the contradictions of neoliberalism is the call by developed nations in Europe and the US for developing countries to adopt neoliberal policies based on opening their domestic markets to international trade while continuing to retain high levels of protection (subsidies) on sectors of their own economy, particularly agriculture (Jones et al., 2005).

In their critique of global tourism that is based on neoliberal practices, Wearing et al. (2019) describe a situation where global tourism has emerged as a consumer culture that has led to unsustainable forms of tourism production and consumption which has begun to have adverse effects on local communities (over-tourism) and the ecosystem. The loss of control of local resources to multinational companies such as Airbnb is a symptom of the impacts of unregulated neoliberal tourism production.

In a critique of neoliberalism Higgins-Desbiolles (2006: 1195) makes the point that "*the mantra of tourism is an 'industry' that is subject only to the rules of the marketplace has been repeated so frequently that to think otherwise is almost viewed as nonsensical.*" In a later critique, Higgins-Desbiolles and Whyte (2015: 106) highlight the pressures that may be placed on communities to enter the market on terms that may be unfair and make them choose between "*ecological integrity, social and community wellbeing*" and economic growth, a consequence of neoliberalism's focus on "*unfettered growth*".

A neo-Marxist perspective

Bianchi (2011: 17) stated that Marxist political economy has "*made a profound contribution*" to how we understand and critique ideas about the capitalist political economy. Given the previous critiques about the dominant role of neoliberalism it is useful to consider the contribution that neo-Marxist scholars can make to our understanding of the operation of global tourism. Observing that there were significant contradictions in laissez-faire economics that led to the exploitation of the working class, Marx postulated that under capitalism, profit was the value added by workers but not paid out in wages. The expropriation of profits by the owners of capital enabled them to gain control over the means of production and exploit workers. The end result of appropriating the surplus value of labour, where workers who are paid less than the full value of their labour, will lead to overproduction, over-accumulation and the ultimate self-destruction of the system. His ideas and those of his colleague Engels were incorporated into Marxist-Leninist philosophies adopted in part by the former USSR. While the experiment failed, many of his insights into the contradictions of capitalism have been adopted by neo-Marxist scholars and continue to provide a useful critique of modern capitalism including neoliberalism.

From a neo-Marxist perspective, the rise of the gig economy where work is capsulized is a classic case of capital exploiting the worker. Removal of workers' rights to benefits such as fixed hours of work, sick leave and annual holidays enables firms to maximise profits without the obligation of returning some of the extra profit gained from productivity increases to workers. In the tourism industry, the trend to casualisation of the workforce is a trend that has been gaining pace for many years.

Bianchi (2011: 23) observed that "*neoliberalism and globalisation have stimulated a resurgence of the power of capital (or 'shareholder value') and thereby accentuated the tendencies towards concentration and centralisation of capital emphasised by Marx (1974: Chapter 25, section 2)*". One example of the concentration of capital cited by Bianchi is the cruise industry where two corporations handle nearly 80% of the Caribbean's passenger capacity.

Neo-Marxist thinkers have described tourism as a form of capitalism (Bianchi, 2011; Fletcher, 2011) where tourism growth provides a mechanism to alleviate the problems of overproduction predicted by Marx. From this perspective, excess capital is able to be reabsorbed into the economy as investment in tourism activity thereby providing a 'fix' to forestall a crisis of overproduction (Harvey, 1989; Fletcher, 2011). Fletcher (2011), building on the views of O'Connor (1988) that capitalism's 'second contradiction' is one of overproduction, argues that sustainable tourism, including ecotourism, provides a means to negate the crisis of capitalism's need to continually expand to survive. By providing a mechanism where finite natural resources can be used in a 'sustainable manner', the danger of natural resource exhaustion through over-extraction can be hopefully averted. From a neo-Marxist perspective, the recent emergence of concerns about over-tourism can be interpreted as an example of overproduction that leads to a crisis, temporarily forestalled by the COVID-19 Pandemic.

Politics of international tourism

International tourism does not simply occur. Instead, a number of conditions must exist. First, one country must recognize the existence of the other. Travel by Israelis to some parts of the Islamic world is impossible because Israel's existence has not been recognized. Second, there must be diplomatic relations between countries. Third, commerce or trade agreements must be signed. While tourism may be a recreational activity pursued by individuals, at a national level, it is a highly sought after trade commodity. Then acceptable entry arrangements must be agreed upon that outline how many people can visit, under what conditions and how long they can stay. These conditions include such features as the need for passports or visas as entry requirements and other entry conditions. They vary from country to country and are negotiated by each country as part of their diplomatic relations. Finally, a range of other barriers must be removed. As Prideaux (2005: 781) writes

> *"The quantum of the flow between country pairs is a function of a matrix of interrelated factors that includes the public and private sector structures supporting flows, diplomatic relations, and economic and noneconomic factors. Countries that have removed barriers to citizens undertaking foreign tourism have limited capability to regulate outbound flows, but still retain significant capacity to increase inbound flows through measures designed to enhance destination competitiveness and encourage citizens to substitute domestic for international tourism. Policies designed to alter the level of inbound and outbound flows through non-regulatory measures may become important where countries facing a balance of payments crisis decide to adopt policies designed to increase exports and reduce imports of tourism."*

It is for this reason that Webster and Ivanov (2016) contend that one of the major barriers to allow people to travel is not the cost but political regulation of borders. Karaman (2016) adds that tourism flows are affected by political, economic and cultural factors whereby security conditions and international relations have a direct influence on mobility.

■ Politics of international trade in services

Most international trade today is governed by rules developed by the World Trade Organization (WTO). Its mandate is to provide the legal ground rules for international commerce that bind governments to keep their trade policies within agreed limits. The international trading system has an enormous effect on tourism also with agreements the procurement of goods and services directly and indirectly (UNWTO/WTO/WTTC, 2018). Direct procurement relates to the provision of hotel and tour operations, while indirect procurement relates to the goods purchased to enable the delivery of services, such as the import of vehicles, staff and the like. It is for this reason that Dwyer (2015) writes about how international bodies such as the WTO and the International Monetary Fund have facilitated globalization by reducing trade barriers.

Currently, about 60 tourism-related agreements and decisions have been developed. While each agreement is complex, they all share the same common, fundamental principles to varying degrees of: removing discrimination between trading partners (most-favoured nation status); treating locally produced and foreign goods equally after they have entered a market (national treatment); gradually working toward freer trade by removing trade barriers; ensuring predictability, stability and transparency through legally binding commitments; promoting fair competition; and aiding economic development (WTO, 2019). WTO agreements cover three broad areas of trade: goods, governed under the General Agreement on Tariffs and Trade (GATT); services governed under the General Agreement on the Trade in Services (GATS); and intellectual property rights, governed by the Trade-Related Aspects on International Property Rights (TRIPS). Each of these agreements states broad principles, while GATT and GATS also include extra agreements and annexes, lists of binding agreements and mechanisms to ensure transparency and resolve disputes

Most aspects of tourism are addressed within the terms and conditions stipulated by the General Agreement on Trade in Services, with one notable exception – services affecting air traffic rights as well as those directly related to the exercise of such rights. At present, more than 133 member countries or territories have made tourism commitments, more than any other type agreement (WTO, 2021). These commitments relate to four aspects of tourism – hotels and restaurants, travel agencies and tour operators, tourist guide services and 'other' (Hoad, 2003).

Tourism has four ways of being traded under the GATS (ITC/UNWTO 2015), even if tourists crossing borders are the most common form of international trade in tourism services. Each of these four modes of supply affects tourism and travel-related services:

- Mode 1 – *Cross-border supply*: travel agents, hotels, and travel guides can be accessed electronically from importing countries.
- Mode 2 – *Consumption abroad*: the mode of supply most typically associated with tourism.
- Mode 3 – *Commercial presence*: chains of hotels and tour operators frequently set up businesses in multiple locations.
- Mode 4 – *Presence of natural persons*: tour operators and hotels often employ foreign staff with skill sets that are not readily available in the country of operation.

Politics of International Air Transport (Bilateral Air Service Agreements)

As mentioned above, air transport is excluded from WTO trade agreements. Yet, if you do not understand how international aviation works, then you did not understand how international tourism works. While it is easy to purchase an airline ticket, the actual ability of airlines to fly between places involves a whole series of bilateral trade agreements signed between the government where the airline is based and every country or territory the airline flies over, lands, drops off or picks up passengers.

First and foremost, it must be appreciated that aviation is a form of international trade. Governments own the airspace above their country or territory and therefore can negotiate permission to fly over or through that airspace (Fayed & Westlake, 2002). Because of this fact, they can control which countries they permit to fly through that space, which airlines can fly, the capacity, costs, and to some extent even prices charged to consumers. Air rights are also highly valuable commodities that can be tied into other trade goods.

The rules for international air travel were set at the 1944 Chicago Convention on Civil Aviation, when 53 governments met to establish an orderly arrangement for post war air travel (Lissitzin, 1964). The convention established the framework for the development of bilateral agreements between countries to ensure the carriage of passengers, mail and freight would normally be restricted to the carriers of these countries. The 1944 Chicago Convention set the legal tone for international aviation to follow. The preamble states that the signing countries have agreed to certain principles and arrangements that will promote the development of international air transport on the basis of equality and opportunity. Article 44 largely eliminated open competition and, instead ensured that government owned carriers could protect their rights, restrict competition and operate profitably (DTRD, 1998).

There were two main outcomes of the Chicago Convention. The first was the assurance that international aviation would be strictly controlled as part of bilateral trade between countries, while the second was the creation of the so-called Freedoms of the Air. Remember a few critical factors when the Chicago Convention was signed. Most international air carriers were owned and operated by governments, not the private sector. Air travel was the domain of a very few modernised countries in the late 1940s. Canada, Australia, the US, the UK and some western European countries dominated international travel. Most other countries were not modernized enough to develop their own international carriers and were therefore dependent on other counties for international air travel. Moreover, aircraft technology limited the distance planes could fly, and so a series of agreements was needed to enable aircraft to land and refuel in order to complete long haul flights.

■ Air Service Agreements

The first international aviation agreement was the Bermuda Agreement, signed in 1946 between the UK and the USA. It set the framework for all air service agreements up to and including the present. Some 3,000 bilateral air service agreements are now in force. They are, or should be lodged with ICAO, the International Civil Aviation Organization, based in Montreal, Canada. In recent years multilateral agreements that involve more than two states have also become popular. Typically, the traditional bilateral aviation agreements included the following elements (Australian Government, 2019; OECD, 2014):

- Routes to be flown, including gateways, which identify which cities are to be flown to and where stopovers are to be included. In this manner, governments can stimulate the expansion or creation of new gateway centres, promote one destination over another, or otherwise attempt to influence the flow of tourists to an area. Often the routes are identified due to market forces. In other instances, governments can lead.
- The capacity to be operated, either in terms of number of flights or number of passengers. Airlines are not obligated to fly that number of flights, instead it represents a maximum number that they can operate. Greater capacity than current demand is usually negotiated, as a means of encouraging growth of certain routes. This capacity is the key factor for the airlines, for it will determine the potential profitability of certain air routes as well as the frequency of flights. The other key factor to consider about capacity is that it includes ramp space or slots at airports, potentially enabling one country and its designated airline to dominate an airport and exclude competition.

- The designation of the air carrier(s). Before an airline can take up rights negotiated under a bilateral agreement, it must first be designated by its government as an international air carrier to the other government concerned. The other government can reject that designation if it is not satisfied that the carrier is substantially owned and effectively controlled by citizens of the designating country.
- How the capacity will be divided between countries. Normally it is split evenly, however, a number of other factors may affect the final division of capacity. Some of these factors relate to the ability of one country to deliver the capacity, while another may be related to trade-offs being made for preferential services elsewhere.
- Tariffs or the prices airlines can charge.
- Other issues including non-scheduled or charter operations, and cabotage. Cabotage is a term that allows foreign carriers to fly domestic routes.

These types of agreements are restrictive. They may work in favour of the country but work against the best interests of the airlines and the flying public for they restrict the market from playing any significant role in aviation. In recent years, pressure has been placed on countries to liberalise their air service agreements or open up their skies to more market forces. Liberalisation does not mean the elimination of air service agreements, for they are still negotiated country to country (EU, 2013). What it does, though, is reduce the artificial constraints that have inhibited the development of the aviation sector for open skies agreements seek to provide broad opportunities for the free market to establish prices, routes and capacity as well as to remove restrictions on the number of carriers that can provide service (OECD, 2014). But, the degree of liberalization is still influenced by the existence of bilateral or multilateral agreements.

Freedoms of the Air are negotiated as part of bilateral trade agreements. These Freedoms include:

- 1st Freedom – the right to over fly
- 2nd Freedom - the right to make a technical stop, for example to refuel
- 3rd Freedom – the right to drop off passengers, for example, for an airline from the the UK to fly to the US and drop off passengers
- 4th Freedom – the right to pick up passengers, for example, an airline from the UK to pick up passengers in the US and and fly them to the UK
- 5th freedom – the right to carry traffic to or from third states, for example, the right of a UK airline to fly to the US, drop off and pick up new passengers and then fly from the US to Mexico.

Politics of visas

Nowhere in international law does the right to enter foreign spaces exist (Neumayer, 2010). Instead, it is a privilege granted by both the generating and receiving country. Either or both can grant easier access to facilitate movements or can create barriers to restrict movements (Liu & McKercher, 2015). The Universal Declaration of Human Rights, in principle, gives individuals the right to exit and enter their own country but not a right to enter foreign spaces. Indeed, it is estimated that approximately two-thirds of the world's population requires a visa prior to departure, 15% can acquire a visa on arrival in country and only 18% are able to enter destinations on a visa free basis (Bangwayo-Skeete & Skeete, 2017).

The issues of entry into different countries and terms of stay have received remarkably little attention within the tourism literature. Yet, they are as fundamental to the understanding of international tourism as bilateral air service agreements for they can encourage or discourage the movement of people. As much as tourism is portrayed as a democratic activity accessible to most, in reality international tourism is highly politicized where certain types of 'undesirables' are actively discouraged from entering a country while 'desirables' are encouraged. Mau et al. (2015) comment that border selectivity is a central reaction to the opportunities and risks stemming from globalization. They continue to note that visa policies are seen as a central instrument in mobility restriction as a form of remote control or preemptive action, by differentiating between 'worthy' or 'trusted' travellers and those considered 'not trusted' or 'risky'. In essence whether governments impose or remove restrictions on cross-border mobility is a reflection of the friendliness between countries and the presence or absence of reciprocal trade agreements (Whyte, 2008). Webster and Ivanov (2016) indicate states generally have policies that favour open borders unless there are states that tend to face some type of real or perceived external threat.

People can enter a country in any number of ways. They may migrate on a permanent basis, come in through various professional and investor migration schemes, enter on student visas and perhaps be allowed to stay afterwards, come in as foreign workers, be identified as quality migrants, come in as part of a family reunion scheme, arrive as digital nomads and many other ways, including visiting as a tourist. The easiest way to gain access is as a tourist, making it also the easiest to abuse. Indeed, the vast majority of illegal immigrants enter a country legally as tourists and then overstay (Neumayer, 2006). It is because of this fact that many countries impose visa requirements as a means of creating a hurdle for visitors, as well as providing a means of tracking them once in the country (Whyte, 2008).

Countries face a difficult balancing act of, on the one hand, encouraging the free flow of people for political, social, economic and other beneficial reasons, while on the other hand, deterring people in order to maintain immigration control and uphold security (Neumayer, 2006). Visa policies fulfil this dual role for they act as a powerful screening mechanism that manages demand (Beenstock et al., 2015). But as Neumayer (2010) notes visas are a crude mechanism. Entry can be denied for a number of reasons, as shown in Table 5.1.

Table 5.1: Reasons why entry to a country can be denied

- Does not meet the entry requirements
- Has committed fraud, deception or misrepresentation in the current or a previous application
- Has a criminal record, has been arrested
- Is a threat to national security
- Does not have a good moral character
- Has previous visa/immigration violations
- Cannot prove to have strong ties to their current country of nationality and/or residence
- Intends to reside or work permanently in the country
- Fails to provide sufficient evidence/documents to prove eligibility for the visa sought after
- Does not have a legitimate reason for the journey
- Has no visible means of sustenance
- Does not have health/travel insurance valid for the destination and the duration of stay
- Is a citizen of a country to which the destination country is hostile
- Has previously visited, or intends to visit, a country to which the destination country is hostile
- Has a communicable disease, such as COVID-19
- Has a passport that expires soon

Four types of visa regimes exist (Karaman, 2016):

- Visa exemption programs (for those seen as being from friendly countries or who pose no threat);
- E-visa applied prior to the trip;
- Visa on arrival, and;
- The formal visa application process that requires attendance or submission of documentation to relevant consulates, as a form of pre-emptive mobility governance (Mau et al., 2015).

Visas typically state the terms of entry (i.e. no work permitted if a tourist), maximum duration of stay and other conditions. To receive one, applicants must often provide proof of a return ticket, evidence of sufficient funds to maintain themselves during the visit, evidence of immunization against certain diseases if needed, confirmation of hotel booking, employer's letter and in some cases, attend an interview in person (Prideaux, 2005). In general, nations with similar economic and political conditions, nearness and membership of international groupings share similar travel documentation requirements (Lee et al., 2010).

The granting of visa waiver conditions for residents of a country signifies that the country is friendly and that its residents are unlikely to pose a security threat, overstay or commit crimes in the country. The waving of visa restrictions further signifies a wish to encourage international trade, foreign investment, enhance political ties or for other reasons. Countries in the developing world may also have an incentive to waive or modify visa requirements in the hope of encouraging foreign investment, while major tourist destinations have an incentive not to impose visa restrictions in order to remain competitive (Mau et al., 2015; Neumayer, 2010).

The system is incredibly unequal, though, and is becoming more unequal over time (Chung et al., 2020). Citizens of high income, politically stable nations enjoy far more liberal access than residents of less developed countries, leading Recchi et al (2021) to conclude the richer a country is, the less its citizens pay for visas to go abroad. Luedtke et al. (2010) identified a range of factors and determined that level of political and civil liberties, per capita income, degree of internal insecurity, presence or absence of health epidemics and geographic location that all play a role. Neumayer (2006) reports that passport holders from 25 countries facing the smallest number visa restrictions come from Western, high income OECD nations with only one or two exceptions. Bangwayo-Skeete and Skeete (2017) found that an increase in the political stability index for a country resulted in an additional 13 countries granting visa free access. They also found that citizens of collaborative multinational organizations, such as the European Union, tend to enjoy greater visa-free access.

By contrast residents of poor, politically unstable countries or countries where the West seems to feel they pose a threat to global security face the most daunting visa restrictions (Webster & Ivanov, 2016). Members of the Francophonie, usually located in Africa, and the Arab League have less visa free travel than most other countries (Bangwayo-Skeete & Skeete, 2017). Neumayer (2006) adds countries that have the least visa-free access have a history of violent political conflict, strict autocratic regimes, are very poor or have some combination of these conditions.

Indeed, global tourism networks have become highly consolidated over time (Chung et al., 2020), highlighting the divide between haves and have nots. As an example, Mau et al. (2015) found citizens from Europe and the Americas have gained more global mobility rights in the last 40 years, while those coming from Oceania and Asia have been excluded from this development. Notably, citizens of politically and economically fragile countries or under-developed states in Africa have even lost mobility rights. Recchi et al. (2020) add, that while Europeans hardly have to work at all for travel permits, visa costs for residents of Sub-Saharan Africa and South Asia often cost the equivalent of several weeks' pay.

A number of studies has documented both the negative and positive impacts of entry restrictions on tourism. Neumayer (2010) suggests the imposition of tight visa requirements can restrict bilateral tourism by up to 60% while others suggest its impact is more in the neighbourhood of a 25% to 35% reduction in arrivals (Atral-Tur et al., 2016; Czaika & Neumayer, 2017; Lawson & Roychodhury, 2010). A number of other studies has shown how the imposition of visa restrictions in response to unique events can deter tourism, at least over the short term (Li & Song, 2013; Song et al., 2012). Interestingly though, Chi et al. (2022) remind us that visa policies alone do not necessarily deter tourists for they also consider the quality of tourism services available. None the less, restrictive visa requirements can act as annoyance, deterring people from visiting some places if others with less restrictive visa policies offer similar experiences.

Liberalizing visa policies can result in significant increases in visitor numbers (Balli et al., 2013), with Lawson and Roychodhury (2013) suggesting that the elimination of visas can result in the tripling of arrivals. Visa-free travel for Koreans visiting Japan produced a 12% increase in arrivals in the first year and 25% increase in the second year (Lee et al., 2010). Likewise, Hu (2013) notes that the implementation of the visa waiver program in the United States has encouraged an additional 1.8 to 2.7 million pleasure or business trips, generating an additional US$6.9 to US$10 billion in direct spending. A joint UN World Tourism Organization and World Travel & Tourism Council study on APEC (Asia-Pacific Economic Cooperation) countries concluded positive visa changes can result in a 5% to 25% increase in arrivals per year over a three-year period (UNWTO, 2013).

Visa issues are becoming more challenging, though, with the rise of the digital nomad. Digital nomadism is increasing in a COVID-adapted world where people are used to working remotely and countries are seeing the benefits or reaching out to these people. Unlike backpackers, digital nomads have a high focus on work during their travels, but still choose their trips based on leisure interest (Chevtaeva & Denizci-Guillet, 2021). In response, governments are enabling access through the creation of special visas that allow the worker (alone, in a couple, or family group) into the local system (Sánchez-Vergara et al., 2023).

Conclusion

This chapter reviewed the various types of political economies under which tourism operates, then entered a more detailed discussion of the politics of international tourism. Both are critical when we consider how tourism will evolve in an uncertain future, when the dominant growth paradigm, the reliance of inbound tourism as a source of foreign currency, is challenged by the need to achieve a net-zero emissions tourism. How we think about tourism may need to change.

■ Discussion Points / Exercises

1. What are the key differences between a capitalist and a Marxist approach to political economies?
2. Why were Keynesian economic approaches replaced by neoliberalist approaches?
3. What factors influence bilateral tourism flows? Select two countries that are seen to be antagonistic to each other and analyse the factors that prohibit international tourism activity.
4. Why is it that some countries impose strict visa requirements on some African and Islamic states?
5. Discuss why airlines want liberalised Air Service Agreements, while some countries may be opposed to such liberalisation.

Points to consider for revision

- A variety of political economy approaches have been adopted to facilitate tourism to greater or lesser success, depending on the system adopted.
- Granting the rights to travel internationally is fundamentally a political decision that is influenced by a range of factors.
- People can enter a country by any number of legal means. Tourist visas are the easiest way to enter a country and, therefore, are also the easiest to exploit.
- Countries impose a range of restrictions on who can enter a country, how long they can stay and set other conditions in order to reduce the risk of illegal immigration.
- International air transport is a form of trade guided by the formation of bilateral air service agreements between states.

Additional readings

Neumayer, E (2010) Visa restrictions and bilateral travel. *The Professional Geographer*, 62(2),171–181.

Prideaux, B. (2005) Factors affecting bilateral tourism flows. *Annals of Tourism Research.* 32(3), 780-801.

Martins, L., Gan, Y. & Ferreira-Lopes, A. (2017) An empirical analysis of the influence of macroeconomic determinants on world tourism demand, *Tourism Management*, 61, 248-260

References

5

Archer, B. (1989), Tourism and island economics: impact and analyse, *Progress in Tourism and Hospitality Management*, 1, 125-134.

Artal-Tur, A., Pallardó-López, V., & Requena-Silvente, F. (2016). Examining the impact of visa restrictions on international tourist flows using panel data, *Estudios de Economía*, 43(2), 265-279.

Australian Government (2019) The Bilateral System—how international air services work. https://www.infrastructure.gov.au/infrastructure-transport-vehicles/aviation/international-aviation/bilateral-air-services-system.

Balli, F., H. O., Balli & K. Cebeci (2013). Impacts of exported Turkish soap operas and visa-free policy on inbound tourism to Turkey. *Tourism Management*, 37, 186-192.

Bangwayo-Skeete, P. & Skeete, R. (2017) Who travels visa-free? insights into tourist hassle-free travel. *Journal of Travel Research* 56(3), 407-418.

Beenstock, M., Felsenstein, D., & Rubin, Z. (2015) Visa waivers, multilateral resistance and international tourism: some evidence from Israel. *Letters in Spatial and Resources Science* 8(3), 357-371.

Bianchi, R. (2011), Tourism, capitalism and Marxist political economy, in Mosedale, J. (ed), *Political Economy of Tourism: A Critical Perspective*, Routledge, London and New York, pp. 17-37.

Boas, T.C., & Gans-Morse, J. (2009) Neoliberalism: From new liberal philosophy to anti-liberal slogan. *Studies in Comparative International Development*. 44, 137-161. https://doi.org/10.1007/s12116-009-9040-5

Bramwell, B. & Lane, B. (2009) Economic cycles, times of change and sustainable tourism, *Journal of Sustainable Tourism*, 17(1), 1-4, DOI: 10.1080/09669580802608061

Chevtaeva, E. & Denizci-Guillet, B., (2021). Digital nomads' lifestyles and coworkation. *Journal of Destination Marketing & Management*, 21, 100633.

Chi, P.Y., Lee, K.C. & Chang, K.I., (2022). Causal effect of tourist visa exemption schemes on international tourist arrivals. *Economic Analysis and Policy*, 75, 427-449.

Chung, M.G., Herzberger, A., Frank, K.A. & Liu, J., (2020). International tourism dynamics in a globalized world: A social network analysis approach. *Journal of Travel Research*, 59(3), 387-403.

Czaika, M. & Neumayer, E., (2017). Visa restrictions and economic globalisation. *Applied Geography*, 84, 75-82.

Demir, E. & Gozgor, G. (2018) Does economic policy uncertainty affect tourism? *Annals of Tourism Research*, 69 (C), 15-17

DTRD (1998) Submission by the Department of Transport and Regional Development to the Inquiry by the Industry Commission into International Air Services, Canberra, March.

Dwyer, L. (2015) Globalization of tourism: Drivers and outcomes, *Tourism Recreation Research*, 40(3), 326-339, DOI: 10.1080/02508281.2015.1075723.

Emmerson, C. & Stockton, I. (2020) How does the size of the UK's fiscal response to Coronavirus compare with other countries? *Institute for Fiscal Studies*, 14 May, https://www.ifs.org.uk/publications/14845

EU (2013). An Overview of the Air Service Agreements Concluded by the EU. European Union. www.europarl.europa.eu/RegData/etudes/note/join/2013/495849/IPOL-TRAN_NT(2013)495849_EN.pdf

Fayed, H., & Westlake, J. (2002). Globalization of Air Transport: The challenges of the GATS. *Tourism Economics*, 8(4), 431-455. doi.org/10.5367/000000002101298214

Fletcher, J., Fyall, A., Gilbert, D. & Wanhill, S. (2018), *Tourism Principles and Practice* 6th edition, Pearson, Harlow.

Fletcher, R. (2011) Sustaining tourism, sustaining capitalism? The tourism industry's role in global capitalistic expansion, *Tourism Geographies*, 13(3), 443-461.

Gozgor, G. & Ongan, S. (2017), Policy uncertainty and tourism demand: empirical evidence from the USA, *International Journal of Tourism Research*, 19, 99–106.

Harvey, D. (1989) *The Condition of Postmodernity: An Inquiry into the Origins of Cultural Change*, Basil Blackwell, Oxford.

Higgins-Desbiolles (2006), More than an 'industry': The forgotten power of tourism as a social force, *Tourism Management*, 27, 1192-1208.

Higgins-Desbiolles, F. & Whyte, K. (2015), Tourism and human rights, in Hall, CM., Gössling, S. & Scott, D. (eds), *The Routledge Handbook of Tourism and Sustainability*, Routledge, London and New York.

Hoad, D. (2003) The General Agreement on Trade in Services and the Impact of Trade Liberalisation on Tourism and Sustainability. *Tourism and Hospitality Research*, 4(3), 213-227.

Hu, X (2013) Economic benefits associated with the visa waiver program – a difference-indifference approach. *Global Journal of Business Research*. 7(1), 81-89.

ITC/UNWTO (2015), *Tourism and trade: A global agenda for sustainable development*, ITC, Geneva: International Trade Centre and World Tourism Organization www.intracen.org/uploadedFiles/intracenorg/Content/Publications/Tourism_and_Trade__low%20res_2014-2015-335.pdf

Jones, C., Parker, M. & Ten Bos, R. (2005), *Business Ethics, a Critical Approach*, Routledge, Oxon.

Karaman, A. (2016) The pernicious impact of visa restrictions on inbound tourism: the case of Turkey, *Turkish Studies*, 17(3): 502-524. DOI: 10.1080/14683849.2016.1170602.

Keynes, J.M. (1935), *The General Theory of Interest, Employment and Money*, MacMillan, London.

Lawson, R., & Roychoudhury, S. (2013) Do travel visa requirements impede tourist travel? *Journal of Economics and Finance.* 40(4), 817-828.

Lee, C., Song, H., & Bendel, L. (2010) The impact of visa-free entry on outbound tourism: a case study of South Korean travelers visiting Japan. *Tourism Geographies* 12 (2), 302–323.

Li S., & Song, H. (2013) Economic impacts of visa restrictions on tourism: A case of two events in China. *Annals of Tourism Research.* 43, 251 – 271.

Li, S., Liu, A. & Song, H. (2019) Does tourism support supply-side structural reform in China? *Tourism Management*, 71, 305-314.

Lin, J. (2012), *New Structural Economics: A Framework for Rethinking Development and Policy*, World Bank Researcher Observer, 26, *World Bank*, Washington, DC.

Lissitzyn, O. (1964) Bilateral agreements on air transport, *Journal of Air Law and Commerce* 30(3), 248-263.

Liu, A. & B. McKercher (2015) The impact of visa liberalization on tourist behaviors: The case of China outbound market visiting Hong Kong. *Journal of Travel Research.* 55(5), 603-611. DOI: 10.1177/0047287514564599.

Luedtke, A., Byrd, D. G. & Alexander, K. P. (2010) The politics of visas. *The Whitehead Journal of Diplomatic and International Relations* 11(1), 147-160

Marx, K. (1974) *Capital: A Critical Analysis of Capitalist Production*, Volume 1, Lawrence and Wishart, London (First English edition published 1887).

Mau, S., Gülzau, F., Laube, L. & Zaun, N. (2015) The global mobility divide: how visa policies have evolved over time, *Journal of Ethnic and Migration Studies*, 41(8), 1192-1213. DOI: 10.1080/1369183X.2015.1005007.

Milhaupt, C. & Zheng, W. (2015) Beyond ownership: State capitalism and the Chinese firm, *Georgetown Law Journal* (665), http://scholarship.law.ufl.edu/facultypub/69

Neumaeyer, E. (2006) Unequal access to foreign spaces: how states use visa restrictions to regulate mobility in a globalised world. *Transactions of the British Institute of Geographers.* 31(1), 72-84.

Neumayer, E (2010) Visa restrictions and bilateral travel. *The Professional Geographer*, 62(2),171–181.

O'Conner, J. (1988), Capitalism, nature, socialism: A theoretical introduction, *Capitalism, Nature, Socialism*, 1(1), 11-38.

OECD (2014) Air Service Agreement Liberalisation and Airline Alliances https://www.itf-oecd.org/sites/default/files/docs/14airserviceagreements.pdf

Palley, T. (2004) From Keynesianism to Neoliberalism: Shifting Paradigms in Economics. *Foreign Policy in Focus*, May, https://fpif.org/from_keynesianism_to_neoliberalism_shifting_paradigms_in_economics/

Peck, J., Brenner, N., & Theodore, N. (2018) Actually existing neoliberalism. In D. Cahill, M. Cooper, M. Konings, & D. Primrose (eds), *The Sage Handbook of Neoliberalism* (pp. 3–15). London: Sage.

Prideaux, B. (2005) Factors affecting bilateral tourism flows. *Annals of Tourism Research.* 32(3), 780-801.

Prideaux, B., Thompson, M & Pabel, A. (2020) Lessons from COVID-19 can prepare global tourism for the economic transformation needed to combat climate change, *Tourism Geographies*, 22(3), 667-678.

Recchi, E., Deutschmann, E., Gabrielli, L. & Kholmatova, N., (2021). The global visa cost divide: How and why the price for travel permits varies worldwide. *Political Geography*, 86, 102350.

Rusu, S. (2011) Tourism multiplier effect, *Journal of Economics and Business Research*, 1, 70-76.

Sánchez-Vergara, J.I., Orel, M. & Capdevila, I., (2023). "Home office is the here and now." Digital nomad visa systems and remote work-focused leisure policies. *World Leisure Journal*, 65(2), 236-255.

Song, H., Gartner, W., & Tasci, A. (2012) Visa restrictions and their adverse economic and marketing implications – Evidence from China. *Tourism Management*. 33(2), 397-412.

UNWTO (2013) *The Impact of Visa Facilitation in APEC Economies*, UN World Tourism Organization and World Travel & Tourism Council. UNWTO, Madrid. www.e-unwto.org/doi/pdf/10.18111/9789284415960

UNWTO/WTO/WTTC (2018) *Tourism, Trade and the WTO: A Joint Communication from UNWTO, WTO, ITC and WTTC*. UN World Tourism Organization. http://www2.unwto.org/ru/node/51274 .

Wearing, S., McDonald, M., Taylor, G. & Ronen, T. (2019) Neoliberalism and global tourism in Timothy, D. (ed) *Handbook of Globalisation and Tourism*, Elgar, pp. 27-43.

Webster, C. & Ivanov, S. (2016) The ideologies of national security and tourist visa restrictions. *International Journal of Tourism Policy*, 6(3-4), 171-190.

Whyte, B. (2008): Visa-free travel privileges: an exploratory geographical analysis, *Tourism Geographies*, 10(2), 127-149.

Williams, A. (2004) Towards a political economy of tourism, in Lew, A., Hall, CM., & Williams, A. (eds) *A Companion to Tourism*, Blackwell, Oxford, pp. 61-73.

WTO (2019) *General Agreement on Trade in Services*. World Trade Organization. www.wto.org/english/docs_e/legal_e/26-gats_01_e.htm

WTO (2021) *Tourism and travel-related service*. World Trade Organization. www.wto.org/english/tratop_e/serv_e/tourism_e/tourism_e.htm.

Part 3 – Evolution of Tourism

Tourism changes tourism! The question is how? This part of the book examines the three major evolutionary/revolutionary and lifecycle models that have shaped our thinking of how tourism, in general, and destinations, in particular, change over time.

Chapter 6 compares and contrasts Butler's and Plog's complementary models and engages in a discussion about which one better reflects how destinations change. Butler adopts a supply-side perspective with carrying capacity as the deterministic factor that influences the likelihood of a destination going into decline. Plog, on the other hand, adopts a demand-side perspective and argues that destinations attract a progressively softer type of tourist. If not managed carefully, destinations are at risk of sowing the seeds of their own destruction.

Chapter 7 takes a different perspective and introduces the concept of complexity theory to describe how destinations evolve rapidly over time. In spite of its broad acceptance in other fields of study, complexity theory has drawn relatively little attention in tourism studies, even though it better reflects the real world situation.

6 Lifecycle Models

By the end of this chapter, the reader will be able to:

- Describe the biological basis of destination lifecycle models
- Evaluate the validity of Butler's and Plog's lifecycle models
- Analyse the strengths and limitations of each model
- Compare and contrast the two models and describe how they integrate closely
- Explain the role of carrying capacity

Introduction

Two tourism lifecycle models developed by Butler (1980) and Plog (1974, 2001) more than 40 years ago have dominated the conceptualisation of this topic. But, while the Plog and Butler models are well known, they are not necessarily known well, for the original sources are somewhat dated, and in the case of Butler's work, obscure. Indeed, one wonders how many people have actually read the original papers? Instead, they are so common that knowledge generally comes from secondary sources which invariably summarize the manuscripts into a few sentences or short paragraphs, and in doing so, simplify the complex and subtle issues raised about the nature of destinations and how they change over time.

Each had humble origins as each was written first as a conference paper and only later converted into a journal article. They are also a product of their times. Tourism in the mid to late 1970s was undergoing rapid expansion as the development of jet aircraft led to the massing of tourism for the first time. Many remote economies identified tourism as their only economic development option and pursued a path of large scale development under the guise that the economic benefits would outweigh any adverse social and environmental impacts. They were unaware of many of the realities of tourism and so, it took authors such as Budowksi (1976), Doxey (1975), Jafari (1974), Wenkman (1975) and many, many others to write cautionary tales about spontaneous development and the inherent risks involved. Butler, in particular, focused on the social and physical environmental impacts of tourism when developing his lifecycle model, and used carrying capacity as the deterministic variable to define the health of destinations. Plog took a demand-side perspective and wanted to understand reluctance to fly commercially.

The basic strengths and weaknesses of each model must be recognised before they are discussed in detail. The devotion to the product metaphor is the main reason why Butler's work is simultaneously praised and criticized – praised because of its elegance, simplicity and intuitive logic, and criticized because it is difficult to prove empirically (Tooman, 1997; Weaver, 2000) and is deficient as a predictive tool (Agarwal, 2002; Hovinen, 2002, Moss et al., 2003). Likewise Plog's work has been both praised and criticized – praised as an early attempt to segment the tourist market by psychographic profile (Chandler & Costello, 2002; Harrill & Potts, 2002; McGuiggan, 2003) and criticized for not accurately depicting lifecycles (McKercher, 2005a).

Biological origins

Both Butler's and Plog's models, have their origins in one or more biological theories (Osland, 1991). The first and simplest analogy is the life cycle that all higher organisms progress through (Crawford, 1984). As Tellis and Crawford (1981) note, individual biological specimens live predictable patterns of birth, growth, maturity, decline and death. Polli and Cook (1969) cite the usefulness of this analogy, for the identification of each stage permits the evaluation of a series of tactical and strategic considerations that can be taken to either progress to the next stage or stem decline. However, this idea is overly simplistic for the biological model applies only to individual specimens, such as a person, a frog or a leaf (Crawford, 1984). It does not apply when groups of things are considered.

Instead, Osland (1991) suggests we need to consider studies of populations of species. This modification forms the second model. Here Osland (1991), summarising the literature, notes the collective shape of the lifecycle curve is influenced by a range of biological factors, including food supply, competitors and the carrying capacity of the environment where the entity lives. Polli and Cook (1969) found it related well to the theory of the diffusion or products, innovation and the overarching lifecycle collections like entities evolve through. As they note, a product finds initial resistance to widespread acceptance and is purchased by only a limited segment of the buying population. Later, a larger segment of buyers adopt it, and sales begin to increase but eventually, the rate of growth decreases as the proportion of adopters gets closer and closer to a maximum. The rate of adoption remains constant throughout the maturity phase and diminishes in the decline phase. This approach makes more intuitive sense, but has been often omitted in favour of the more simplistic individual lifecycle model.

However, both the specimen and population models are static. Tellis and Crawford (1981) support a third, alternative evolutionary lifecycle model. As they state *"in biology the evolutionary cycle is used to explain the phenomenon of the growth and proliferation of species, groups of biological specimens that can breed within but not outside each group. Indeed, while the life-death cycle is a rigid, highly predictable one telling the story of each biological specimen (the individual horse, dog, etc.), the evolutionary cycle is a dynamic, open-ended one telling an even more exciting story of the origin, growth, and proliferation of entire species"*(Tellis & Crawford, 1981: 127). Later Crawford (1984:6, 7) argues and evolutionary approach is more suitable to consider for it contains four features of products. They are:

- *Cumulative*, where each change in the object under study builds on previous changes. This condition has been well documented in the Darwinian concept of evolution;
- *Directional*, where each change moves the object to greater complexity, efficiency and diversity. First, the species becomes more diverse, with more variations in the group. Second, the variations permit greater efficiency wherever the group needs it and third, the evolution introduces greater complexity. The evolution of the telephone, the car, even the cake mix, documents all three of these aspects of direction in change;
- *Motivated*, which refers to the source of change or what is causing it. In biology there are three such sources. The first is internal to the specimens or its gene sets. Second is the environment where living conditions favour certain mutations or certain behaviour changes. Third, there is mediation, which is unique to humans where we try to eliminate undesirable species (the measles virus), develop useful ones (hybrids of corn), and to control or maintain rare ones (zoo animals);
- *Patterned*, in that there is a clear pattern of change that closely corresponds to what we can call the traditional lifecycle pattern of: divergence or the start of new clans or species; development through growth through increased numbers; differentiation by increasing variation, including subspecies; stabilization with long periods of stability or stagnation, when there is no force for change, and perhaps, demise of the loss of a species.

Butler's model

Butler's model has become the industry standard (Wang et al., 2016). It is often quoted, but not particularly accurately. In particular, the model has often been described as a 'lifecycle' model that is assumed to borrow heavily from the product lifecycle concept. But, right from the start, there is a misconception, for Butler's original paper is titled *The Concept of a Tourist Area Cycle of Evolution* and not as it has been summarised recently as the TALC (tourism area life cycle) model. The difference is profound. Lifecycle models imply a finite cycle with a beginning, middle and end. Evolution, on the other is an ongoing process of continuous change from a simple to a more complex form with change occurring in response to environmental conditions by natural selection of randomly appearing mutations. Evolution, therefore, implies an open ended, infinite process where change is part of the normal order of events. Moreover, it relates to how entire systems collectively interact to influence change. Murphy (1985) was one of the first to liken destinations to an ecosystem in his discussion of community tourism. Over time, evolution can produce systems that are almost completely different from those found during the initial stages, although one can logically map the progression or change over time.

6

The opening paragraph of his paper states unequivocally

> *"There can be little doubt that tourist areas are dynamic, that they evolve and change over time. This evolution is brought about by a variety of factors including changes in the preferences and needs of visitors, the gradual deterioration and possible replacement of physical plant and facilities, and the change (or even disappearance) of the original natural and cultural attractions which were responsible for the initial popularity of the area. In some cases, while these attractions remain, they may be utilized for different purposes or come to be regarded as less significant in comparison with imported attractions"* (Butler 1980:5).

So, he argues from the outset that tourism changes tourism and that destinations evolve, rather than have a classic fixed life cycle. Where the 'lifecycle' confusion comes in, is that he argues the evolution can be mapped along the classic product lifecycle curve for convenience sake, more so than for absolute accuracy. But, note from Figure 6.1, the phases intentionally do not correspond with product phases.

His model is predicated on a number of assumptions:

- Demand changes as destinations evolve, with the type of tourist becoming increasingly 'softer' as destinations mature;
- Supply of products and services changes, usually in response to changing consumer demand, but occasionally in a proactive manner to lead demand change. Again, the change is to the creation of 'softer' products – better quality accommodation, better services and more built attractions that

develop the extrinsic appeal of the destination rather than relying on the intrinsic appeal of its original cultural and heritage landscape

- the limits of growth are determined by the destination's carrying capacity and once it is reached or exceeded, the destination is at risk of entering decline. At risk, but not pre-determined to do so!

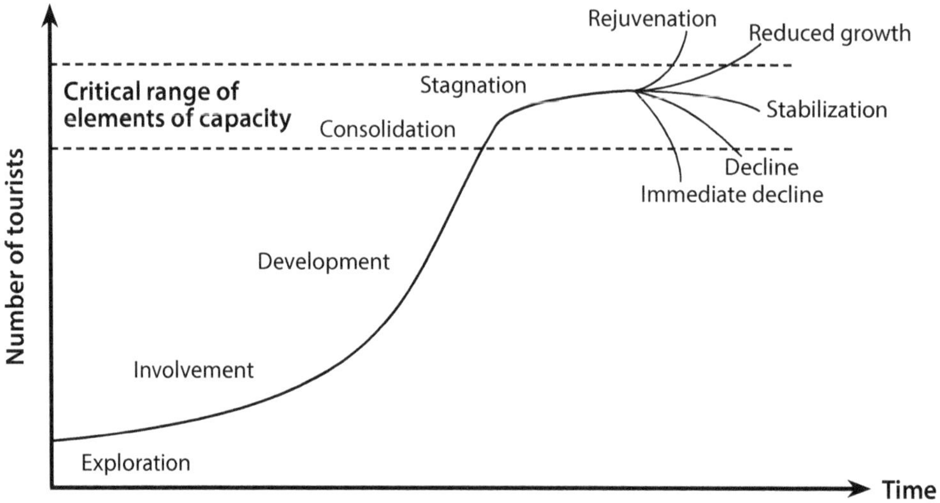

Figure 6.1: Butler's lifecycle. Source: Butler, 1980

The model has five main stages before a series of post-stagnation options emerge. The key features are shown in Table 6.1 below and summarised briefly below.

Table 6.1: Stages in Butler's Lifecycle

Stage	Characteristic
Exploration	Few explorers or drifters visiting sites with no public facilities Visitors attracted to the destination by a natural physical feature Specific visitor type of a select nature
Involvement	Limited interaction between local residents and the developing tourism industry leads to the provision of basic services Increased advertising induces a definable pattern of seasonal variation Definite markets are begins to emerge
Development	Development of additional tourist facilities and increased promotional efforts Greater control of the tourist trade by outsiders Number of tourists at peak periods exceeds the resident population, inducing rising antagonism by the latter towards the former
Consolidation	Tourism has become a major part of the local economy, but growth rates have begun to level off A well-delineated business district has taken shape Some of the older deteriorating facilities becomes less attractive to visitors Local efforts are made to extend the tourist season

Stagnation	Peak numbers of tourists and capacity levels are reached The destination has a well-established image, but it is no longer in fashion
Post-stagnation	Five possibilities, reflecting a range of options that may be followed, depending partly on the success of local management decisions. At either extreme are rejuvenation and decline

Exploration represents the first stage for a destination. Small numbers of independent tourists arrive. There is little existing tourism infrastructure and visitors tend to be attracted by the natural or cultural assets of that destination. They use local facilities. Tourism tends to be small scale and locally owned. Although relationships between hosts and guests may be intense, overall impacts are small because numbers are small. Usually leakages from the local economic system are small, since most of the tourists' needs are produced locally.

As the number of visitors increases, the destination enters the *Involvement* stage. At this point, a formal tourism industry begins to appear, dominated almost exclusively by local residents. A distinct tourist season may begin to appear and also some formal travel arrangements for tour operations can emerge. Contact between local residents and tourists remains high. Relationships between hosts and guests, although a little more formalized, are still personal and the scale of tourism and its growth rate are still manageable. Economic benefits per tourist can be quite large, since leakages remain small.

The *Development* stage reflects the period where large scale tourism begins. The area begins to develop an identity as a tourism destination, large numbers of tourists begin to arrive, and large scale development begins. Local involvement in tourism declines and outside investment increases. Purpose-built attractions begin to replace natural or cultural attractions. The type of tourist will change from someone seeking to explore the destination to someone looking for more of a rest and relaxation holiday. It is during this stage that adverse social and environmental impacts tend to be noted first, as local control over tourism diminishes and residents must begin to adjust their lifestyles to cater to the needs of the growing numbers of tourists. Congestion, traffic problems, increased competition for scarce resources, inflation and disruption of local ways of life are noted.

Consolidation occurs when the growth rate slows, but the total number of tourists continues to increase. Tourism is now a major economic activity with the industry dominated by national or multinational organizations. Butler suggests the destination takes on the characteristics of a product which is branded and marketed as a single entity. Because of the sheer number of tourists, much money enters the economy, however, leakages are also large as amenities these tourists look for are likely to be imported. Antipathy to tourism may reach its peak because of serious social and environmental impacts, while the economic benefits may be dispersed in an uneven manner.

Stagnation occurs once the peak number of arrivals has been reached and parts of the physical, social, psychological and ecological environment have reached or exceeded their carrying capacity. Tourist enclaves may emerge whereby tourists and locals are separated, mostly because locals have been pushed out of tourism development zones. As a result, contact between tourists and hosts is highly institutionalised. The area is now well known, but has fallen out of favour and instead of generating new arrivals will become increasingly reliant on repeat visitors. Local involvement in tourism starts to return as national and multi-national firms begin to abandon the place.

After stagnation comes decline. Right? Wrong! In classic product lifecycle literature, decline normally follows stagnation, but not here. Butler identifies five possible options, of which only two relate to decline. These options are:

- *Decline* (slow or rapid) where the area enters a rapid or slow period of decline. The destination area falls out of favour with tourists.
- *Stability* where the absolute number of tourists stays relatively stable or declines only slowly over time.
- *Rejuvenation* (slow or rapid) where the destination area reinvents itself.

Yes, three options are quite positive. On the one hand, the destination can maintain a level of maturity and remain stable indefinitely. But, equally, as well, Butler illustrates that it can enter a period of renewed growth. Admittedly, rejuvenation will involve a conscious strategic decision on behalf of the local community, either by adding new and different built attractions and activities to target new markets, or by taking advantage or previously untapped natural resources, again to attract new visitors.

His model is not without its critics as well as supporters, as acknowledged by Butler himself (2011) in a reflective commentary. Many have argued that it lacks utility because it is not deterministic and predictive. One cannot show when one phase ends and the next begins. Butler (2011:6) counters this criticism by stating "*the purpose of the model was to draw attention to the dynamic nature of destinations and propose a generalized process of development and potential decline which could be avoided by appropriate interventions (of planning, management and development), or as suggested in the title of the article, the management of resources.*"

Carrying capacity

Perhaps, though, the biggest criticism of Butler's model is the idea that carrying capacity is a fixed concept. Butler argues that the limits of growth are defined by the carrying capacity of the resort area and once exceeded, a range of negative and sometimes irreversible impacts are likely to occur (Butler 1999). The UN World Tourism Organization (2018:3) defines tourist carrying capacity as "*the maximum number of people that may visit a tourist destination at the same time, without causing destruction of the physical, economic and sociocultural environment and an unacceptable decrease in the quality of visitors' satisfaction*." The idea has been broadened in the literature to include many more concepts, such as: experiential/psychological carrying capacity - the lowest degree of enjoyment tourists are prepared to accept before they start seeking alternative sites or destination; economic carrying capacity the ability to absorb tourist functions without squeezing out desirable local activities; political carrying capacity at which point political support for tourism diminishes and; social carrying capacity the tolerance of the host community for tourism.

The concept of carrying capacity was recognized well when the original paper was written, but since then has been subjected to a great deal of criticism since the real world does not fit well into an idealized hypothetical world (McCool & Lime, 2001). Butler himself even acknowledges this point (Butler, 2011). The biggest challenge is the focus on numbers as a determinant of carrying capacity, with the assumption that large numbers equate to negative impacts (Koens et al., 2018). The second issue is that somehow it is an objective measure. But, McCool and Lime (2001) ask the question of from whose perspective is carrying capacity measured – the local or the tourist? Tourists have a much higher threshold than locals. Likewise, Jurado et al. (2013) highlight the challenge of which aspect of carrying capacity to measure: social, physical, psychological, economic or political? They may all be quite different. A third issue is the inherent seasonality of tourism, and with it comes the cost decision of what level of infrastructure is needed. Overcrowding may be an issue during part of the year in highly seasonal destinations but may not be a problem at other times. This observation leads to the fourth concern about carrying capacity and that is, from a social perspective, it tends to forget that people are resilient and adjust to change. Social disruption it is transitional whereby, eventually communities adapt to these changes over time (Perdue et al., 1999).

Perhaps, the biggest issue is that carrying capacity is a malleable, subjective and complex concept that is more of a social construct than an ecological absolute (Lindberg et al., 1997; Weaver, 2000). As such, it can be altered through new product development and site hardening. Romeril (1989) feels that while carrying capacity is simple in theory, the management application is complicated by strong value dimensions. Some users will adopt a liberal attitude, accepting significant modification of the environment as an acceptable outcome for greater use of a resource, while others will perceive that any modification as representing unacceptable degradation.

Sites can be hardened and the carrying capacity increased. Indeed, changes in carrying capacity are implicit in the development of the Butler framework, for the process of touristification is the process of carrying capacity enhancement. Building small, locally owned properties enhances the carrying capacity of a destination somewhat, while building larger resorts and hotel complexes increases it a lot. Likewise, the shift from natural and cultural attractions to built attractions is, in essence, an exercise in carrying capacity expansion. The same can be said for most infrastructure development from water and sewage provision to electricity provision and enhancement of the transport infrastructure.

The unrecognized subtlety of Butler's model

Most people tend to look at the S-shaped curve as a reflection of the entire evolutionary process of the destination. They fail to appreciate the most subtle and exciting condition in Butler's lifecycle. It is open ended. The *x* axis, reflecting time, and the *y* axis, reflecting the number of tourists, have arrows at the end implying that destination evolution is not bounded by either time or the volume of visitors (McKercher, 2005b). Instead it is an open ended process that extends beyond the narrow confines of the framework shown. The open-ended nature of the life cycle model is also reflected in what happens post stagnation, where the outcome is also open ended. Indeed, some years ago Crawford (1984) illustrated that the fixed and predictable lifecycle can only be applied to individual specimens and not to groups of things, including complex system.

In reality, the framework shown by Butler does not reflect the destination's total life cycle from beginning to end, but rather individual phases in an ongoing life cycle pattern. The recognition of an evolutionary biological model reminds us that the only constant in tourism is change. As destinations mature and evolve, they by definition, will diversify their product offerings and broaden their market bases. In doing so, the character of a destination will change. But, that does not necessarily mean that ongoing growth is inevitable. Instead, using a biological analogy, destinations like other ecosystems, may endure a prolonged period of stability before being subjected to shocks that change their trajectory (Ibanez et al., 2017). They may also display a range of evolutionary patterns. Choy (1992) recognized this pattern explicitly in his work on the island destinations, while Prideaux's (2000) model discussed elsewhere recognizes implicitly. It is also explicit in the complexity theory approach to destination change.

Figure 6.2 shows how each of these individual lifecycle phases can be aggregated to reflect the overall evolution of a destination. Destinations may undergo multiple stages of involvement, development consolidation and stagnation as the nature of the destination changes. Moreover, the critical range of capacity that defines the consolidation and stagnation phase is malleable and can change as the nature of the destination changes. The trend may be upward, as shown in Figure 6.2, downward, in the case of destinations in long term decline, or highly variable in destinations undergoing rapid evolution (McKercher & Wong, 2024). Not all destinations necessarily evolve through all of Butler's stages. Some will go directly from post stagnation to rebirth, while others will enter a period of decline before entering a recovery period.

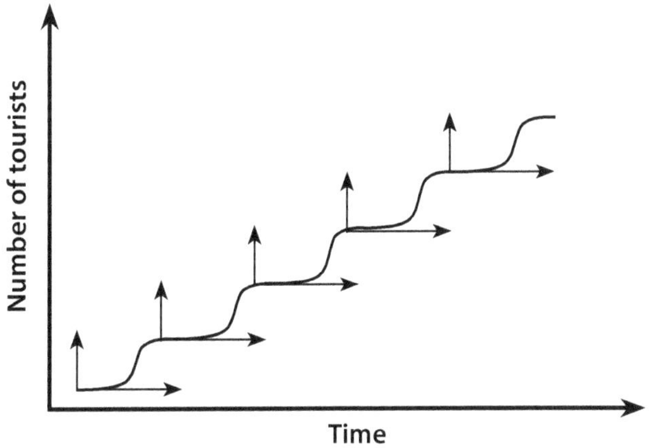

Figure 6.2: Cumulative lifecycle phases

Plog's model

The other tourism model referred to most in the literature was developed by the social psychologist Stanley Plog. Plog's lifecycle model is based on the underlying assumption that as a destination changes it attracts a progressively less adventuresome tourist. In doing so, these tourists require more and more entertainment oriented attractions that reflect, increasingly, their home environment. As he states (Plog 1974:55) *"any approach to understanding why destination or resort areas rise and fall in popularity inevitably must involve an understanding of the psychology of people who travel."*

Plog's work as a lifecycle model has been largely forgotten in recent years. Instead the main contribution is seen as being one of the first attempts to define market segments by the tourists' underlying psychological profile. But, as with Butler, this reinterpretation of his work is a misconception, for the original paper was titled *Why Destination Areas Rise and Fall in Popularity.* His work looks at how destinations change as they attract initially the adventuresome tourist who discovers the place, through growth as the market base broadens, and then finally to decline and demise as the market base shrinks.

His paper states unequivocally (Plog 1974: 58) *"the introduction of each new type of traveler has meant a larger population base from which to draw travelers...[but] when the appeal of the resort passes the magic mid-point in the population curve of travelers... it begins to draw on a smaller number of travelers."* Elsewhere he adds (Plog 1974: 58) *"thus, we can visualize a destination moving across a spectrum, however gradually or slowly, but far too often inexorably towards the potential of its own demise. Destinations carry with them the potential seeds of their own destruction as they allow themselves to be more commercialized and lose qualities which originally attracted tourists."* The only flaw in this logic is that markets are additive and not subtractive (Prideaux, 2000). Rather than displacing existing markets, destination evolution tends to broaden the market base by attracting a more diverse set of users.

His model is shown in Figure 6.3. Two features make this model unique. First, it is read from right to left. While it may be confusing to some, the reason is actually very simple, for the groupings are based on responses to a series of Likert scaled questions. People who were more likely to answer '7' tend to be allocentric in nature, whereas those who answer closer to the '1' end of the scale are more psychocentric. The second feature is that the categories are quantitatively based. Plog feels all populations are normally distributed, including the tourist population. As such, they can be segmented based on deviation from the norm. The midcentric groups capture all those people whose responses are within one standard deviation of the average score, while the other groups are based on scores of between one and two standard deviations and greater than two standard deviations from the mean, respectively. It is for this reason that the mid group represents just under two-thirds of the population, while the two 'Near' groups represent about one-third of the population and the two groups at the extreme end represent the rest.

The labels have changed over time as first he tried to soften the negative connotations applied to people at the psychocentric end of the spectrum and later as the model was commercialized (see www.besttripchoices.com). Table 6.2 summarizes some of the key features of each group, along with the different labels attached to them, while each is described briefly below.

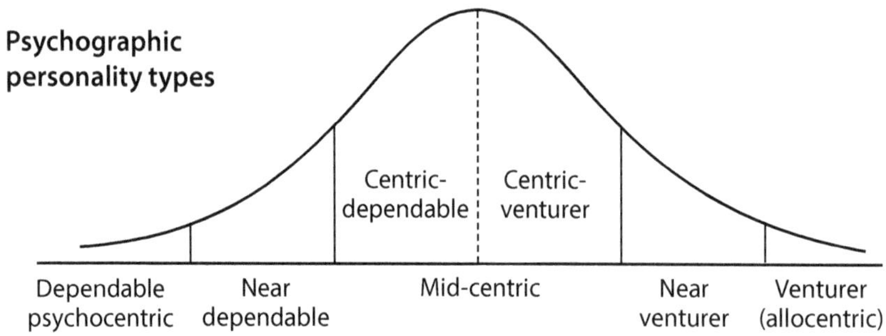

Figure 6.3: Plog's Lifecycle model. Source: Plog 2001

Plog feels Allocentrics or Venturers are the first type of tourist attracted to a destination. The term Allocentric refers to the root word 'allo' which means very informed, or a person with varied interests and 'centric' meaning the focus of his/her patterns. An Allocentric tourist is somebody who is interested in a sense of discovery and likes to be immersed in new activities.

Table 6.2: Plog's Classification

Current segment name*	**Traditionals**	**Sightseers**	**Journeyers**	**Voyagers**	**Pioneers**	**Venturers**
Initial name	**Pyschocentrics**	**Near Psychocentric**	**Midcentric**	**Midcentric**	**Near Allocentrics**	**Allocentric**
Other names	Authentic/dependables	Near Authentic				
Approx. size of the group	About 3%	17%	30%	30%	17%	about 4%
Importance and frequency of travel	Travel unimportant Take fewer trips and stay away for fewer nights than all others	Travel not very important and travel less than others	Travel somewhat important and travel an average amount	Travel important and travel an average amount	Travel important and travel often	Very important part of their life and travel the most
Preferred types of travel and destinations	Nearby well established destinations Major hotel chains and entertainment facilities not available at home Prefer domestic travel and familiar atmosphere Like group tours	Prefer popular places with lots of sunshine Repeat visitor Prefer domestic travel Cruises are popular Package tours if travelling internationally	Like crowds Travel for fun, recreation, and to spend time with family Prefer domestic travel Cruises are popular Package tours if travelling internationally	Prefer a modest selection of good hotels Repeat visitors to favourite places Prefer independent travel but will buy packages Prefer to be active Wanderlust	Like to travel to foreign destinations and seek new experiences Plan trips and set an itinerary of places they want to visit Wanderlust Soft adventure Travel internationally to culturally different places	Like to visit relatively unknown, uncommon and non-tourist destinations before others Prefer longer duration trips Want to explore and discover Unique cultures are especially attractive Meet people of different backgrounds Prefer high activity and hard adventure Happy to tough it

* as used at www.besttripchoices.com

6

As they begin to talk to their friends about the vacations that they have recently discovered, their friends are attracted to destinations. Members of the Near Allocentric/Pioneer sector are not discoverers, but are tourists who can still withstand a fair amount of strangeness. However, they do require hotels, restaurants and tourist services. The advent of this group really represents the beginning of the tourism life cycle.

As destinations become more popular, the Midcentric market begins to arrive. Plog initially identified only one group of Midcentrics who represented two-thirds of all tourists. He has since divided them into two groups based on whether their inclination is to the Allocentric or Psychocentric end of the spectrum. The Voyagers share more in common with Allocentrics, in that they are somewhat adventuresome and want to engage in wanderlust activities. The Journeyers share more in common with the Pyschocentrics. They still enjoy travel but are more cautious. Both groups represent the typical tourist who requires a large number of services and facilities and who travels primarily for rest and recreation. An influx of Midcentric tourists is thought to displace the other groups. The destination has now reached its maximum potential because it is an attracting the broadest audience possible.

As the appeal of the place passes a certain point, it begins to attract Sightseers / Near Psychocentrics/Near Authentics who are rather constrained in their travel. They travel less and prefer at home style of accommodations and experiences. The Psychocentric / Dependable / Authentic / Traditionals tourist is the opposite of the Allocentric. The root of the word Psychocentric is psyche, which means self. These tourists are self-focussed, fearful of change and the outside world. They travel little and when they do prefer as little change from their home environment as possible. Plog argues that destinations move along the continuum from Allocentric to Pyschocentrics, and in doing so, inevitably must enter a period of decline.

■ Questioning Plog

McKercher's (2005a) critique highlights three assumptions that form the basis of the model that may not be valid.

1) Can destinations exist simultaneously at different places in the continuum?

The model defines lifecycle stage from the perspective of a single market. In the original case lifecycle seemed to be assessed from the perspective of a New Yorker. Thus, Asian and African destinations were seen as appealing to the Allocentric or Venturer tourist. But, from the perspective of someone living in Asia, a place like Japan may not be all that foreign and may be seen as lying on the mid-point of the scale. How a destination is classified, therefore, depends on the perspective of the market. In other words, perceived lifecycle stage is market specific, rather than destination specific. Each market has a unique perception shaped by geographical proximity, political orientation, stage of economic development, similarities or differences in culture, language and shared histories and a host of other factors.

A few researchers have acknowledged the destinations can be seen to exist at multiple stages in their life cycle simultaneously. Prideaux's (2000 and 2004) resort development spectrum model identifies a series of growth phases that resorts can progress through as their market base broadens. The shift from one stage of development to another (i.e. from regional to national or national to international) is induced by a broadening in demand producing or resulting in a concomitant broadening of supply. If a destination can exist simultaneously at multiple points along its lifecycle continuum, depending on the perspective of different markets visiting it, then it cannot be said any destination is at a definitive lifecycle stage.

2) Can markets evolve fully through their lifecycles?

Plog posits that destination lifecycle is similar to the diffusion model of products. It begins with innovators (Allocentric / Venturer segment) discovering a place, experiences growth as early adaptors (Near Allocentrics / Pioneers) visit and shows signs of maturity as Midcentrics visit. It then enters terminal decline when the destination loses favour with all but the rump group of laggard or Psychocentric / Dependable / Authentic / Traditionals tourists.

This proposition is predicated on the validity of the assumption that each market can evolve fully through this life cycle pattern. To do so, each market must be a source of Allocentric / Venturer, Midcentric and Psychocentric/Dependable tourists. This proposition ignores the significant mitigating impacts that physical and cultural distance will have on the type of tourist who is willing to visit. Smith and Xie (2003) illustrate the impact of cultural distance in their work examining tourist flows from America to Canada. Their comparisons of the volume of visitors entering Canada as opposed to travelling within the United States showed that the act of crossing an international border added the equivalent of an average of 1,650 km to a journey and that crossing from English-speaking Vermont to French-speaking Quebec added the equivalent of 3,700 km. Therefore, not all markets can evolve. Some must start as Allocentric and can never progress beyond Midcentric. Others start as Psychocentric.

3) Must psychocentric destinations be in decline?

Lastly, the assertion that Psychocentric destinations must be in decline needs to be questioned. This model suggests that tourism destroys tourism and yet there is very little hard evidence to support this proposition. Instead, tourism changes tourism. Many so-called Psychocentric destinations located in the pleasure periphery of large cities have undergone a renaissance as destination marketing organizations and local business people have responded to the growing market demand for short break holidays, spa vacations, and the growth in the meeting and convention market. True, the nature of tourism demand may have changed as these destinations evolve to cope with cyclical demand, peaking on weekends and troughing during the week, but in doing so, many have also become four season resort areas.

Does Plog work empirically?

In theory, Plog's ideas make a lot of sense. In theory! Reality, though is different, for many tests show Plog's model is good at identifying tourists' ideal destinations, but not where they actually travel. Figurc 6.4, shows the difference between the ideal destination and the actual destination visited. In reality, more people will visit so called Psychocentric or Traditional destinations, regardless of their profile. Table 6.3 shows how even Venturers (Allocentrics) prefer Traditional (Psychocentric) destinations. Litvin (2006) revealed that ideal destinations fit nicely along Plog's proposed continuum but that actually, places visited leaned heavily toward the Psychocentric end of the continuum. A follow-up study by Litvin and Smith (2016), as shown in Table 6.3, revealed that almost 90% of all travel could be considered to Psychocentric destinations instead of the predicted 20% (including both Psychocentric and Near Psychocentric destinations). Moreover, instead of two-thirds of travel being to Midcentric destinations, the actual figure was closer to 10%. Likewise, little travel was taken to Allocentric destinations.

Table 6.3: Destination choice by type of tourist

Destination type	Psychocentric / (Traditional / Sightseers)	Midcentric (Journeyers / Voyagers)	Allocentric (Pioneers / Allocentrics)	All tourists
Allocentric	0.4%	0.9%	1.6%	0.9%
Midcentric	7.8%	10.9%	15.6%	11.2%
Psychocentric	91.8%	88.2%	82.8%	87.9%

Source: Litvin and Smith (2016)

Look closely at the 'most recent trip' line shown in Figure 6.4. It looks an awful lot like the classic distance decay curve, further reinforcing the fact that distance decay seems to be a dominant law in tourism. Why is it that actual trip patterns reflect a distance decay curve and not the predicted Plog-like line? Time availability, cost travel party size, motivation and a range of other factors dictate that most travel is to nearby destinations. There is another factor as well. Plog notes that the Allocentric tourist is a very active tourist, taking many trips a year, while those who lean towards the Allocentric side of the Midcentric continuum are also active travellers. It is unrealistic to assume that every one of them will take a typical Allocentric type of trip, but as active travellers they will likely take more short haul trips to nearby destinations, as well as the possibility of a longer haul trips that reflects their underlying personality traits. This observation further challenges the belief that Psychocentric destinations must be in decline.

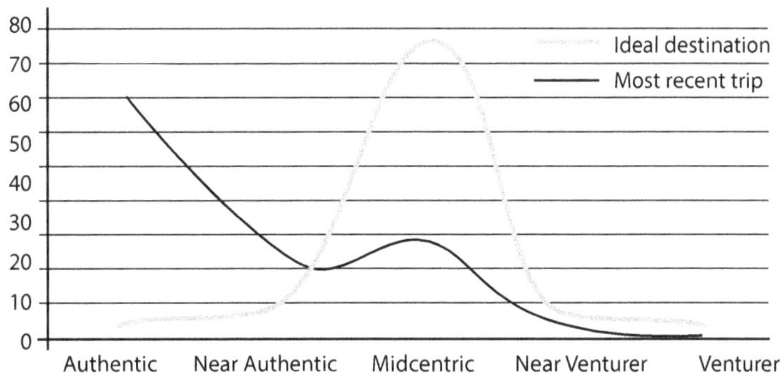

Figure 6.4: Source: from Litvin 2006

These findings raise a further two key questions about Plog's work. First, he does not consider the role differing motivation may play in individual trip choice. Instead, he gives blanket descriptors and assumes all travel by that person fits the generic descriptor. But, one's underlying psychological makeup is not necessarily a deterministic factor that defines the motivations for a trip. You may have an Allocentric personality, but simply want to get away from your job for a weekend, or spend some time with family and young children. Your motives may dictate that you take a Pyschocentric or Midcentric type of trip. Having said this though, it is unlikely that someone with a Psychocentric personality would want to climb Mt Everest.

In fact, an alternative interpretation of Plog is that personality traits determine the upper limit of the type of trip an individual is comfortable with, unless a range of special considerations is made to reduce the discomfort level. This relationship is shown in Figure 6.5. Much like as seen with distance decay, where anybody who can travel can travel a short distance, the same situation applies to Plog. Anybody who can travel can choose a Psychocentric destination or type of holiday and probably be quite satisfied with that experience if that type of trip reflects their desired motives, regardless of their underlying personality characteristics. A Midcentric type of personality would likely be satisfied with a type of trip that reflects a Psychocentric or Midcentric experience if that was their motive, and the Allocentric tourist would be satisfied with all types of trips up to and including an Allocentric trip.

But, because of his or her underlying personality traits, the Psychocentric type of tourist would be unlikely to take a more adventuresome type of trip unless a huge environmental bubble was created, as is often done when either travelling to visit friends (high degree of familiarity) or travelling in a package tour (high degree of safety). The Midcentric tourist would also likely not be interested in a more venturesome type of trip, unless significant measures were taken to reduce the level of strangeness to an acceptable level.

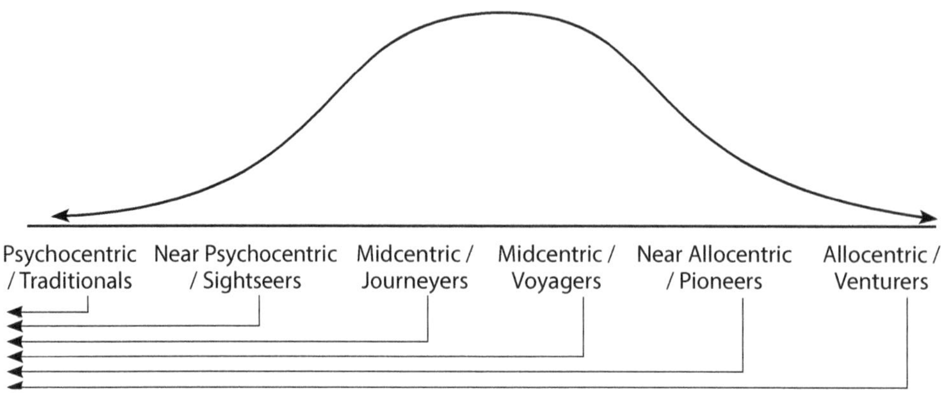

Figure 6.5: Plog's psychographic characteristics determine the 'maximum' type of trip

The second feature is that Plog assumes destination populations are exclusive instead of cumulative. Here, he suggests that 'softer' tourist progressively displaces the hardier tourist until only a rump of the population remains. But Prideaux (2000) suggests otherwise.

The complementarity of Butler and Plog

Butler and Plog are essentially looking at destination lifecycles from different sides of the same coin. Butler adopts the supply-side development process in response to changing demand and, rarely, in anticipation of changing demand, while Plog adopts a demand-side perspective and examines how the supply of facilities and services for tourists changes as the profile of tourists change. Indeed, Butler (1980) makes a number of specific references to Plog indicating that the Allocentric tourist can be found during the exploration stage of the life cycle, the Midcentric tourist is typical in the development stage and the Psychocentric tourist can be found in the stagnation stage of the life cycle.

Table 6.4 compares the features of Butler's and Plog's life cycle stages to illustrate just how complementary they both are. Small numbers of tourists who are satisfied with limited facilities, want intense contact with the local community and are attracted initially by the intrinsic natural and cultural assets of the destination area reflect the desires of the prototypical Allocentric tourist. The Near Allocentric tourist shares much in common with the Allocentric tourist but is not quite as hardy and, therefore, wants some better services and facilities. This segment is up to six times larger than the Allocentric segment, which then stimulates the initial development of purpose-built tourist facilities, often by local entrepreneurs who see a business opportunity to cater to this growing market. It is at this stage of the life cycle that tourism development begins to occur rapidly.

Table 6.4: Complementarity of Butler and Plog

Butler	Plog
Exploration Small numbers initially Restricted by lack of access, facilities Limited local knowledge Use of local facilities Intense contact with local residents Appeal rests with natural and cultural attractions	**Allocentric / Venturer** Prefer non-touristy areas First to 'discover' new places Can cope with limited facilities Prefer heavy interaction with local residents Independent tourists Travel a high priority
Involvement Visitor numbers begin to grow Initial development of tourist facilities Contact with locals high Emergence of a tourist season	**Near Allocentric / Pioneer** Popularise places Want a comfortable bed Want tourist facilities including resorts, hotels and activities Travel a high priority
Development Well defined tourist area Local involvement in development declines as outside organisations establish facilities Development of built attractions Formalisation and commercialisation of natural and cultural attractions Peak season crowding begins to emerge	**Midcentric / Voyageurs & Journeyers** Rapid growth continues Tourists want better quality and variety in facilities, accommodation and activities Want to be isolated from local community and happy to be in tourist enclaves
Consolidation Growth continues but rate of increase slows National and multinational firms are evident Less growth in new developments Well defined tourism nodes	**Near Pyschocentrics / Sightseers** Want well developed facilities Prefer built attractions Want known brands Prefer domestic travel Travel not a high priority
Stagnation Peak arrivals reached Area well known but beginning to lose its appeal Artificial, built attractions dominate Heavy reliance on repeat visitation	**Pyschocentrics / Traditionals** Prefer familiar destinations Heavy repeat travel Prefer highly developed places Prefer familiar environment including known brands Low activity level Travel a low priority

The rapid growth development phase corresponds with the Midcentric tourist discovering and popularizing the destination area. Remember that according to Plog's quantitative measures, Midcentrics represent two-thirds of the travelling population. The Midcentrics who lean towards Allocentrism want to explore and discover but they want to do so from the comfort of good facilities and services. Those who begin to lean towards the Psychocentric end of the spectrum are starting to become less active and more demanding of facilities that represent the extrinsic appeal of the destination. Because of these factors, well-defined tourist areas need to be developed and the essential character of the destination changes.

The consolidation and stagnation stages correspond closely to the Psychocentric end of the spectrum where the destination area has lost its unique character and broad appeal and instead begins to attract a heavily repeat market of low activity visitors. The Near Psychocentric tourist is not particularly active, wants well-developed facilities and known attractions while the Psychocentric tourist is the least active of all groups of tourists.

Conclusion

These two models still resonate today, for they continue to be referred to frequently in the literature. But, neither asks the key question of "*Do destinations have lifecycles or simply progress through a series of lifecycle phases*?" We know destinations evolve. Plog says they evolve from a demand perspective as an increasingly softer group of tourists force change to the destination. Butler takes a more supply side approach and also sees how the provision of goods and services changes as destinations evolve. Both assert that in a single point in time it is possible to depict the destination as existing at some point on their various continua.

But, neither answers of questions of whether destinations have lifecycles or simply progress through a series of lifecycle phases. Plog in fact asserts their lifecycles are finite, while Butler's work is interpreted by many as also depicting a finite process, in spite of the model's open-ended nature. Indeed, Plog (2001) says a destination needs an earthquake to revise its fortunes. Butler (1980) argues places can be revitalised, but also presents a rather pessimistic view that revitalisation involves targeting a specialist interest group of tourists who may provide a short term period of growth before the place loses its competitiveness once more.

Where is the empirical evidence of such a finite lifecycle process? Few if any destinations have truly lost their flavour and disappeared, unless they have been victims of suburban encroachment or some natural disaster. Indeed, we may need a formal rethink of the entire lifecycle model. Food for thought.

Discussion points / Exercises

1. Butler and Plog adopt different biological models when developing their lifecycles. Butler supports an evolutionary model whereas Plog supports a diffusion model. Both have gained acceptance in the literature. Which one do you support and why?
2. People seem to support either Butler's or Plog's idea, but rarely both. Yet, each model has its merits. Select which model you support and then argue why the other model also has merits.
3. What factors would propel a destination into a new lifecycle phase as suggested in Figure 6.2?
4. Each of these models has its strengths and its limitations. Can you think of another model that captures the dynamic nature of destinations better?
5. What would cause an Allocentric tourist to select a Psychocentric destination?
6. Visit www.besttripchoices.com, take the personality quiz (http://besttripchoices.com/travel-personalities/quiz/) and see what type of holiday suits you. Do you agree with this assessment?

Points to consider for revision

- Two competing lifecycle models have dominated the literature over the past 40 plus years. Both have strong biological roots.
- Butler's model is predicated on the assumption that carrying capacity is fixed and once the carrying capacity of a destination is exceeded, it could enter a period of decline, but is not mandated to do so.
- The subtlety of Butler's model, and in particular the recognition that it is open ended suggests destinations can evolve through multiple lifecycles.
- Plog's model is based on the diffusion or innovation of products.
- Plog posits destinations tend to have a closed lifecycle and once they pass the magic midpoint of their market curve, the seeds of their demise are sown.
- Empirical testing of Plog's model does not support the absoluteness of his lifecycle, primarily because he does not consider motivations as a key determinant of a trip.
- It is likely that Plog's model represents the maximum extent of change each segment is willing to undergo unless they wrap themselves in a bubble of familiarity.
- Butler's and Plog's models are complementary and overlap in many places.

Suggested additional readings

Butler R. W. (1980) The concept of a tourist area cycle of evolution: implications for management of resources *Canadian Geographer* 24(1), 5-12.

Butler, R. W. (2011) *Tourism Area Lifecycle: Contemporary Tourism Reviews. Oxford:* Goodfellow Publishers.

Litvin, S. & Smith, W. (2016) A new perspective on Plog psychographic system. *Journal of Vacation Marketing* 22(2), 89-97.

Plog, S. (1974) Why destination areas rise and fall in popularity. *Cornell HRA Quarterly* 14(4), 55-58.

References

Agarwal, S. (2002) Restructuring seaside tourism: The resort lifecycle *Annals of Tourism Research* 29(1), 25-55.

Budowski, G. (1976) Tourism and conservation: Conflict, coexistence or symbiosis? *Parks*, 4, 3-6.

Butler R. W. (1980) The concept of a tourist area cycle of evolution: implications for management of resources *Canadian Geographer* 24(1), 5-12.

Butler, R. W. (1999) Sustainable tourism: A state-of-the-art review, *Tourism Geographies*, 1(1), 7-25, DOI: 10.1080/14616689908721291

Butler, R. W. (2011) *Tourism Area Lifecycle: Contemporary Tourism Reviews. Oxford:* Goodfellow Publishers.

Chandler, J. & Costello, C. (2002) a profile of visitors as heritage tourism destinations in East Tennessee according to Plog's lifestyle and activity level preferences model. *Journal of Travel Research* 41, 161-166.

Choy D. (1992) Life cycle models for Pacific Island destinations *Journal of Travel Research* 31(3), 26-31

Crawford, M. (1984) Business took the wrong life cycle from biology. *Journal of Consumer Marketing* 1(3), 5-11.

Doxey, G. V. (1975) A causation theory of visitor-resident irritants: methodology and research inferences. In *Travel and Tourism Research Associations Sixth Annual Conference Proceedings* (195-198), San Diego, September.

Harrill, R. & Potts, T. (2002) Social psychological theories of tourism motivation: exploration, debate and transition. *Tourism Analysis* 7(2), 105-114.

Hovinen, G. (2002) Revisiting the destination lifecycle model *Annals of Tourism Research* 29(1), 209-230

Jafari, J. (1974) The socio-economic costs of tourism to developing countries. *Annals of Tourism Research* 1(7), 227-262.

Jurado, N., Damian, I., & Fernandez-Morales, A. (2013) Carrying capacity model applied in coastal destinations. *Annals of Tourism Research,* 43,1-19.

Koens, K., Postma, A., & Papp, B. (2018) Is Overtourism Overused? Understanding the Impact of Tourism in a City Context. *Sustainability* 10(12), 15 https://www.mdpi.com/2071-1050/10/12/4384 .

Lindberg, K., McCool, S., &Stankey, G. (1997) Rethinking carrying capacity. *Annals of Tourism Research* 461-465.

Litvin, S. (2006) Revisiting Plog's model of allocentricity and psychocentricity...one more time. *Cornell Hotel and Restaurant Administration Quarterly.* 245-253.

Litvin, S. & Smith, W. (2016) A new perspective on Plog psychographic system. *Journal of Vacation Marketing* 22(2), 89-97.

McCool, S. & Lime, D. (2001) Tourism carrying capacity: tempting fantasy or useful reality? *Journal of Sustainable Tourism,* 9(5), 372-388, DOI: 10.1080/09669580108667409

McGuiggan, R. (2003) A vacation choice model incorporating personality and leisure constraints theory *Tourism Analysis* 8(2): 183 – 186.

McKercher, B. (2005a) Are psychographics predictors of destination lifecycles? *Journal of Travel and Tourism Marketing* 19(1), 49-55

McKercher, B. (2005b) Destinations as products? A reflection on Butler's lifecycle *Tourism Recreation Research* 30(3).

McKercher, B., & Wong I (2024) Can destinations have multiple lifecycles? In Butler R. (Ed) *Tourism Area Lifecycle: Review, Relevance and Revision.* Channelview Publications

Moss S., Ryan C., & Wagoner, C. (2003) An empirical test of Butler's resort product life cycle: Forecasting casino winning. *Journal of Travel Research* 41, 393-399.

Murphy, P. (1985) *Tourism: A Community Approach,* Routledge, London.

Osland, G. (1991), Origins and development of the product life cycle concept. *Scholarship and Professional Work - Business* 237. http://digitalcommons.butler.edu/cob_papers/237

Perdue, R., Long, P., & Kang, Y. (1999). Boomtown tourism and resident quality of life: The marketing of gaming to host community residents. *Journal of Business Research,* 44(3), 165-167.

Plog, S. (1974) Why destination areas rise and fall in popularity. *Cornell HRA Quarterly* 14(4), 55-58.

Plog S (2001) Why destination areas rise and fall in Popularity: An update of a Cornell Quarterly classic *Cornell Hotel and Restaurant Quarterly* 42(3), 13-24.

Polli, R. & Cook, V. (1969) Validity of the product life cycle. *The Journal of Business,* 42(4), 385-400.

Prideaux, B. (2000) The resort development spectrum – A new approach to modeling resort development. *Tourism Management* 21:225 – 240.

Prideaux, B. (2004) The resort development spectrum: The case of the Gold Coast, Australia *Tourism Geographies* 6(1), 26-58.

Romeril, M. (1989) Tourism and the environment: accord or discord? *Tourism Management,* September, 204 - 208.

Smith S. L. & P. F. Xie (2003) Estimating the distance equivalence of the Canada-U.S. Border on U.S.-to-Canada visitor flows. *Journal of Travel Research* 42(2), 191-194.

Tellis, G., & Crawford, C. M. (1981) An evolutionary approach to product growth theory. *Journal of Marketing*, 45(4), 125-132.

Tooman L. A. (1997) Applications of the tourism life-cycle model. *Annals of Tourism Research* 24(1), 214-234.

UNWTO (2018) *'Overtourism'? Understanding and Managing Urban Tourism Growth Beyond Perceptions: Executive Summary*. Madrid: UN World Tourism Organization.

Wang, X., Weaver, D., Li, R., & Zhang, Y. (2016) In Butler (1980) we trust? Typology of citer motivations. *Annals of Tourism Research* 61, 216-218.

Weaver, D. (2000) A broad context model of destination development scenarios *Tourism Management* 21, 217-224.

Wenkman, R. (1975) The Pacific tourism blight. *Annals of Tourism Research* 3(2), 68-77.

7 Tourism as a Complex System

By the end of the chapter, the reader will be able to:

- Describe why existing lifecycle models are deficient
- Appreciate that tourism, by its very nature, is inherently volatile
- Explain how and why tourism functions in a nonlinear manner
- Explain the key features of a complex system.

Introduction

The last chapter examined Butler's and Plog's lifecycle models. They suggest that tourism works as a linear, predictable manner, moving logically through a pre-determined lifecycle. While the speed of change may be variable, progression through the stages up to, and possibly including, decline seems inevitable. These models and the models identified earlier in this book share a number of features in common. They recognise that any tourism system begins with the tourist and that any system needs some destination features and a linkage between the tourist and the destination. The models all have a number of advantages helping us understand the constituent parts of tourism and some of the simple linkages between elements. They are also deficient in a number of areas, though, for they do not work in practice.

This chapter takes an alternative view to the evolution of tourism by looking at complexity theory. In the last 20 years, complexity theory has made strong inroads into management disciplines, but has only recently gained interest in the tourism sector (Altinay & Kozak, 2021; Baggio, 2008; Hartman, 2023; Sainaghi & Baggio, 2017; Scott et al., 2008).

Existing tourism models

The idea of examining tourism from the perspective of complexity theory was mooted first by Faulkner and Russell (1997, 1998) and McKercher (1999) who questioned the utility of existing models to explain how destinations change over time. They argued that most models proposed a reductionist approach to tourism where one could understand how it worked by disaggregating it into its component parts, identifying the relationships between parts and then re-aggregating it. Such models make it easy to study tourism's components, but do not necessarily explain how tourism works. Moreover, they tend to be written from the perspective of an overarching public sector tourism organization with the implicit understanding that somehow tourism can be controlled from above by planners, as the World Tourism Organisation asserted 30 years ago (WTO, 1994) and continues to assert.

And yet, they do not explain why it is so hard to control the genie of tourism once the development process starts and why, as Pearce (1989) stated, spontaneous, catalytic and extensive development are the most common types of development observed, especially in emerging economies. It also does not explain why, after almost 70 years of concern about and research into the social, cultural and environmental impacts of tourism, over-tourism continues to remain a vexing problem today. If, indeed, tourism can be controlled, then one would have expected these issues to have been resolved a long time ago.

■ Why existing models do not work

Why do existing models not work? Models represent simplified versions of reality in order to help make sense of complicated ideas. By design, they are selective as to which elements they include and therefore exclude. They tend to focus narrowly on selected destination variables and argue that simple cause and effect relationships exist between these variables. As a result, most of the models cannot appreciate the independent, yet complex interrelationships that exist among the multitude of players involved in tourism, where any tourism business must both coexist with and compete fiercely against other businesses to survive.

The models further fail to reflect the dynamic nature of tourism, where hundreds or thousands of businesses, depending on the size of the destination, enter into and exit the marketplace, change ownership or reposition themselves radically each year. And so, instead of representing a stable, closed system, the tourism system can be seen as an open system subject to constant dynamic interactions with a whole series of internal and external agents (McDonald, 2009; Brouder & Eriksson, 2013). Hartman (2023) adds we must consider tourism simultaneously at multiple levels, including the individual, emergent structures and agency interface that all influence the process of adaption, especially in times of change.

As a result, a degree of instability is both inherent and essential. Indeed, tourism destinations are in a constant state of change to a greater or lesser extent (Harman, 2023), influenced by changing consumer demand, government actions (Altinay & Kozak, 2023) and product innovations (Yeager et al., 2023). From time to time apparently stable systems are thrown into a period of tremendous instability (Farrell & Twining-Ward, 2004) that rattle them to their core only to re-emerge. How a system reacts, the extent of the change and the length of time involved to respond fully to that change, depend on the inherent resilience of a place (Holladay, 2018). In some cases, destinations may emerge unscathed, while in other cases, they may be shaken to their core and be forced to embark on a new path. In extreme cases, decline may occur that could ultimately lead to the virtual abandonment of a destination (Baum, 1998). These perturbations can be caused by external factors such as COVID-19, war, terrorism or natural disaster, internal events involving new developments or financial failure of key players or by technological innovation affecting transportation linkages in the tourism distribution system. Let's not forget how fickle tourists are as well, as they tend to overreact positively to the latest fad or negatively to bad news.

Most existing models focus on stability of systems or their orderly, linear change. They are predicated on predictability and central tendencies and the threat posed by outliers. Yet, outliers are often precursors of change. The models, thus, cannot accommodate the ongoing turbulence that is inherent even in mature, stable tourism communities. Nor can they explain the apparent organic resilience seen in many systems (Baggio, 2008; Cochrane, 2010). Some 40 years ago, Crawford (1984), writing about biological evolution of systems, identified the differences between cumulative change and directional change. Cumulative change refers to the gradual but continuous evolution of systems, built on previous stages, as ascribed to by most existing lifecycle models. Directional change, on the other hand, forsakes gradual change and instead rapidly moves an object into greater complexity and diversity. It is for this reason that de Larrea et al. (2021) suggest destinations need to stop focussing on a small number of issues and instead embrace complexity theory to better understand change and innovation.

In short, at their heart, existing models do not appreciate that tourism operates in a non-linear manner that is akin more to a living ecosystem than to a machine (McKercher & Wong, 2024), with a clearly defined hierarchy of dominant and subservient players and clear inter-relationships between entities. All other tourism activity revolves around these features, including the demand for accommodation, secondary attractions, amenities, services, shopping, other activities, improved access, and the interest in the travel trade to bring tourists to the area. Farrell and Twining-Ward (2004: 277) are more blunt when they state "*the central problem is that tourism researchers schooled in a tradition of linear, specialized, predictable, deterministic, cause-and-effect science, are working in an area of study that is largely nonlinear, integrative, generally unpredictable, qualitative, and characterized by causes giving rise to multiple outcomes, quite out of proportion to initial input."*

A different approach – complexity

Traditional scientific thought, or the Cartesian-Newtonian approach, views systems as machines (Faulkner & Russell, 1997) that may be closed or open ended. To understand how a machine works, all you have to do is break it into its component parts, understand how each part works and how they fit together. Moreover, any machine should work in a predictable manner, where a certain input produces the same output time and time again. Faulkner and Russell (1997, 1998) highlight some other features of the Cartesian-Newtonian approach. To begin, it is assumed that the failure to understand how a system works is due to lack of information, and if enough information can be gathered, then any system can be understood fully. Systems function by fairly simple scientific laws, whereby understanding the laws means that you can control the system. They are deterministic. They tend toward equilibrium and are driven by negative feedback, where individual differences, externalities or other changes noted are seen as being an exception or noise.

Complex systems are dramatically different. Chaos theory, and its companion model, complexity (Lewin, 1993) emerged from the realization that many systems operate in a complex, non-linear, non-probabilistic, non-deterministic and dynamic manner (Gleick, 1987; Kellert, 1993 as per Overman, 1996). Instead of operating in a machine-like manner, these have been likened to living systems where a vast number of elements form an array of rich, subtle, complex and varied relationships. Baggio (2008:3) defines complexity as:

> *"structures which depend only in part on the nature of their constituents and whose global behaviours and characteristics cannot be deduced from the knowledge of their elementary building blocks. Complexity concerns the unforeseen adaptive capacities and the emergence of new properties in systems that arise as the quantity and the quality of the connections among individuals and organisations increase... A system is considered complex if its parts interact in a nonlinear manner. Simple cause and effect relationships among the elements rarely exist and instead a very little stimulus may cause unpredictably large effects or no effect at al."*

Sedarti, Santos and Pintassilgo (2018) assert destinations act as complex systems with numerous interactions between the sectors operating within destinations and multiple stakeholders with varying and conflicting interests. Hartman (2023) adds the interplay between autonomous, external pressures, including disasters and crises, changes in demographics and other factors may trigger actors that affect how destinations develop and evolve over time. Indeed, he states that destinations are in a persistent state of *becoming*. Both argue that the complexity of the problems that emerge in tourism systems and the dynamic interrelationships between and among stakeholders mean that traditional linear thinking has to be discouraged.

At first glance, relationships in such systems may appear to be haphazard and random, but at a deeper level recognizable patterns emerge that help explain the system's functioning as a whole. Because the relationships are open and so complex and because this complexity engenders an innate level of instability, it is extremely difficult to predict accurately the future movement of the system. Indeed, one of the features of complex systems is their inherent adaptability (Stevenson et al., 2009), which is influenced by and in turn influences its environment.

Scott, Baggio and Cooper (2008:176) identify the key features of complex adaptive systems, including:

- *Nondeterminism* – it is impossible to anticipate precisely the behaviour of such a system even knowing the function of all of its elements.
- *Presence of feedback cycles* both positive and negative that influence the overall behaviour of the system.
- *Distributed nature* where many properties and functions cannot be precisely localized for in many cases there are redundancies and overlaps.
- *Emergence and self-organization* where the system exhibits emergent properties that are not directly identifiable or predictable from the knowledge of single components meaning the system continues to revolve in its complexity up to the next self-organization level.
- *Limited decomposability* where the dynamic structure of the system has to be studied as a whole and it is difficult if not impossible to understand how it works by decomposing it into its component parts.
- *Self-similarity* whereby the system will look like itself at different scales if magnified or made smaller. Thus complexity can be used to understand global tourism, as well as tourism as a national, regional, local and even enterprise level.

Rodriguez-Giron and Vanneste (2019) further add that seven features make tourism a complex system, including:

- People and organizations are subject to diverse motivations and conditions.
- Each visitor interacts with a multiplicity of facilitators, within commercial and non-commercial contexts, for the realization of the desired experience (thus operating in their own unique tourism system).
- The tourism phenomenon itself is quite varied and can include independent travel, customised itineraries and prearranged packaged tours.
- Tourism is multidimensional involving both social nature and institutional players (thus necessitating the need to consider both technical and heuristic considerations)

- Tourism is characterized by its multiscalarity
- Tourism is vulnerable to externalities
- Tourism involves intangible personal experiences.

Some other features must also be considered. Complex systems work on the principle of the 'strange attractor'. In business, strange attractors have been likened to a common vision, sense of meaning, strategy or value system that drives people to achieve a common goal (Svyantek & DeShon, 1993; Covey, 1994; Zahra & Ryan, 2007). Altinay and Kozak (2021) indicate that government authorities, and especially health authorities acted as strange attractors during COVID-19. As a result, the system will manage itself, often in an unknowing manner, towards a common goal. But because the system is non-probabilistic, because the relationships between elements are so rich, and because the system is open to external stimuli, it is impossible to predict accurately the future position over time. Instead, the rules will dictate where the system is likely to move within broad parameters. All systems evolve and self-organise to ensure the survival of their components (McDonald, 2009).

The butterfly effect is probably the best known element of complex systems. It states that a butterfly flapping its wings in one location could trigger a weather event in another, distant location. Of course, like most homilies, this one is overly simplistic. Indeed, the butterfly effect is a misnomer. The correct term is "*sensitivity to initial conditions.*" This idea recognizes that small and apparently insignificant changes can precipitate a chain reaction that culminates in a fundamental shift in the structure of the system in one place, but the same action may have no effect somewhere else. Small changes may initiate positive feedback loops that feed on themselves, spinning more and more benefits. Thus it is possible for the initiative of a single entrepreneur to precipitate a major shift in the evolution of a destination, with changes reverberating throughout the entire system.

The corollary of the butterfly effect is the lock-in effect. Inertia within a system is common, whereby an accident of history or initiative produces a strong network of mutually reinforcing relationships that may endure long after the initial conditions have been superseded. The classic example is the computer keyboard. The keyboard we all use was designed for manual typewriters, with the letters organized in such a way that the keys would not jam. The advent of electronic keyboards overcomes the problem of jamming. Yet, the keyboard design remains unchanged.

Both the butterfly and lock-in effect relate to the feedback cycle of complexity. Linear systems theory is predicated on negative feedback where systems move to a stable state (Stevenson et al., 2009). Tourism functions as much by positive feedback as it does by negative feedback. There is a reason why the strong seem to get stronger and the weak seem to get weaker, and that is the role of positive feedback.

Mass tourism destinations, such as Waikiki in the US state of Hawai'i, are classic examples of how positive feedback can create a lock-in effect giving them an apparent insurmountable competitive advantage. The first hotels were built here in the late 1800s and early 1900s. The advent of jet aircraft, coupled with Hawai'i's statehood and the discovery of the area by celebrities, spurred massive development in the 1960s. Waikiki has now about 170 properties offering more than 30,000 units of accommodation, making it by far the largest resort region in the state. For a variety of reasons this position is unassailable.

Phase shift and the associated idea of the edge of chaos recognize that from time to time otherwise stable systems receive such profound shocks that they are shifted from one state to another. Chaotic systems may have experienced prolonged stages of relative stasis, interrupted by (r)evolutionary change or phase shift. A feature of large, interactive, dynamical systems is that they evolve naturally toward the edge of chaos (Lewin, 1993; Altinay & Kozak, 2021). We are all familiar with how COVID-19 had such a devastating impact on the tourism industry. What some do not realise, though, it that is also created the need to re-establish stability and a sense of order in a new tourism system.

This observation leads to another feature of complex system, *bifurcation*. Bifurcation can represent the flashpoint of change (Altinay & Kozak, 2021) where a system's character and structure are disrupted, forcing it to move in possibly different directions than in the past. Zahra and Ryan (2007: 855) note that contrary to the usual representation of the destination life cycle as presented by Butler's diagrammatic use of the S-shaped curve, tourism development is often characterized by nonlinear spurts and interventions that are significant. Like any open system, the tourism system can be exposed to exogenous shocks.

Because complex systems are self-organizing, they have tremendous adaptive ability and are remarkably resilient. It has been observed that living systems, when confronted with change have the capacity to fall apart so that they can reorganize themselves to be better adapted to their current environment (Flower, 1993). Within a tourism context, Faulkner and Russell (1997) note otherwise apparently stable systems may be upended from time to time and completely reorganised, resulting in a new, stronger and more resilient system. Cochrane (2010) comments on how quickly a system can change depends on its adaptive capacity, as a function of its ability to absorb disturbance and reorganise itself. It may emerge as fundamentally the same type of entity or re-emerge in a substantially different form. Indeed, self-organising systems may result in the emergence of radical forms, structure and procedures, fundamentally changing the existing world (Altinay & Kozak, 2021).

Components of a complex system

McKercher (1999) proposed an open-ended complex tourism system comprised of nine different sets of actors. It is reproduced in Figure 7.1. While the model is inelegant, it does identify key component and their roles. The actors include, and are described briefly below:

- The *Traveller*, who is the essential player in tourism, for without people travelling no tourism would occur;
- The *Communication* vectors used to connect the traveller to the destination;
- The *Considerations* or factors that influence the effectiveness of the Communication vectors used;
- The *Destination* or *Internal tourism community* consisting of all businesses involved in tourism at the destination;
- *External tourism agencies* (public and private sector) that try to influence tourism;
- *Other tourism-related externalities*, such as alternative tourism destinations that affect a destination's ability to attract travellers;
- *Non-tourism related externalities*, or macro-environmental forces, such as changing political, economic or social conditions, war, pr natural disaster, that affect people's ability to travel;
- *Outputs* from the system – both desired and undesired, and;
- *Rogues* or *Chaos makers* who can push a system to the edge of chaos.

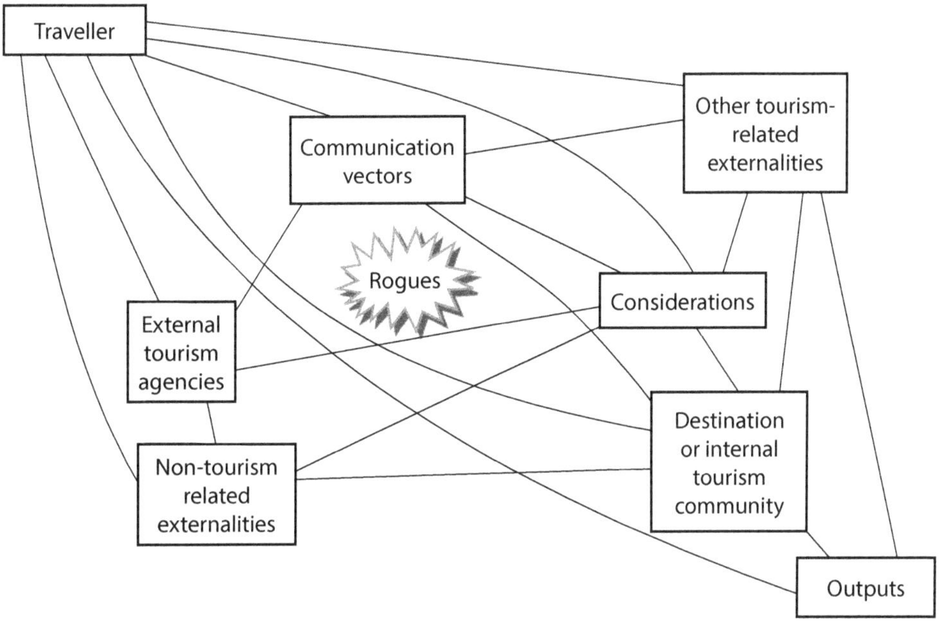

Figure 7.1: A complex tourism system. Source: McKercher 1999

Essentially, tourism is about people travelling, with the business of tourism about linking tourists to destinations. For this reason, the traveller or tourist must be the starting point in the consideration of any tourism model. Markets are dynamic, erratic and non-linear (Faulkner, 1997; Pike, 2005), with tourism markets, in particular, being noted by their great volatility (Var et al., 2006). If the key input does not behave in a rational manner, then the entire system cannot function in a deterministic manner.

Communication vectors connect the traveller with the destination and include both formal marketing activities as well as the dissemination of markers. Communication channels are exceedingly complex (Pearce et al., 2004), involving all levels of the formal travel trade, other tourism retailers, the efforts of public sector marketing and promotion agencies and direct communication between businesses and the travelling public.

However, the effectiveness of different communications activities is influenced by a number of considerations. Considerations from a consumer's perspective, relate to getting the right information, while from the destination's perspective they relate to resource availability and expertise. Other factors such as distance, travel party composition and willingness/ability to respond positively also affect the likelihood of connecting the tourist to the destination.

The destination or the internal tourism community describes the businesses that exist within any destination area. The word 'community' is used in its ecological sense, where a community is described as a living, dynamic, self-organizing system with a group of organisms interacting with one another, cycling and recycling matter through the system. Anything crossing the boundary of that community is considered as either being an input or an output. Importantly, the community as a whole is reliant on the ongoing health of its keystone species.

External tourism agencies, those public and private sector agencies that function outside the commercial tourism world, can have an effect on most of the elements within this model. No product, destination, region or country operates in isolation. Its success depends on how well it competes against other tourism products and how well they compete against it.

Tourism initiatives that occur elsewhere can have a fundamental impact on the competitiveness of a destination, just as initiatives within the system can affect the competitiveness of other destinations. For this reason, a variety of other tourism related externalities need to be included in any tourism model.

Other tourism related externalities can have a dramatic impact on tourism activity. Lifu Island, Vanuatu has benefited from the development of South Pacific cruises originating in Australia. A vibrant community-based tourism sector has been developed to cater to day trippers from the cruise ship. Sometimes the net result can be negative, where the development of new destination areas may lead to the decline in other areas, especially if market access has changed.

Two other elements have the potential to plunge tourism into chaos: one is external to the tourism system; the other is internal. The role of non-tourism related externalities is well recognized in the tourism literature. These occur rapidly and without warning. Previously stable tourism communities are plunged into a state of chaos, usually caused by a dramatic fall in visitors. Certainly, one does not need to look further than the COVID-19 outbreak that devastated global tourism.

Tourism is not immune from global events, such as war, terrorism, natural disasters or disease. A vast array of non-tourism related externalities can dramatically impact the sector. The COVID-19 outbreak posed the greatest threat to the global economy in almost 80 years as it brought tourism to a standstill.

What is less well recognized is the role of chaos makers, or rogues operating within the tourism system (Faulkner and Russell, 1998. Russell & Faulkner, 1999). These individuals can single-handedly transform an organization, destination, region or country.

Chaos makers or rogues can singlehandedly transform tourism. Global travel changed in 1841, when Thomas Cook organized a day trip by train for 570 people to travel in England. By 1845 he was organising trips on a fully commercial basis and in 1855 took the first guided tour to Europe.Walt Disney invented the theme park in the 1950s. Conrad Hilton created the world's first global hotel chain. Freddie Laker introduced the idea of low cost carriers in the 1960s and shook up trans-Atlantic travel. When Tony Fernandes took over the struggling Air Asia, he created Asia's first international low cost carrier network that has revolutionised travel in this part of the world.

Lastly, this model considers the outputs of tourism as an intrinsic aspect of any tourism system, for the impacts or perceived impacts of tourism at the host destination or other destinations are increasingly affecting its development. Governments support tourism out of the belief that it will provide a variety of positive outcomes, such as employment, enhancement of the tax base or the provision of foreign currency. In addition, tourism is seen as a powerful tool to help modernize societies. On the other hand, opponents of tourism citing its legacy of adverse impacts have successfully blocked tourism development proposals. Positive or negative, the impacts of tourism are playing a greater role in development decisions.

■ What does it mean?

Complexity theory helps explain the evolution/revolution of destinations better than existing models. It also helps identify policy avenues that can help resolve apparently intractable tourism issues.

The adoption of complexity theory means that destinations do not have lifecycles per se, nor do they evolve in a predictable manner. Instead they tend to go through periods of relative stability before they are fundamentally shaken and transformed. Stability may include an increase, decrease or no change in arrivals. Destinations may also enjoy a period of development in response to current market situations, but, at some point this period of stability ends, caused by either a conscious choice on behalf of the destination's managers and governing bodies, or by one or more external events beyond the control of the destination. The resulting evolutionary pattern as identified by Prideaux (2000). He integrated elements of Butler, Plog and chaos into his model of resort development (Figure 7.2).

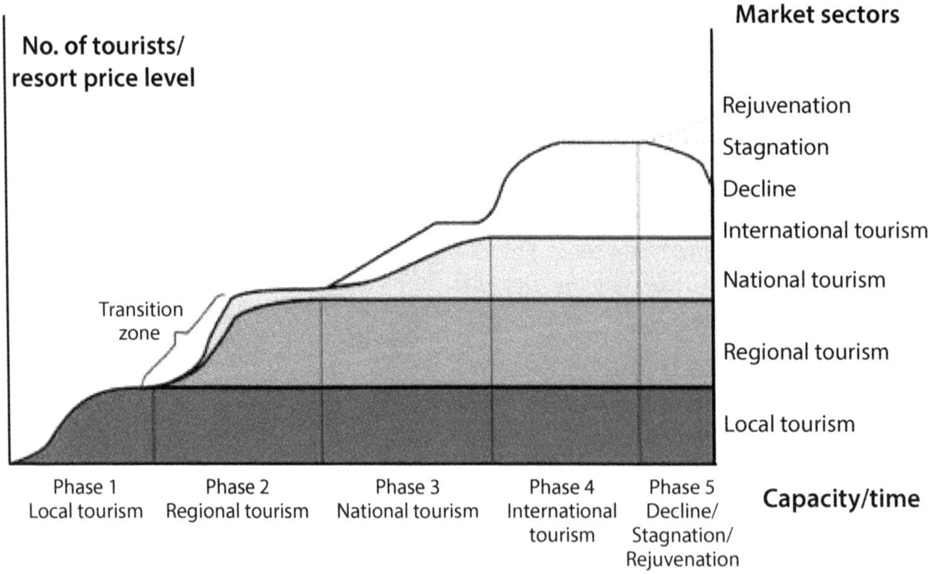

Figure 7.2: Prideaux's Lifecycle Model. Source: Prideaux 2000

Whereas Butler and Plog regard destinations or a destination's markets as single entities, respectively, Prideaux recognizes their complexity. He argues that catalyst to destination evolution is the emergence of new markets that stimulate new development and in turn, transform the destination. He also identifies a number of elements of change that push destinations to the edge of criticality and initiate phase shift.

Surprisingly little research has been conducted examining this topic, mostly because it is difficult to conduct a sufficiently long duration study to document the impacts and their resultant consequences. It must be recognised, though, that the revolutionary events that push places to and beyond the edge of chaos are fundamentally different than the occasional shocks that disrupt systems over the short term, but leave no long lasting impact. Natural disasters and crises can rattle systems, but their impact is usually transitory. Instead, what we are talking about is transformative events that completely change the tourism system.

Conclusion

Much has changed in our thinking of how tourism has evolved over the last 40 years. Systems models and the Cartesian-Newtonian approach that try to make things ordered nicely have largely been questioned. Instead, we now appreciate that tourism is a far more complex activity involving a myriad of stakeholders at all levels who interact in a manner that is hard to model and predict accurately. The complexity model proposes an alternate structure for tourism that suggests we need to consider externalities as well as internalities when looking at the structure and evolution of tourism. It also suggests that the nature of destination change is more revolutionary than evolutionary.

■ Discussion points / Exercises

1. Compare and contrast Plog's and Butler's models with the complexity model described in this chapter. Where are they similar and where do they diverge?
2. Identify different chaos makers or rogues and discuss how they have shaken the tourism system.
3. Complexity looks at longer term evolution of a destination. Can one-off crises alter a destination's lifecycle? If so, how?
4. Plog suggests the less robust tourist displaces the more robust one. Prideaux suggests the results are cumulative. Argue which case you feel is more applicable in tourism.

Points to consider for revision

- Tourism is a fundamentally unstable activity that is guided by the rules of complexity and chaos theory
- Tourism systems function much like living ecosystems
- Phase shift, the lock in effect and the butterfly effect are particularly relevant to tourism

Additional readings

Baggio, R. (2008) Symptoms of complexity in a tourism system. *Tourism Analysis* 13(1), 1-20.

Faulkner, B. & Russell, R. (1997) Chaos and complexity in tourism: in search of a new perspective. *Pacific Tourism Review* 1, 93-102

Prideaux, B. (2000) The resort development spectrum – A new approach to modeling resort development. *Tourism Management* 21, 225-240.

References

Altinay, L. & Kozak, M., (2021). Revisiting destination competitiveness through chaos theory: The butterfly competitiveness model. *Journal of Hospitality and Tourism Management*, 49, 331-340.

Baggio, R. (2008) Symptoms of complexity in a tourism system. *Tourism Analysis* 13(1), 1-20.

Baum, T. (1998). Cases from the periphery, in Faulkner B., Laws, E. & Moscardo, G. (Eds) *Embracing and Managing Change in Tourism: International case studies*, Routledge. pp. 117–138.

Brouder, P., & Eriksson, R. (2013) Tourism evolution: On the synergies of Tourism Studies and Evolutionary Economic Geography. *Annals of Tourism Research*, 43, 370-389 http://dx.doi.org/10.1016/j.annals.2013.07.001

Cochrane, J. (2010) The sphere of tourism resilience, *Tourism Recreation Research*, 35(2), 173-185, DOI: 10.1080/02508281.2010.11081632

Covey S. R. (1994) The strange attractor. *Executive Excellence*. 11(8), 5-6.

Crawford, M. (1984) Business took the wrong life cycle from biology. *Journal of Consumer Marketing* 1(3), 5-11.

de Larrea, G.L., Altin, M., Koseoglu, M.A. & Okumus, F., (2021). An integrative systematic review of innovation research in hospitality and tourism. *Tourism Management Perspectives*, 37, 100789.

Farrell, B., & Twining-Ward, L (2004) Reconceptualizing tourism. *Annals of Tourism Research*, 31(2), 274-295.

Faulkner, B. & Russell, R. (1997) Chaos and complexity in tourism: in search of a new perspective *Pacific Tourism Review* 1, 93-102.

Faulkner, B. & Russell, R. (1998) *Movers and shakers: The chaos makers on the Gold Coast*. Paper presented at the 1998 Australian National Tourism and Hospitality Research Conference, Gold Coast.

Flower, J. (1993) The Power of Chaos *Healthcare Forum*. 36(5), 48-55.

Gleick, J. (1987) *Chaos: Making a New Science* Abacus Books, London

Hartman, S. (2023). Destination governance in times of change: a complex adaptive systems perspective to improve tourism destination development. *Journal of Tourism Futures*, 9(2), 267-278.

Holladay, P. J. (2018). Destination resilience and sustainable tourism development. *Tourism Review International*, 22(3-4), 251-261.

Lewin, R. (1993) *Complexity: Life on the Edge of Chaos* Phoenix, London.

McDonald, J. (2009) Complexity science: an alternative world view for understanding sustainable tourism development. *Journal of Sustainable Tourism* 17(4), 455-471.

McKercher, B. (1999) A chaos approach to tourism. *Management* 20(4), 425-434.

McKercher, B., & Wong I (2024) Can destinations have multiple lifecycles? In Butler R. (Ed) *Tourism Area Lifecycle: Review, Relevance and Revision.* Channelview Publications

Overman E. S (1996) The new science of administration: chaos and quantum theory *Public Administration Review*. 56(5), 487-491.

Pearce, D. (1989) *Tourist Development 2nd Ed.* Harlow: Longman.

Pearce, D., Tan, R., & Schott, C. (2004) Tourism distribution channels in Wellington, New Zealand *International Journal of Tourism Research* 6(6), 397-410.

Pike, S (2005) Tourism destination branding complexity. *Journal of Product & Brand Management*, 14(4), 258-259. https://doi.org/10.1108/10610420510609267.

Prideaux B (2000) The resort development spectrum – A new approach to modeling resort development. *Tourism Management* 21, 225-240.

Rodriguez-Giron, S. & Vanneste, D., 2019. Tourism systems thinking: towards an integrated framework to guide the study of the tourism phenomenon. *Tourism Culture & Communication*, 19(1), 1-16.

Russell, R. & Faulkner, B. (1999) Movers and shakers: chaos makers in tourism development, *Tourism Management* 20(4), 411-423.

Sainaghi, R., & Baggio, R (2017) Complexity traits and dynamics of tourism destinations. *Tourism Management*, 63, 368-382

Scott, N., Baggio, R., & Cooper, C. (2008) *Network Analysis and Tourism: From Theory to Practice.* Clevedon: Channel View Publications

Sedarati, P. Santos, S & Pintassilgo, P., (2018): System dynamics in tourism planning and development, *Tourism Planning & Development*, DOI:10.1080/21568316.2018.1436586

Stevenson, N., Airey, D., & Miller, G. (2009) Complexity theory and tourism policy research. *International Journal of Tourism Policy* 3(2), 206-220.

Svyantek D. J. and DeShon R. P (1993) Organisational attractors: A chaos theory explanation of why cultural change efforts often fail. *Public Administration Quarterly*, 17(3), 339-355.

Var, T., Kaplan, M. & Yurt, O. (2006) Challenges and opportunities of developing wine tourism in a small community in Turkey. *e-Review of Tourism Research,* 4(1), 9-18.

WTO (1994) *National and Regional Tourism Planning: Methodologies and Case Studies,* Routledge, London.

Yeager, E., Boley, B.B. & Goetcheus, C., (2023). Conceptualizing peer-to-peer accommodations as disruptions in the urban tourism system. *Journal of Sustainable Tourism,* 31(2), 504-519.

Zahra, A., & Ryan, C. (2007) From chaos to cohesion—Complexity in tourism structures: An analysis of New Zealand's regional tourism organizations. *Tourism Management* 28(3), 854-862.

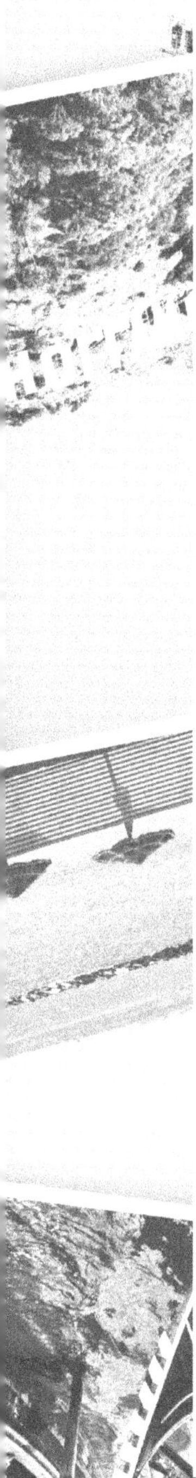

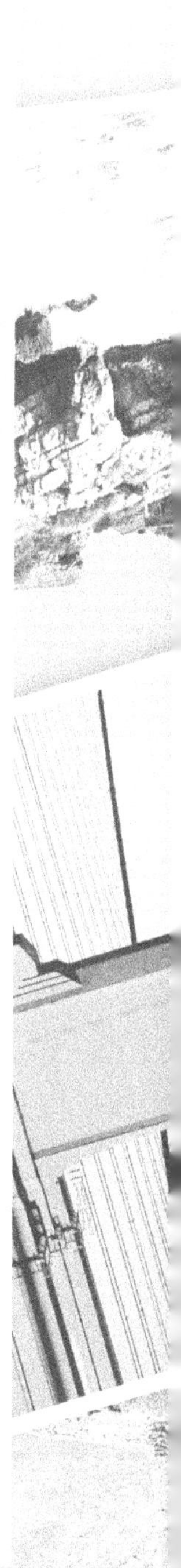

Part 4 – The Tourist

So far, the book has looked at defining tourism, understanding its constituent parts to see how tourism works and examining different lifecycle models. Now it is time to consider the central component of tourism – the tourist. The tourist has been the subject of much research by sociologists, anthropologists, psychologists, members of the leisure and recreation community and others for close to 60 years. The next few chapters review their thoughts.

Chapter 8 asks the important question of why do people travel? This chapter explores a range of progressively sophisticated models and theories that try to explain tourist motivation. It begins with a review of Maslow's hierarchy and concludes with a discussion of Pearce's current Travel Career Pattern.

Chapter 9 considers the thoughts of sociologists and anthropologists. It begins by delving into ideas that were prevalent in the 1960s and then considers such core concepts as the environmental bubble, tourism as a quest and the liminal nature of travel.

Chapter 10 focuses more on how the leisure and recreation community views tourism in general and issues relating to propensity and intensity in particular. This chapter examines how the evolving literature on barriers to travel translates into different attitudes toward travel and subsequent travel patterns, the confounding role of the leisure paradox and how generational cohort theory influences travel.

Chapter 11 focusses on how tourists consume destinations. Tourism occurs within a local destination area. That is where the attractions, activities and experiences are located and where the positive and negative impacts occur. This chapter begins by conceptualising what a local destination area is and then explores a number of hitherto under-examined theories and models explaining why tourists behave the way they do in destinations.

8 Why Do People Travel?

By the end of this chapter, the reader will be able to:

- Describe the strengths and weaknesses of Maslow's hierarchy of needs
- Analyse Dann's push-pull model
- Apply Crompton's motives
- Critique Iso-Ahola's optimal arousal model
- Apply Pearce's evolving travel career ladder/pattern.

Introduction

Understanding why people travel for pleasure has been the focus of tourism research since the earliest days of tourism academia. Iso-Ahola (1983:50) states "*while many questions about recreational travel remain to be raised and answered, perhaps the most fundamental of them all are the ones that deal with motivation.*" As Song and Bae (2018) note in their review of the literature, motivation influences destination choice, perceived benefits, satisfaction and on-site experiences.

A number of increasingly sophisticated models has been developed over the years that has attempted to develop a comprehensive theory of tourist motives. Initially models were adapted from consumer behaviour literature. Recently, though, tourism specific theories have been developed. The task has been challenging for the underlying reasons to travel are covert in that they reflect individual's private needs and wants (Pearce, 2005). Many individuals have difficulty articulating the reasons why they travel (Crompton, 1979), potentially leading to method and response bias if the reasons are posited in a closed ended questionnaire. As a result, many of the earliest attempts to examine the issue conflated motives with activities pursued and trip purpose (see Lundberg 1971), while other popular models tended to identify unmet needs as motives, even though they are different (Tasci & Ko, 2017).

Another challenge with trying to identify motives is that different disciplines adopt different approaches to explore this concept. Jamal and Lee (2003) argue, for example, that social psychology and sociology represent the two major discipline areas that have influenced this area of research. Sub-disciplines within each host discipline, in turn direct studies. As a result, the study of motives is typical of Smith et al.'s (2013) type two theory where many legitimate, but competing models and/or theories exist. For example, Dann (1981), Page (2019) and Jamal and Lee (2003) identify a range of approaches that have been adopted over the years, including:

- Travel as a response to what is lacking yet desired;
- Destination pull in response to motivational push;
- Needs based approaches where pleasure travel is a means to satisfy unmet needs;
- Intrinsic motivations as reflected by a desire to obtain some reward for travelling;
- Values based approaches;
- Disequilibrium and the need for a break;
- Motivation as fantasy and a liminal break;
- Expectancy based approaches which look at achieving outcomes as a prime motivation;
- Quest for authenticity and a search for one's centre;
- Motivation as a trip purpose, etc.

An additional challenge in developing motivation theory was highlighted by Pearce (2011:43) who stated

> *"The criteria for a sound motivational framework highlight some of the unique attributes of tourism ... its episodic, dynamic, relationship-dependent, future-oriented and varied character ... Good motivation theory [thus] needs to be multi-motive, dynamic, measurable and relatively easy to communicate. In addition, it is most essential that the approach functions as an organiser and synthesizer of existing information, with a desirable characteristic that it has a future-oriented or predictive capacity."*

This chapter explores the evolving literature on motives from Maslow through to the Travel Career Pattern proposed by Pearce and his colleagues.

Before beginning though, it is important to define terms. Dann (1981) explains that etymologically motivation is derived from the Latin word '*movere*' to move. Pearce (2005:54) defines motives as the forces that drive travel behaviour. They can be biological and socio-cultural. Iso-Ahola (1982:257) defines them as internal factors that arouse, direct and integrate a person's behaviour. Moutinho (1987:16) identifies motives as a state of need, a condition that exerts a push on the individuals toward certain types of action that are likely to bring satisfaction.

In a similar manner Mill and Morrison (2012) suggest a distinction is needed between needs and wants. A need is an amorphous concept. People may only be partially aware of their needs. Some people say it is a latent, or hidden element. A want is the concrete definition or transformation of a need. The difference between a need and a want is awareness.

Maslow's Hierarchy of Needs

In many ways, Abraham Maslow is the father of motivation research, with his work playing a significant role in tourism motivation research. He sought to understand why people are driven to act, by trying to understand how particular needs can be met at particular times (Kotler et al., 2003). He posited that human needs can be classified in a hierarchical manner (Maslow, 1943) from the most pressing to the least pressing. His theory suggests lower order needs had to be met before higher order needs can be considered, but once met, they fade into the background as we strive to have higher order needs satisfied. Maslow (1943) also asserts that for most people the hierarchical nature of needs is relatively fixed. He does recognise, though, that needs are related to each other, being arranged in a hierarchy of prepotency. As he explains "*the most prepotent goal will monopolize consciousness and will tend of itself to organize the recruitment of the various capacities of the organism. The less prepotent needs are minimized, even forgotten or denied* (Maslow, 1943: 394, 395)."

His model (Figure 8.1) can be divided into deficiency needs and growth needs (McLeod, 2018), with the first four levels referring to deficiencies that need to be fulfilled before the top level growth need of self-actualization can be met. It is typically portrayed as a pyramid, with its dimensions described below:

- *Physiological* needs of food, water, sex, air, sleep clothing;
- *Safety and security* needs encompassing freedom from threat or danger or a secure, orderly and predictable environment;
- *Social* needs like love and belonging needs that reflect feelings of connection to and affectionate relationships with family, friends, teammates or one's community;
- *Esteem* needs of self-respect, accomplishment, reputation, recognition and prestige; and
- *Self-actualisation* needs that fulfil one's potential

(Hsu et al. 2008).

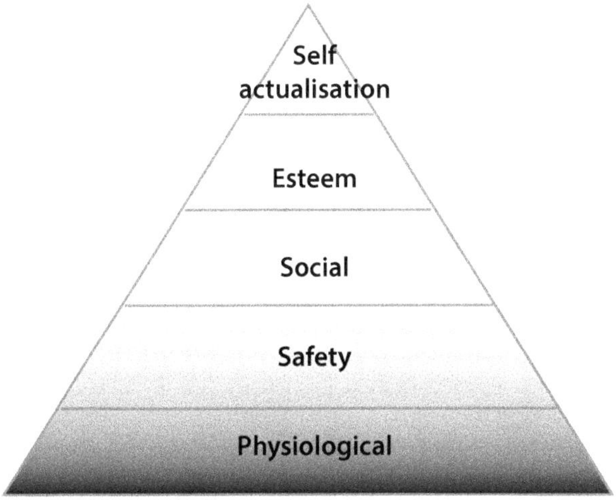

Figure 8.1: Maslow's Hierarchy of Needs

The model is intuitively appealing, in part because of its parsimoniousness and in part because we can all relate to its dimensions. It has been used so widely in tourism that a *Google.Scholar* search using the keywords *Maslow* and *Tourism* yielded over 36,000 results and more than 13,000 since 2020, covering every conceivable topic from destination marketing to food tourism consumption, motives for volunteer tourism, seniors' travel, impacts and residents' perceptions, tourism planning, quality of life, dental tourism and many, many more topics. Virtually every tourism marketing and consumer behaviour text, along with most introductory tourism texts refer to it. Yousaf et al. (2018) and others have also provided a range of examples of how destinations can provide goods and services that can satisfy each tier of needs.

There is only one problem with Maslow's Model. There is very little evidence to support it in a tourism context. It is particularly difficult to apply Maslow's hierarchy in understanding why people travel, although it may be useful in understanding how tourism experiences are delivered. A number of criticisms of the theory have been raised in both the tourism and mainstream psychology literature. To begin, there is scant empirical evidence to support the conclusions (Kaur, 2013; Kay, 2003). The reason is that needs are not necessarily hierarchical nor are they singular (Denning, 2012). You may be able to seek social interaction without necessarily having physiological and security needs met. Indeed, that is the very basis of much commercial adventure tourism, where social interaction is a driver of participation and yet people willingly place themselves in some physical risk. Maslow suggests such actions are incompatible.

The observation that needs are not singular is also challenging. Again, using commercial adventure tourism as an example, the drivers of participation may be physiological, social and esteem related, and in some cases self-actualisation may come into play. All may be equally powerful in the participation decision, with no single item standing out. This issue is something that has vexed Pearce in the development of his travel career pattern. More on this later.

Moreover, from a tourism standpoint, the same action may satisfy many needs simultaneously, defying simple categorisation. Eating may satisfy a physiological need if the person is hungry, a relationship need if dining with friends and relatives, a self-esteem need if eating at a fine dining restaurant, or perhaps even a self-actualisation need if the person is a culinary tourist and has always wanted to eat a certain type of food in a certain destination.

One of the challenges in applying Maslow's hierarchy in a tourism context is that the same action may satisfy different needs. Here in Vienna, for example, sampling local food migh satisfy physiological needs of hunger, social needs of dining with friends and family, esteem needs of consuming something unique in a foreign country or perhaps even self actualisation needs if you always wanted to travel to Vienna and partake in its tangible and intangible culture. Which need is being satisfied, then depends on the context of the activity.

Likewise, some core tourism motives are not covered by the hierarchy or can be subsumed into multiple tiers. Pearce and Lee (2005), for example, identify novelty and escape/relaxation as two of the four motivational backbones behind all travel decisions, yet they do not easily fit into the traditional hierarchy of needs. Finally, the causal link between need and behaviour is not as absolute as desired.

Indeed, Pearce and Packer (2013: 389) conclude that "*in some sixty years after its initial formation, it can be suggested that the value of Maslow's work lies not in the hierarchical system associated with his studies, but resides more simply in the diversity of motive forces he identified.*"

Plog

The same observation can be made about Plog's (1974, 2001) models. Good ideas, but not applicable in practice! Plog's ideas have been discussed in detail in Chapter 6 and need not be repeated here again, other than to acknowledge that his Allocentric/Psychocentric continuum has been used as a proxy to understand tourist motives. Critics of Plog's model as a motivational theory state that it neither explains tourist motivations nor predicts tourist behavior (Huang & Hsu, 2009). Litvin and Smith (2016), reviewing the literature, note many people have questioned how such a simplistic model that gives a person an absolute label can predict travel behaviour when tourists travel with different motivations on different occasions. Yoo et al. (2018) also show little empirical support for his work. One other criticism is that he lumps two-thirds of the travelling public into the same Midcentric category, which proves to be of little utility when trying to segment the market.

What his works do, though, is recognise that tourists are not a homogeneous group and that a variety of factors influence their likelihood of travelling and the type of holiday they may prefer ideally. He also argues effectively that one's underlying social-psychological make-up may set the upper limits of the type of travel one is comfortable with unless a large environmental bubble is created. It must be placed within the historical context of the lack of tourist motivation theory some almost 50 years ago. Early researchers, such as Lundberg (1971) were among the first to write about why people travel. But in hindsight, this work is rather simplistic, for it conflates purpose (sports, adventure tourism), with activities (such as shopping), motives (ego enhancement, escape, personal growth) and other factors (socially acceptable behavior, acquiescing to cultural norms). Plog's work represented a paradigm shift in that he tried to associate one's underlying personality traits to travel preferences. It is just that this work has been superseded by more sophisticated research.

Dann's push–pull model

The late 1970s and early 1980s saw a flourishing of research on motives by such scholars as Dann (1977), Crompton (1979) and Iso-Ahola (1982). Much of this work is still relevant today.

Dann's (1977) push-pull model is arguably the best known and most ubiquitous one relating to motivations (Crompton, 1979. Uysal et al., 2008. Prayag & Ryan, 2011). Its popularity lies in part because it provides a simple and intuitive approach to define tourist behaviour (Klenosky, 2002) and in part because it attempts to show a link between drivers of travel and destination choice (Yousaf et al., 2018). It also builds on and/or informs many other motivational works, including Maslow's hierarchy, Crompton's work and to some extent Pearce's travel career approach

(Pearce 2011, Prayag & Ryan, 2011). Moreover, Klenosky (2002) states push and pull factors have generally been characterized as relating to two separate decisions made at two separate points in time—one focusing on whether to go, the other on where to go. Below is a summary of the key points of this important work.

Essentially, Dann (1977) argues that push factors are those factors that predispose tourists to travel, while pull factors are the features of the destination that attract tourists to certain places. He illustrates that most of the research conducted to date (remember 1977) focussed on the pull factors at destinations, in much the same way that marketing theory on the time focussed on the pull factors of products (the 'if you build it they will come' syndrome). Push factors were largely ignored. Yet, he argues that while destinations may provide a range of products and services, the actual decision to visit is predicated on a prior need to travel. Push factors then are the antecedent considerations before pull or destination factors can be considered. Thus, he states the question of what makes people travel can only be answered by considering push factors, whereas the question of why they choose certain destinations may involve both push and pull considerations.

Push factors drive people to travel, while pull features are features of the destination that serve to attract people and satisfy their needs. Graham Dann argued two factors, ego enhancement and anomie drive tourism. Ego enhancement, coupled with the desire to learn about exotic lost civilisations, drives many people to visit places like Petra in Jordan. Alternately, anomie, or the desire to get away from it all is a powerful push factor to explore remote regions like Qatar. In other instances, anomie may be the driving force behind seeking a sun, sand and sea resort holiday, where people simply need a short break from their everyday life.

He argues that the motivation to travel lies in the twin factors of anomie and ego-enhancement. Anomie refers to a state of disequilibrium from an anomic or socially disoriented society. Thus, one push factor is the desire to get away from it all or transcend the isolation of everyday life. Tourism provides an opportunity for stronger and more influential social interaction for those living in an isolated world, or conversely social isolation if living in a world where privacy is at a premium. Both needs can only be fulfilled when away from the normal environment.

Ego-enhancement too derives from personal needs, where from time to time we need to have our ego boosted. One strategy is to travel where a tourist can go to a place where his/her social position is unknown and, consequently, can feel superior on that basis on such anonymity. Moreover, status can be gained when the person returns from the trip.

These two push factors are quite limited and do not explain all types of pleasure travel. Both Crompton's (1979) and Pearce's (2011) various studies expand the range to a far more comprehensive list.

Combined, Dann asserts that travel can represent the fulfilment of basic needs, while offering an alternative world to that in which the tourist lives. It also lets the tourist behave in a way that is normally circumvented in the daily life, by enabling subterranean values to be released and allowing the person to reside, temporarily in a fantasy world. Such a fantasy world relieves the humdrum work of the anomic life, while providing the opportunity for ego-enhancement by acting out in an alien personality.

Pull factors, on the other hand, represent those attributes of the destination that appeal to tourists. Fyall et al. (2019) characterise them as encompassing social and cultural features of the destinations, its physical characteristics and the range of attractions that exist to satisfy needs. Indeed, Uysal et al. (2008) indicate the number of pull factors is limited only by the scope of the destination. It is therefore, easier to define and measure pull factors. They suggest that the attitude of a tourist toward a vacation destination may be a measure of that destination's ability to pull or attract the tourist. But, as Pearce (2011) indicates, pull factors are variable and as such are hard to isolate as key features that induce travel to a destination. He illustrates that one activity or destination attribute can be valued by different travelers for very different reasons. Nikjoo and Ketabi (2015) found pull factors varied depending on the motive of the tourist. Those who were driven by anomie and wanted to escape tended to select resort style vacations, while those who were travelling for ego enhancement chose more culturally oriented holidays.

It was believed initially that the push and pull factors occurred sequentially (Dann, 1977), in that people were pushed to travel and then assessed the pull features of destinations to satisfy their needs (Lewis et al., 2010). It is for this reason that Mutinda & Mayaka (2012) argue that push factors are more significant than pull features as determinants of destination choice. Others, though, suggest that the process is interrelated and interdependent. People may be pushed by their own internal forces and simultaneously pulled by the external forces of the destination and its attributes (Klenosky, 2002). Indeed, the link between push and pull forces is more artificial than real (Uysal et al., 2009) for people may not be able to articulate their push reasons explicitly, while searching for destinations that can satisfy an amorphous need.

More recent research suggests either push or pull factors or both may play a role, depending on the nature of the experience desired. Pull features seem to play a more critical role in nature based tourism than push features, as Arowosafe et al. (2022) found in the appeal of waterfalls and rock climbing in Nigeria, and Valencia et al. (2023) noted in farm stay travel. In some instances, the pull features associated with last chance tourism (Salim et al., 2022), new and emerging destinations (Tojib, et al., 2023), must-visit destinations (Alrawadieh et al., 2023) or unusual places such as North Korea (Shi et al., 2023) that enable people to escape their cultural practices at home (Michael et al., 2022) are powerful draw cards. In other instances, more prosaic pull factors may be at play, such as the chance to participate in sports (Mishra et al., 2022) or to visit attractions that are unavailable in one's home community, such as casinos (Carvalho et at., 2023).

It is difficult to separate pull features of destinations from associated push motives, though, for the type of pull features identified above are likely a co-function of push features related to escape (nature-based and farm tourism) or ego-enhancement (last chance tourism, must-visit places and new destinations). It is for this reason that Dancausa et al. (2023) state the pull factors of a ghost tour can be fostered by various push factors, and that thee push motives of the quest for novelty and emotional experience may match the pull motive of the ghost tours to attract participants.

In other instances, push motives are clearly the dominant driving force, although, the type, influence and weighting of push factors is culturally dependent (Soldatenko et al., 2023). Matiza and Kruger (2022) note domestic tourists are motivated primarily by relaxation and self fulfilment needs, while Smith et al. (2022) identified a range of push factors of cultural tourists that played a more important role than the pull of the attraction. Interestingly, Chinese LGBT tourists who travel to Thailand are pushed by their desire and ability to express their sexual identity in a more welcoming environment (Subyubon et al., 2023). Somewhat ironically, Liu (2023) also identified government policy as a powerful push factor in nations with a large trade surplus that are facing significant external pressures to reduce trade imbalances.

Crompton's Motives

John Crompton's (1979) work titled *Motivations for Pleasure Vacation* is one of the most influential motivational studies published in the last 40 years. It has received over 6500 citations and again is referred to in most texts. As Crompton notes (1979: 409), motivations are only one of the variables that may explain tourist behavior, they are considered to be the crucial variable because they are the impelling and compelling force behind all behavior. This relatively simple, though elegant study involved 39 unstructured interviews with a convenience sample of adults. While the study explored a range of socio-psychological variables, motives were a primary focus of the research. Interviews were recorded and content analyzed to identify patterns of responses which provided the basis of the conceptual issues identified in the study. The key features of the study are summarized below.

Essentially, Crompton's work is another form of push study. He began by identifying a four stage continuum in the travel decision process. Like Dann before him, and as strongly inferred by Jafari (1987) and Cohen (1979), the process begins with the recognition of a state of disequilibrium, which could best be satisfied by the second stage, the recognition of the need for a break from existing routine. Disequilibrium could be classified into two categories, short term, and long term. Short term disequilibrium occurs when a particular set of circumstances or events triggers disruption to homeostasis. It demands immediate attention and could be satisfied by a single vacation. In many ways, short term disequilibrium explains the popularity of short break vacations. Long term disequilibrium has deeper causes. It is considered ever present but postponable and can only be resolved by a series of pleasure vacations. In some ways, as well, short term disequilibrium is reflected in Cohen's Recreational tourist who travels to re-create and rejuvenate, while long term disequilibrium may be reflected by the type of vacation taken by Cohen's Diversionary tourist who travels simply to sooth a troubled soul.

Uysal et al. (2008) pointed out that disequilibrium in an individual's cultural, social, and psychological needs can be a primary motivation for travel. Interestingly, Crompton notes a break from routine does not necessarily mean a change to doing different kinds of things or to a change in lifestyle. It may simply involve a change in location where things occur. He does suggest some form of inversion as Graburn (1982, 2018) suggested, whereby the break from routine may occur in a different geographic or social setting. Activities could also be pursued at differing levels of intensity.

The third phase in the continuum is the consideration of behavioral alternatives, where people can choose to stay at home, travel for pleasure, to visit friends and relatives or go on a business trip. The fourth stage examines the range of motives that can help determine the nature of the trip and the choice of destination.

A total of nine motives emerged from this work. Seven could be classified as socio-psychological and two as cultural. Socio-psychological motives were overtly identified rarely by respondents and proved difficult to articulate. Crompton argues that such motives reflect the hidden agenda that drives travel. The cultural motives were more concerned with the destination's attributes than with the social or psychological status of the individual. Importantly, they are not exclusive. Instead, they operate in tandem or in combination with other motives. Crompton thus concludes that a destination can attract customers whose motives are not homogeneous or necessarily compatible. The nine motives are summarised in Table 8.1.

Table 8.1: Crompton's Motives

Socio-psychological motives

1 Escape from perceived mundane environment
- Temporary escape
- No single optimum environment
- Anticipation of the event can outlast the event

2 Exploration and evaluation of self
- Re-evaluate and rediscover more about oneself
- Liminal state aided self-discovery

3 Relaxation
- More mental relaxation than physical relaxation
- Pursue activities of interest

4 Prestige
- Rarely stated overtly

5 Regression
- Do things that were inconceivable in their normal lives, like adolescent or childlike behaviour

6 Kinship

7 Social interaction
- Meet new people
- Most likely to interact with tourists and not locals

Cultural motives

8 Novelty
- Curiosity, adventure, new and different experiences

9 Education
- Develop a rounded individual

Crompton also made another important observation. He determined that particular pull attributes of destinations are relatively unimportant if people traveled to have broadly based social psychological needs of regression, escape and relationship development met. Thus, almost any destination could satisfy these types of needs. This situation is especially true in multi-product urban destinations that represent a dynamic agglomeration of attractions, culture, events, landscapes and services that provide tourism space for consumption (Framke, 2002).

Crompton (1979) developed a broader set of push factors that drive people to travel, including relaxation and regression. Summer homes are very popular in Canada, enabling people to relax in an informal setting, spend quality time with close family members and engage in forms of regression reminding them of happy childhood experiences they otherwise cannot have in their busy lives.

Iso-Ahola optimal arousal theory

Seppo Iso-Ahola (1982) argued that proponents of push-pull theories in general and sociologists like Dann in particular did not pay enough attention to psychological theory when considering motivation (Moscardo, 2014). He also suggested that deep insights into travel motivations could arise from examining the vast array of leisure research, since pleasure tourism is an extension of leisure (Iso-Ahola, 1983). He argued people pursue leisure as a potential satisfaction-producer for two main reasons: to provide certain intrinsic rewards such as feelings of mastery and competence and to leave the routing environment behind (Iso-Ahola 1982: 258). He categorised these two elements as an 'approach' or 'seeking', and 'avoidance' or 'escapist' dimensions. Both are present in leisure motivation in general and in tourism motivation in particular, although he does suggest the unique nature of tourism many places greater emphasis on the escapist motive.

People may be pushed to escape the personal world, such as personal problems, difficulties or the interpersonal world of co-workers, family members and relatives. And they may seek personal rewards such as feelings of mastery, learning about other cultures and the like, or interpersonal awards, such as deeper social interaction with family and friends, meeting new friends or socialising with members of a travel party. Together these four elements can be organised in a two by two matrix, where tourists can be placed in one of four quadrants (Figure 8.2). One group may travel primarily to escape their personal environment and seek personal rewards, while other groups may escape their personal environment and seek interpersonal rewards; escape their interpersonal environment and seek personal rewards; or escape their interpersonal environment and seek interpersonal rewards. Interestingly, while Iso-Ahola criticises Dann's work, Pearce (2011) asserts both the escaping and seeking dimensions represent a form of push motive.

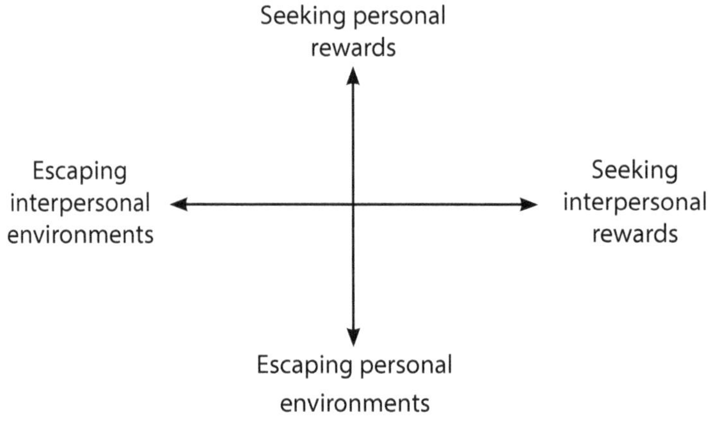

Figure 8.2: A social psychological model of tourism motivations. Source: Mannel & Iso-Ahola, 1987

The optimal arousal theory (Iso-Ahola 1983) that evolves from these two considerations suggests that tourists must balance between novelty and security. The extent of escaping from and degree of seeking to represents the desired state of stimulation that tourists can accept (Pearce, 2011; Prayag & Ryan, 2011). As Iso-Ahola (1983: 53,54) states

> *"when the social environment provides too many inputs and stimuli for the individual to handle (i.e., the situation is too incongruous in relation to the stored information), the individual tends to withdraw from the environment. On the other hand, when the environmental inputs are overly similar to the coded and stored information and experience, the individual becomes bored and seeks situations which offer more incongruity, uncertainty, novelty, and complexity. Thus, this tendency to seek the optimal incongruity and arousal channels the individual's leisure behavior."*

Pearce from travel career ladder to travel career pattern

The current state of the art for tourist motivation is Pearce's Travel Career Pattern model. It had been an evolving framework for more than 30 years until his untimely death just a few years ago. What began as a travel career ladder, based on Maslow's hierarchy, has since undergone a number of iterations, modifications and revisions until today the travel career pattern model bears little resemblance to the initial piece of work.

Pearce posited initially that people have travel careers. As he summarised in an early paper "*people have a career in their tourist behaviour. Like a career at work, people may start at different levels, they are likely to change levels during their life-cycle and they can be prevented from moving by money, health and other people. They may also retire from their career or not take holidays at all and therefore not be part of the system*" (Pearce 1992: 125). The initial iteration proposed a travel career ladder based on Maslow's hierarchy. Like Maslow, the needs of travelers were seen as organized into a hierarchy or ladder, with relaxation needs being at the lowest level, followed in order by safety/security needs, relationships needs, self-esteem and development needs, and finally, at the highest level, fulfilment needs (Pearce & Lee, 2005: 227). The premise was that people's motivations would change as their level of experience grew. And, as they grew they would seek to have higher needs met (Pearce, 2022). While he recognised that people's motives may be found at multiple levels, the initial model suggested one set of needs would dominate, thus enabling someone to be positioned neatly at one place on the ladder.

8

Great idea, intuitively appealing, but like Maslow, not practical in practice. Ryan (1998) wrote an insightful critique of the model arguing it is too simplistic and not supported empirically. He also noted that empirical research conducted to that point had been cross-sectional and, like a lot or work on lifecycle theory, could identify a situation at a single point in time, but not necessarily show change over time. He then went on to raise six questions about the concept, including its lack of philosophical congruence with Maslow, the validity of measures used and the methods adopted in different studies, and possible dissonance between expressions of needs and the ability to satisfy needs. In addition, laterally, subsequent work by Pearce and Lee (2005) and McKercher and du Cros (2008) suggest the pyramid may actually be inverted, for inexperienced tourists consistently express views that their first trips reflect a degree of self-actualisation, but as tourists gain experience, their needs tend to move down the pyramid to have lower order needs of relationship or safety met.

And so, a rethink of the Travel Career Pattern was undertaken (Pearce, 2022). Rather than progressing up or down a 'career' ladder, what Pearce and his students found was that the mix of motives changed as people gained travel experience. Most of us are still driven by core motives of escape, relaxation and building close relationships which have been labelled the backbone or skeleton of all travel (Pearce & Lee, 2005). But they identified two outer layers that also play a varying role (Figure 8.3). The middle layer is more focused on the desire to learn, fulfilment and building relationships with local communities. This layer has been further subdivided into more extrinsic factors, such as host site interaction and more intrinsic factors like self actualization and personal development. In the outer layer, several peripheral and less important motives such as nostalgia, the quest for autonomy and romance have been identified (Pearce, 2022). Importantly, this work recognizes that a pattern of and mix of motives is evident in most travel and not necessarily a single dominant one as believed previously (Pearce, 2011).

To a large extent, then, core motives are generic, while middle and outer layer motives become more specific. His framework identifies 14 different motive factors, which comprise more than six dozen specific motivate variables, as summarized in Table 8.2. His various studies have analysed the relationship at a factor level, more so than at an individual motive level.

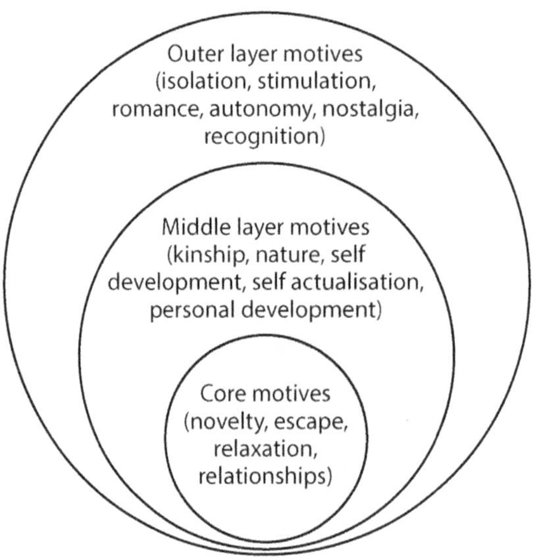

Figure 8.3: Pearce's travel career pattern. From Pearce 2011

This model has been tested in a number of situations and has been largely supported (Song & Bae, 2018), although, again cultural differences do come into play (Oktadiana et al., 2017, Wannasan & Jianvittayakit, 2022). Ward (2014), for example found different segments of the seniors' market had multiple motives and interests which influenced their destination choice as well as frequency of travel. Zhang & Peng (2014) concluded that using the TCP model could aid in the design of packages for Chinese tourists visiting Cairns in Far North Queensland, Australia.

Table 8.2: Disaggregating constituent motive variables from Pearce's travel career pattern

Factor	**Constituent variables**
Motive level: Core	
Novelty	Experiencing something different Feeling special atmosphere of the vacation destination Having fun Visiting places related to my personal interests
Escape/relax	Being away from daily routine Getting away from everyday physical stress/pressure Getting away from the usual demands of life Getting away from everyday psychological stress/pressure Giving my mind a rest Not worrying about time Resting and relaxing
Relationship	Being with others who enjoy the same things as I do Contacting with family/friends who live elsewhere Doing something with my family/friends Doing things with my companions Strengthening relationships with family/friends Strengthening relationships with my companions
Motive level: Middle	
Kinship (relationship-security)	Being near considerate people Being with others if I need them Being with respectful people Feeling personally safe and secure Feeling that I belong Meeting people with similar interests/values
Self-development through host-site involvement	Developing knowledge of the area Experiencing different cultures Following current events Learning new things Meeting new and varied people Meeting the locals Observing other people in the area
Nature	Being close to nature Being harmonious with nature Getting a better appreciation of nature Viewing scenery

Self-actualization	Being creative Experiencing the things I have always wanted to do Feeling inner harmony/peace Gaining a new perspective on life Understanding more about myself Working on my personal/spiritual values
Self-development – personal development	Develop my personal interests Developing my skills and abilities Gaining a sense of accomplishment Gaining a sense of self confidence Knowing what I am capable of Using my talents and skills
Motive level: Outer	
Isolation	Avoiding interpersonal stress and pressure Being away from crowds Enjoying isolation Experiencing open space Experiencing peace and calm
Stimulation	Being spontaneous Experiencing the risk involved Experiencing thrills Exploring the unknown Feeling excitement Having daring/adventuresome experiences Having unpredictable experiences
Romance	Being with people of the opposite sex Having romantic relationships
Autonomy	Being independent Being obligated to no-one Doing things my own way
Nostalgia	Reflecting on past memories Thinking about good times I've had in the past
Recognition	Being recognized by other people Having others know I have been there Leading others Sharing skill and knowledge with others Showing others I can do it

Source: Pearce 2005

These studies have largely confirmed that middle level motives were more important to experienced travellers than outer motives, while those with limited travel experience value all motives were equally important (Pearce, 2011), while

Wu et al. (2019) found middle layer motives were of particular importance to inexperienced tourists. Agarwal et al. (2023) studied repeat visitors to India and found general support for the model. They conclude that the core motives remain equally important for all tourists irrespective of the number of visits they made to India, while the middle layer motives of repeat tourists revealed differences between those who had visited India twice compared to those who visited three times or more. They, along with outer layer motives, became more important the more frequently one visited. This finding more or less corroborates the finding of other research that suggests people who have the 'itch' to travel are driven by a range of motives, with core motives dominating (Gizem et al., 2022).

It has also been suggested that the model can be used to segment the market (Song & Bae, 2018). A study examined behaviours of tourists in Bali (McKercher et al., 2021) noted differences depending on which core, middle and outer layer motives were identified. People travelling for novelty and escape were much more active than those travelling for relationship reasons. They tended to participate more in resort-based leisure activities. Likewise differences were also observed among the groups who identified a mix of core and intrinsic or extrinsic second tier motives, second tier motives only or whose travel decisions were dominated by third tier motives.

Conclusion

Do we understand why people travel? The answer is an equivocal "yes", for travel involves a series of personal decisions. Each one of us has a slightly different reason to travel and thus, it is impossible ever to come up with a definitive model, unless it contains a virtually unlimited set of inter-linked variables.

But, do we have a better theoretical understanding of travel? The answer here is an unequivocal "yes!" As this chapter illustrates, the last 50 years has seen a dramatic improvement in both the sophistication in understanding and theoretical development of ideas about why we travel. Early work tended to conflate ideas or borrow intuitively appealing, but ultimately rather simplistic, ideas that did not fit the complex nature of tourism. Moreover, these early models tended to place people in absolute categories that failed to appreciate that the factors that drove trips could vary from trip to trip. The early 1980s saw a well-spring of ideas that have currency today, especially the identification of the push-pull model and the beginning of efforts to isolate individual motives of escape, relaxation, relationship and the like. Work since then, culminating in the travel career pattern, has added additional dimensions of both complexity and sophistication by recognising that we are driven to travel by a basket of motives and that the mix of motives and importance placed on each one varies by trip.

Discussion points / Exercises

1. Four different motivational models and theories were identified in this chapter. Identify common features that apply to all or most of them.
2. Discuss how motivational studies have evolved over the past 50 years.
3. Can you map push-pull onto Pearce's travel career pattern model?

Points to consider for revision

- Models of increasing complexity have been developed to understand why people travel.
- Sociology and psychology are the dominant disciplines that have contributed to the understanding of the field.
- Motivational studies really began with Maslow, but for a variety of reasons, the model does not apply to tourism. Instead, it may reflect what people do, rather than why they travel.
- Push-pull theory represents the father of tourism motivation models.
- Crompton's nine motives encompass the main reasons why people travel
- Pearce's evolving travel career pattern builds on Maslow's, Dann's and Crompton's works but importantly, represents the multi-faceted nature of motives and the different roles core, middle and outer layer motives play in pleasure travel.

Suggested additional readings

Crompton, J. L. (1979). Motivations for pleasure vacations. *Annals of Tourism Research*, 6(4), 164-182.

Dann, G. (1977) Anomie, ego enhancement and tourism. *Annals of Tourism Research*, 4(4), 184-194.

Pearce, P. & Lee, U. (2005) Developing the travel career approach to tourist motivation. *Journal of Travel Research* 43: 226 -237.

References

Alrawadieh, Z., Prayag, G. & Alrawadieh, Z., (2023). A cognitive appraisal perspective of emotional accessibility at heritage sites: empirical evidence from the UNESCO World Heritage Site of Petra. *Journal of Heritage Tourism*, 18(2), 145-163.

Arowosafe, F., Akinwotu, O., Tunde-Ajayi, O., Omosehin, O. & Osabuohien, E., (2022). Push and pull motivation factors: A panacea for tourism development challenges in Oluminrin waterfalls, Nigeria. *Journal of Policy Research in Tourism, Leisure and Events*, 14(1), 63-74.

Carvalho, N., Rodrigues, H. & Brochado, A., (2023). Double or nothing: push and pull factors of casinos in Europe. *GeoJournal*, 88(2), 2233-2246.

Cohen, E. (1979) A phenomenology of tourist experiences, *Sociology*, 3, 170 -201.

Crompton, J. L. (1979). Motivations for pleasure vacations. *Annals of Tourism Research*, 6(4), 164-182.

Dancausa, G., Hernández, R.D. & Pérez, L.M., (2023). Motivations and constraints for the ghost tourism: A case study in Spain. *Leisure Sciences*, 45(2), 156-177.

Dann, G. (1977) Anomie, ego enhancement and tourism. *Annals of Tourism Research*, 4(4), 184-194.

Dann, G. (1981). Tourist motivation: An appraisal. *Annals of Tourism Research* 8(2), 187-219.

Denning, S. (2012) What Maslow Missed. *Forbes*, March. https://www.forbes.com/sites/stevedenning/2012/03/29/what-maslow-missed/#fb41ede661b5

Framke, F (2002) The destination as a concept: a discussion of the business-related perspective versus the sociocultural approach in tourism theory. *Scandinavian Journal of Hospitality and Tourism*, 2(2), 92-108

Fyall, A., Legoherel, P., Frochot, I, & Wang, Y (2019) *Marketing for Tourism and Hospitality: Collaboration, Technology, Experiences.* Oxon: Routledge.

Gizem, C., Tuğba, P. & Derya, T. (2022). Travel itch as a new phenomenon: travel motives and attributes of itchers. *Tourism and Hospitality Management*, 28(1), 83-99.

Graburn, N. (1983) The anthropology of tourism. *Annals of Tourism Research*, 10, 9-33.

Graburn, N. (2018) Secular oilgrimage: A general theory of tourism. In Gmelch, S. & Kaul, A. (eds) *Tourists and Tourism: A general reader 3rd Ed.* Long Grove: Waveland, 17–28.

Hsu, C., Killion, L., Brown, G., Gross, M., & Huang, S. (2008) *Tourism Marketing: An Asia-Pacific perspective*. Milton: John Wiley & Sons Australia.

Huang, S. & Hsu, C. (2009) Travel motivation: Linking theory to practice. *International Journal of Tourism, Culture and Hospitality Research*, 3(4), 287-295.

Iso-Ahola, S. (1982). Toward a social psychological theory of tourism motivation: A rejoinder. *Annals of Tourism Research* 9(2), 256-262.

Iso-Ahola, S. (1983) Towards a social psychology of recreational travel, *Leisure Studies*, 2(1), 45-56, DOI: 10.1080/02614368300390041

Jafari, J. (1987) Tourism models: the sociocultural aspects. *Tourism Management*, 8(2), 151-159

Jamal, T. & Lee, J (2003). Integrating macro and micro approaches to tourist motivations: Toward an interdisciplinary theory. *Tourism Analysis* 8(1), 47-59.

Kaur, A. (2013) Maslow's need hierarchy theory: Applications and criticisms. *Global Journal of Management and Business Studies* 3(10), 1061-1064.

Kay, P. (2003) Consumer motivation in a tourism context : continuing the work of Maslow, Rokeach, Vroom, Deci, Haley and others, in *Proceedings of the*

2003 Australian and New Zealand Marketing Academy conference, ANZMAC, Adelaide, S. Aust., pp. 600-614. https://dro.deakin.edu.au/eserv/DU:30036658/kay-consumermotivation-2003.pdf

Klenosky, D (2002) The 'pull' of tourism destinations: a means-end investigation. *Journal of Travel Research*. 40, 385-395.

Kotler, P., Bowens, J. & Makens, J. (2003) *Marketing for Hospitality and Tourism, 3rd edition*. Upper Saddle River: Prentice Hall.

Lewis, C., Kerr, G., & Pomering, A. (2010) Destination choice by young Australian travellers: a theoretical explanation to a practitioner problem. *Proceedings of the 20th Annual Conference of the Council for Australian University Tourism and Hospitality Education (CAUTHE)* (pp. 1-16). https://ro.uow.edu.au/commpapers/1513.

Litvin, S. & Smith, W. (2016) A new perspective on Plog psychographic system. *Journal of Vacation Marketing* 22(2), 89-97.

Liu, X., (2023). Cause analysis of the development of Japan's tourism industry based on the push-pull theory. *Geographical Research Bulletin, 2*, 2-21.

Lundberg, D. (1971) Why tourists travel. *Cornell Hotel and Restaurant Quarterly*, 11(4), 75-81. https://doi.org/10.1177/001088047101100413

Maslow, A. H. (1943). A theory of human motivation. *Psychological Review*, 50*(4)*, 370-96.

Matiza, T. & Kruger, M., (2022). Profiling the South African recreational domestic tourist in the Era of COVID-19. *Leisure Sciences*, 1-22.

McKercher, B & du Cros, H. (2008) A comparison of international travel between an emerging and a mature source market. *Asia Pacific Journal of Tourism Research*. 13(3), 265-280.

McKercher, B. Tolkach, D., Mahadewi E., & Byomatara, D. (2021) The relationship between motive and in-destination behaviour. *Journal of Hospitality and Tourism Management*. 46, 432-439.

McLeod, S. (2018) Maslow's Hierarchy of Needs. Simplypyschology.org. https://www.simplypsychology.org/maslow.html.

Michael, N., King, B. & Fotiadis, A. (2022). Escaping from cultural practices at home: an exploration of push and pull. *Tourism Culture & Communication, 22*(1), 13-29.

Mill R. C., & Morrison A. M. (2012) *The Tourism System: An Introductory Text* Prentice Hall, NJ.

Mishra, S., Malhotra, G., Johann, M. & Tiwari, S.R., (2022). Motivations for participation in active sports tourism: a cross-national study. *International Journal of Event and Festival Management, 13*(1), 70-91.

Moscardo, G. (2014) Do tourists travel for the discovery of 'self' or search for the 'other'? *Tourism Recreation Research* 39(1), 2-10.

Moutinho, L. (1987) Consumer behaviour in tourism. *European Journal of Marketing*, 21(10), 3-44.

Mutinda, R. & Mayaka, M., (2012). Application of destination choice model: Factors influencing domestic tourists destination choice among residents of

Nairobi, Kenya. *Tourism Management*, 33(6), 1593-1597.

Nikjoo, A. & Ketabi, M. (2015) The role of push and pull factors in the way tourists choose their destination, *Anatolia*, 26(4), 588-597, DOI: 10.1080/13032917.2015.1041145

Oktadiana, H., Pearce, P.L., Pusiran, A.K. & Agarwal, M., (2017). Travel career patterns: The motivations of Indonesian and Malaysian Muslim tourists. *Tourism Culture & Communication, 17*(4), pp.231-248.

Page, S. (2019) *Tourism Management 6th Ed* Oxon: Routledge.

Pearce, P. (1992) Fundamentals of tourist motivation. In D. Pearce & R. Butler (Eds), *Tourism Research: Critiques and challenges* (pp. 85-105). London: Routledge and Kegan Paul.

Pearce, P. (2005) *Tourist Behaviour: Themes and Conceptual Schemes.* Clevedon: Channel View Publications.

Pearce, P. (2011) Travel motivation, benefits and constraints to destinations. In Wang, Y & Pizam A. (Eds) *Destination Marketing and Management.* CABI, pp. 39-52

Pearce, P.L., (2022). Consolidating the travel career pattern approach to tourist motivation. In Sharpley, R (Ed) *Routledge Handbook of the Tourist Experience*, pp.169 -184.

Pearce, P. & Lee, U. (2005) Developing the travel career approach to tourist motivation. *Journal of Travel Research* 43, 226-237.

Pearce, P. & Packer, J. (2013). Minds on the move: new links from psychology to tourism. *Annals of Tourism Research* 40, 386-411.

Plog. S, (1974) Why destination areas rise and fall in popularity, *Cornell Hotel and Restaurant Quarterly* 14(4), 55-58.

Plog. S, (2001) Why destination areas rise and fall in popularity: an update of a Cornell Quarterly Classic, *Cornell Hotel and Restaurant Quarterly* 42(3), 13-24.

Prayag, G. & Ryan, C. (2011) The relationship between the 'push' and 'pull' factors of a tourist destination: the role of nationality – an analytical qualitative research approach, *Current Issues in Tourism*, 14(2), 121-143, DOI: 10.1080/13683501003623802

Ryan, C. (1998) The travel career ladder: An appraisal. *Annals of Tourism Research* 25(4), 936-957.

Salim, E., Mayer, M., Sacher, P. & Ravanel, L. (2023). Visitors' motivations to engage in glacier tourism in the European Alps: comparison of six sites in France, Switzerland, and Austria. *Journal of Sustainable Tourism*, 31(6), 1373-1393.

Shi, D., Yi, B., Zhang, B. & Yu, D. (2023). Travelling to an unusual destination: Chinese tourists' preferences for visiting North Korea. *Current Issues in Tourism, 26*(13), 2128-2144.

Smith, M., Pinke-Sziva, I., Berezvai, Z. & Buczkowska-Gołąbek, K. (2022). The changing nature of the cultural tourist: motivations, profiles and experiences of cultural tourists in Budapest. *Journal of Tourism and Cultural Change*, 20(1-2), 1-19.

Smith, S., Xiao, H., Nunkoo, R., & E. Tukamushaba (2013) Theory in Hospitality, Tourism and Leisure Studies. *Journal of Hospitality Marketing and Management* 22, 875-894.

Soldatenko, D., Zentveld, E. & Morgan, D. (2023). An examination of tourists' pre-trip motivational model using push–pull theory: Melbourne as a case study. *International Journal of Tourism Cities.*

Song, H. & Bae, S. (2018) Understanding the travel motivation and patterns of international students in Korea: using the theory of travel career pattern, *Asia Pacific Journal of Tourism Research,* 23(2), 133-145, DOI: 10.1080/10941665.2017.1410193.

Subyubon, P., Thongrom, P., Tiwasing, A. & Chaya, P., (2022). Chinese LGBT tourist motivation traveling to Bangkok. *International Journal of Early Childhood Special Education,* 14(1), 1933-1942.

Tasci, A. & Ko, J. (2017) Travel needs revisited. *Journal of Vacation Marketing,* 23(1), 20-36

Tojib, D., Tsarenko, Y., Hin Ho, T., Tuteja, G. & Rahayu, S. (2022). The role of perceived fit in the tourist destination choice. *Tourism Analysis, 27*(1), 63-76.

Uysal, M., Li, X., & Sirakaya-Turk, E. (2008) Push–pull dynamics in travel decisions. In Oh, H. (Ed) *Handbook of Hospitality Marketing Management.* London: Routledge. pp 412-440.

Valencia, J.P., Cerio, C.T. & Biares, R.R. (2022). Tourists' motives and activity preferences to farm tourism sites in the Philippines: application of push and pull theory. *Cogent Social Sciences,* 8(1), 2104706.

Wannasan, S. & Jianvittayakit, L. (2022). Travel motivations of Generation Y Chinese and Taiwanese tourists visiting Thailand: The application of the travel career pattern approach. *International Journal of Multidisciplinary in Management and Tourism,* 6(1), 72-87.

Ward, A. (2014). Segmenting the senior tourism market in Ireland based on travel motivations. *Journal of Vacation Marketing,* 20(3), 267-277.

Wu, J., Law, R., Fong, D.K.C. & Liu, J. (2019). Rethinking travel life cycle with trave\ career patterns. *Tourism Recreation Research,* 44(2), 272-277.

Yoo, C., Yoon, D & Park, E. (2018) Tourist motivation: an integral approach to destination choices. *Tourism Review* 73(2), 169-185.

Yousaf, A. Amin, I., & Santos, J (2018) Tourists motivations to travel: A theoretical perspective on existing literature. *Tourism and Hospitality Management,* 24(1), 197-211.

Zhang, Y., & Peng, Y. (2014) Understanding travel motivations of Chinese tourists visiting Cairns, Australia. *Journal of Hospitality and Tourism Management.* 21, 44-53.

9 Sociological and Anthropological Concepts in Tourism

At the end of this chapter, the reader will be able to:

- Understand tourism as a quest for contrived reality
- Evaluate how tourism is a quest for authenticity
- Define the environmental bubble and critique how it forms the basis of commercial tourism
- Evaluate Cohen's five roles of tourists
- Analyse Cohen's quest for the other
- Assess and be able to critique the concept of liminality
- Identify aspects of risk and how they affect travel behaviour.

Introduction

Tourism is both a business and a social phenomenon. To understand how it works, then, one must understand both the business and human sides of the phenomenon. Ever since large scale tourism began, people have been examining it and trying to place it in a sociological perspective. These efforts have involved an examination of why people travel, the social reasons for travel and the type of tourist experience they are either getting or seeking. This chapter examines how the founding sociologists and anthropologists looked at tourism and how their work still shapes our thinking.

Social perspectives on tourism

Three dominant themes in the sociology of tourism have emerged over the last 60 years. They are:

- Boorstin's essentially negative 1960s view that tourism represents contrived reality;
- MacCannell's 1970s more positive view that tourism represents a quest for authenticity, and;
- Cohen's more neutral views that tourism represents a quest for one's centre.

In reviewing these themes, it is important to understand a few points. One is that most take a very American-centric view of the world. The second is to remember that these works were written over a nearly 30-year period and that they are heavily influenced by the dominant social paradigm at the time of their writing.

Boorstin's Contrived Reality

In the early 1960s, sociologists were becoming increasingly concerned about the contrived and illusory nature of the human experience in American society. They were concerned that people were no longer experiencing reality in their lives, instead they were being presented a series of pseudo-events. As Boorstin (1964) described them, they were immersed in a thicket of unreality.

Tourism, and especially mass tourism, was seen as being just another example of how American life had become overpowered by pseudo-events and contrived experiences. Boorstin felt that (in 1964, remember) the modern tourist was simply a passive onlooker who was isolated from the host environment and the local residents. As a result, tourists chose to disregard the real world around them, instead choosing the isolation of tourist ghettos. Both the tourist experience and the type of tourism development had to become more contrived, in which the tourist moved in a closed self-perpetuating environment that was isolated from the rest of the world. This was both caused by and reflective of the nature of modern (1964) American society, of which tourism was only a part.

There is much to what Boorstin discusses that is still relevant today. Mass tourists can only withstand a certain amount of strangeness. They like to witness strangeness of the host environment from the security of their own surroundings. Too much strangeness becomes threatening. In addition, he introduced the idea of the 'tourist ghetto' that is isolated from the real world. Destination areas like the Gold Coast in Australia, Disneyworld, Waikiki in Hawaii and many places in the Mediterranean are prime examples of tourist ghettos. On a smaller scale, many destination resorts are physically removed from the outside world. They have been constructed as discrete areas that, by design, inhibit any contact with the real world. The only types of contact that tourists are likely to have with locals are either in a servant-master relationship, in a commercial relationship, or through the window of a self-enclosed, air conditioned bus. No one would describe these places as being real except possibly as real tourist destinations.

But, as Cohen (1972) argues, Boorstin's concept has a number of fundamental weaknesses. To begin, he labels all tourists as being the same. He fails to recognize that there are a large number of tourists that do not fit the mass tourist model, and even among mass tourist's significant differences exist. Moreover, his work is editorial and observational in nature, rather than scientifically rigorous.

Boorstin felt the modern tourist was simply a passive onlooker who was confined largely to tourist ghettos or enclaves and did not want to experience the destination in any meaningful way, nor was he or she able to do so. He felt these enclaves divorced from the real world, as shown in this resort image in Da Nang, Vietnam. Tourists are actively encouraged to stay in the resort area, and equally discouraged from engaging the broader community in a meaningful manner. Some of his arguments are valid, but his core thesis fails to appreciate that this is exactly the type of experience many tourists seek.

■ MacCannell (1973) - quest for authenticity

Some ten years later Dean MacCannell (1973) came up with a revised view of tourism and tourist motivations. Indeed, he was rather dismissive of Boorstin's ideas. He argued that the modern tourist is not a victim of a contrived and illusory culture, but instead is on a quest for authenticity that involves paying homage to the symbols of modernity, in this case, tourist attractions. He suggested that the modern mass tourism journey shares many similarities with religious pilgrimages. Tourism centres have become the modern equivalent of pilgrimage centres, where people go to these places for similar reasons that pilgrims go to the Vatican,

Mecca and Lourdes. Whereas pilgrims attempted to visit a place where an event of religious importance actually occurred, tourists present themselves at places of social, historical, and cultural importance (MacCannell, 1973:593). Likewise, while Boorstin suggested that attractions were of little significance, MacCannell adopted a completely different view and sees them as highly significant social symbols.

The quest for authentic tourist attractions, not contrived ones, becomes the central motivation for tourism. What is the difference between an authentic and contrived tourist attraction? Well, the difference is small and often perceptual. If the visitor goes to a destination and perceives it as being 'authentic' then it is. In a reflective piece, MacCannell (2018) commented that he *"sees tourism as a ritual of modern peoples, in which they scour the world looking for some kind of 'authenticity', a quality which many think is lacking in modern social life; and in this intention they are aided by modern hosts who attempt to 'stage' it for them"*. On the other hand, if the visitor perceives an attraction as being contrived, then it is not regarded as being authentic.

This example reflects MacCannell's the concept of a front and back of house continuum. Building on Goffman's (1959, as cited by MacCannell 1973: 598) idea of a front and back region, MacCannell identifies a six stage continuum, that includes:

- Stage 1: Goffman's front region; the kind of social space tourists attempt to overcome, or to get behind.
- Stage 2: a touristic front region that has been decorated to appear, in some of its particulars, like a back region... Functionally, this stage is entirely a front region, and it always has been, but it is cosmetically decorated with reminders of back-region activities: mementos, not taken seriously, called 'atmosphere'.
- Stage 3: a front region that is totally organized to look like a back region... This is a problematical stage because the better the simulation, the more difficult it is to distinguish it from stage 4.
- Stage 4: a back region that is open to outsiders... It is the open characteristic that distinguishes these especially touristic settings (stages 3 and 4) from other back regions. Here access to most non-touristic back regions is somewhat restricted.
- Stage 5: a back region that may be cleaned up or altered a bit because tourists are permitted an occasional glimpse in.
- Stage 6: Goffman's back region; the kind of social space that motivates touristic consciousness.

As the pressure by tourists to see 'real' places (or the back of the house) expands, destination areas are forced to create tourist attractions and present them as being real. In other words, to satisfy tourist demand, they have to stage authenticity.

The creation of inauthentic 'real' tourist attractions is a consequence of tourism development.

MacCannell's work introduced important concepts of tourist space, which is different from tourist ghettos. Like Boorstin's work, his thesis is perceived to adopt a global view of tourism and failed to recognize that different types of tourists existed who are seeking different tourist experiences.

Cohen (1972) - strangeness vs. familiarity, or value adding?

Erik Cohen took Boorstin's initial concept and developed it one step further. Cohen's work examined tourism from the perspective of 'strangeness' versus 'familiarity' by looking at the extent, variety and degree of change tourists seek in their various social contacts. The degree to which strangeness or familiarity prevails in the tourist's role or activities determines the nature of the tourism experience as well as the effects he/she has on the host society.

While all tourists are to some extent strangers in the host community, the extent that the tourist's role is pre-determined will dictate the manner in which tourists interact and the images they will develop of one another. Tourism evolved from a basic change in our view of the world, from isolation to a global perspective. Initially few people travelled, primarily because of ignorance of other people's cultures which made people leery of seeking change. Because our view of the world changed, people became more interested in experiencing strange and novel situations but only to the point that this strangeness remained non-threatening.

People want to experience change from the security of their own environmental bubble. What is an environmental bubble? It is a bubble of familiarity that reduces the level of strangeness or threat to an acceptable level, enabling people to participate. Think of an umbrella and raincoat. You can envelop yourself in a raincoat and use an umbrella to protect you from the rain, enabling you to go out and enjoy the experience.

The idea is predicated on the assumption that people will want to participate in things only up to the point that the perceived risk outweighs the perceived reward. If the reward is high and the risk low, then there is little need for a bubble. But, if the perceived risk is high, then things must be done to reduce it in order to facilitate participation. The environmental bubble exists throughout all aspects of the tourism system. Think of what hotels do to reduce perceived risk. Foreign television channels, multilingual staff, concierges who are familiar with the destination, different types of food on offer, especially at breakfast, in-room safes for security and many more actions are undertaken to be seen to reduce risk. Tour guides play the same roles by interpreting and mediating the destination experience. Airlines provide an environmental bubble through their safety announcements at the start of the flight and request that people keep their seatbelts fastened.

It is also important to appreciate that people have different levels of risk tolerance, and therefore, seek different levels of bubble provision. Some tourists may be more tolerant of risk or accept a greater degree of risk (Chang, 2011) than others (Thapa et al., 2013). These risk tolerant tourists are more courageous and, therefore, are more likely to engage in risk-taking behaviours (Chang, 2011), including going to new or perceived to be unsafe destinations (Karl, 2018). By contrast, those who are more sensitive to risks may be less likely to take any adventures at the destination (Scott & Mowen, 2007). Rittichainuwat and Chakraborty (2009) add that those who perceive risk to be high are more likely to opt for alternate destinations.

Most tourists are not adventuresome risk takers. Instead they are often on a once a year break and looking for relaxation, fun, spending time with family and novelty. They seek strange and exciting environments, only up to the point that they are not too strange and exciting. In other words, the thrill must be perceived as being greater than the risk. If the perceived risk is too great, then most people will choose not to participate or will alter their behaviour. This street market in Dhaka, Bangladesh is one such example where the perceived risk exceeded the perceived benefit of visiting. It is a thrilling place to visit for those looking for an authentic experience, but might be 'too authentic' for others who may feel intimidated by entering. Few tourists feel comfortable enough to venture in. Alternately, a well-managed heritage site, such as Stonehenge is extremely popular, for the benefits of visiting far exceed the risk.

Here, Plog's categorisation is useful. The repressed Psychocentric / Authentic tourist cannot cope with much change. He or she will seek as much familiarity as possible when travelling. But, if placed outside their comfort zone, will require a massive environmental bubble in order to feel comfortable (fully escorted package tour, familiar food, known brands of hotels and restaurants, etc.). The various groups of Midcentric tourists are also risk averse, but not to the same extent. They are more willing to be adventuresome, but still desire a bubble of familiarity, such as branded hotels, availability of local food if desired and perhaps some guided day tours or a clear itinerary they can follow. So while they want to see new sites, they still want to perceive them from the comfort of their own culture. The Allocentric/Venturer tourist, on the other hand is the least risk averse tourist and is most willing to accept the destination on its own merits. He or she eschews an environmental bubble.

Jaakson (2004) and others prefer the more neutral term 'tourist bubble.' It conveys the same idea but removes the discussion from insinuations that it is linked to the natural environment. He suggests the bubble has two dimensions – a physical emphasis of it as a type of place, and a psychological emphasis of it as the tourists' *eigenwelt* of motivations, attitudes, and belief systems. Some tourists, thus place themselves in a secure physical place that is isolated from the rest of the destination and therefore only see a rather sanitized version of the place (Newman, 2002). Emotionally, as well they seek a secured, protected, and normalized environment often in an artificial, segregated environment devoted to consumption and play (Newman, 2002). It is for this reason that Jaakson (2004) has observed a pack mentality among many tourists. Andreu, Kozak, Avci & Cifter (2005) call this pattern the 'herding effect'.

9

The likelihood of isolation is enhanced if substantial language or cultural differences exist between the hosts and the visitors. It is this controlling and standardizing of the tourist product, however, that often provides the images that visitors have of local arts and culture. Unfortunately, many tourism destinations the world over look very similar. Countries lose their individuality as the area's geography and culture are packaged and reduced to bite sized elements for easy consumption. In a sense, mass tourism simply becomes a series of photographs and stamps on passports. Mass tourists themselves rarely walk away from a destination with any greater understanding of the area they are visiting.

Serial monotony is a particularly challenging issue in tourism, especially in destinations that lack unique features and a clear brand personality (Pratt, 2013). Here tourists are overwhelmed by much of a muchness. The first attraction visited may be appealing, but after a while all begin to look the same. The novelty of consuming one place wanes very quickly. Unfortunately, many cultural and heritage attractions are very similar, leading tourists to believe that if they have seen one there is no point in seeing others.

Serial monotony is a challenge for many destinations that try to present their culture as being unique. Dances are one way in which cultural features are portrayed, as in this example of Hanni dancers in rural China. For many pleasure tourists with only a modest interest in learning about indigenous culture, these dances soon meld together. Pretty costumes, traditional instruments, the same theme of the dances (sowing seeds, harvesting, falling in love, etc.) may be interesting the first time, but after a while they become quite similar.

Why has this happened? Urry (1990a and b) identified a number of reasons. First, most travellers are not adventuresome drifters who want to take the destination as it is presented to them. These people exist, but they are a minority. Most people do not want to feel that they are facing real risks and real discomfort in pursuit of their vacation. Second, the market dictates the types of services and amenities demanded. If the vast majority of travellers are looking for comfort and demanding some sort of environmental bubble, then the industry must accommodate them. But, third, people still want to feel like they are experiencing some type of novel experience and want to feel that they have been exposed to a different culture. It is just that they either do not want to or are incapable of experiencing this in any real or meaningful manner. As DeKadt stated (1978) most travellers are not anthropologists or sociologists. They are primarily consumers who are temporarily unemployed and who are seeking a break, an escape from their normal life. Finally, tourism is a commercial activity and as with any commercial activity, efficiencies can be achieved when the product is standardized, packaged, mass produced and easily consumed. Because of its nature, clients must be processed as efficiently and smoothly as possible. Therefore, the tourist's experience must be ordered, predictable and controllable.

Many people decry the standardisation, modification and commodification of tourism as presenting a superficial experience that removes the individual from engaging the destination in a meaningful manner. Such an experience can be found here in a Segway tour of Tel Aviv in Israel. Yet, the need to be able to process tourists quickly, efficiently and profitably, while ensuring they have an enjoyable and often once in a lifetime experience dictates that tourist experiences need to be ordered and as predictable as possible. Moreover, it must be remembered that most people are travelling for leisure, escape and fun. Standardising and commodifying products may remove some spontaneity, but ensure that all tourists will be guaranteed the same quality of experience.

9

Cohen developed a four-fold typology of roles tourists play (Table 9.1). It is divided into two parts, those who follow an institutionalised path and those who he classifies as non-institutionalised tourists. The following section summarises Cohen's work (1972: 167-173).

Table 9.1: Cohen tourist typology

Institutionalised tourist	Organised mass tourist Individual mass tourist
Non-insitutionalised tourist	Explorer Drifter

Institutionalized tourists fall into one of two roles:

- *Organized mass tourists* are the least adventuresome and remain largely confined to their environmental bubble. The guided tour conducted in an air-conditioned bus represents the prototype of the organized mass tourist. The itinerary is fixed in advance and all stops are well prepared and guided. The tourist makes almost no decisions and stays almost exclusively in the microenvironment of the home country. Familiarity is at a maximum, while novelty is at a minimum. Mass tourism, and the commodification of the mass tourism product enables tourists to take in the novelty of the area without any physical or emotional discomfort. In a sense, they can observe it without experiencing it first-hand. One of the main purposes of mass tourism (remember the operative word is *mass*) is to see attractions, whether the attractions are genuine or contrived. Even if the attraction is genuine, it often has to be transferred and somehow controlled to make it more suitable for mass consumption.
- The *individual mass tourist* is similar to the organized mass tourist, except the tour is not entirely preplanned. However all the major arrangements are still made through a travel agent. Tourists experience the destination from within the environmental bubble of their home country and venture out of it only occasionally.

He then identifies two types of non-institutionalized tourist:

- The *explorer* tries to get off the beaten track but is nevertheless comfortable in accommodations and reliable means of transport. This person dares to leave the environmental bubble more than the other two but is still careful to be able to step back when the going becomes too rough. Though novelty dominates, the tourist is not immersed completely in the host society.
- The *drifter* ventures furthest away from the beaten track and from the customer's way of life.

Value adding as a synonym for the environmental bubble

Cohen's idea of the environmental bubble is brilliant but is not developed further to explain the commercial nature of tourism, for the entire business model of tourism is devoted to environmental bubble creation. Only the industry uses the term 'value adding'! The entire basis for the commercial delivery of tourism is all about bubble creation, right from the travel trade through transport providers, accommodation houses, in-destination transport and services and attractions. They exist to make travel more accessible by removing barriers and creating a sense of familiarity.

Think about what you are paying for. In reality, you are paying for the bubble. If you wanted, you could sleep at low cost hostels. But, most of us want the comfort and convenience of better quality accommodation. We could travel throughout a destination on public transit and explore sites on our own. But, commercially provided tours are popular because they remove the risk and highlight the thrill. We could buy and cook our food, but restaurants are preferred for they remove the uncertainty of dining. The list goes on. What we pay for is less the tangible product on offer and more the augmented service that reflects the creation of a bubble.

The commercial delivery of tourism is all about environmental bubble creation as this image of a group of tourists on a day trip of the Colosseum in Rome illustrates. The tour product on offer includes a tour guide, interpretation, bus transport and access to the site. Tourists could have undertaken this trip themselves at a much lower cost, but may have had to line up for a long time before they could enter the Colosseum. Why would anyone pay for such a tour when you could do it for almost nothing? The reason is the value adding offered by the tour company, or in sociological terms, the environmental bubble created by the operator. The guide offers some level of expertise, interpretation and significantly a safe, mediated experience. Tourists are taken to highlight sites and therefore do not feel they miss out on anything of importance. Additionally, they did not have to worry about getting lost. In other words, they were willing to pay a price premium for the environmental bubble and the operator added value (and additional income) to the experience through bubble creation.

There are added benefits as well. To begin, bubble creation broadens the market, by reducing the skill level needed to participate, optimizing the thrill and providing needed security to convince people that risk has been reduced to a sufficient level. Reisinger and Mavondo (2005:213), reviewing the consumer behavior literature, identify seven types of risk: financial (losing or wasting money), functional or performance (not performing, not meeting the customers' needs and requirements), physical (injury or illness), social (losing personal and social status), psychological (damaging self-image and/or reflecting poorly on personality), satisfaction (not delivering satisfaction), and time (not performing on time, taking too much time, and/or wasting time). Wickens (2002) argues the need for an environmental bubble relates primarily to risk reduction. Most tourists seek to balance the excitement of visiting a destination against the possible risks involved in doing so. If the risk exceeds the reward, tourists will need to enmesh themselves in a bubble of familiarity to reduce risk to an acceptable level.

Finally, satisfaction can be enhanced through the modification, standardization, and commodification of the product. Most tourists, and especially long haul tourists may only visit a destination once or may only visit an attraction in a destination once. They want to have the best possible experience during this visit. The best way to do so for as many tourists as possible is to provide a highly commodified and standardized experience that seeks to offer a consistently high quality opportunity for all who visit. Within the cultural tourism space, for example, purpose-built cultural theme parks represent extreme environmental bubble formation that offer a safe, controlled and controllable environment (Craik, 1997), while Hughes (1998) also observes the arts industries have standardized their product in order to satisfy consumer demand and enable effective signification to the consumer.

■ Cohen (1979) and the 'Centre and the Other'

Cohen also built on MacCannell's ideas of tourism as a quest. His later work (Cohen, 1979) sees individual's alienation from society as lying along a five scaled continuum. He tied his thesis into the contemporary sociological premise of 'the Centre' and the 'Other'. Basically, every society possesses a centre, which is its ultimate moral value. The Centre represents the paramount symbols of a society, be they social, cultural or religious.

For most people, the Centre is located within the confines of that society. Leisure and recreation activities, and by extension, tourism offer a temporary release from the Centre that serves a number of beneficial purposes. If a person is closely tied to his or her Centre, then the type of tourism experience he or she will select may have little regard for authenticity in their tourist experience. Why? Because they are already connected to their Centre and thus don't need the reinforcement of a tourism experience. But, most people in modern and post-modern societies are not centred, and indeed, are alienated to some extent. For them, tourism has a different meaning. It becomes an opportunity to look for their Centre, to reconnect with the Centre they have become alienated from, or simply to escape from a Centre.

Cohen identified five different roles tourists could play, depending on their circumstances at the time. While his categories resemble Plog's to some extent, he argues that each trip is context specific. So, the same tourist may be on a quest in one trip, looking to re-create on another or simply want to sooth his or her soul on a third trip. Each one has different needs and each one perceives authenticity in a different manner. The five roles are:

- Recreational;
- Diversionary;
- Experiential;
- Experimental, and;
- Existential.

Recreational tourists are basically content with their lives. For them, tourism is a move away from the Centre that eventually reinforces it. The goal of travel, as the label implies, is to re-create, to regenerate and to restore mental and physical well-being. For them, their vacation is simply an opportunity to recharge their batteries. It has no deeper meaning. Travel is not undertaken for self-realization or for self-learning. Essentially, these people are already well centred in their home culture, but need a break. As a consequence, they seek mass entertainment and often seek what some would call shallow, trivial and easy to consume tourism experiences. They seek recreation, want to be entertained and are willing to accept make believe or pseudo-events on their own terms.

Diversionary tourists are alienated from their Centre, but are not seeking other Centres. They, as the name implies, are simply seeking a temporary escape from their boring and meaningless routine. Their goal is to heal the body and to sooth the spirit but not to re-create or to reconnect with their core values. Authenticity has no meaning for them. Instead they are simply seeking something that will entertain them for the short term. These people are most like the type of traveler Boorstin identified in his work. They are reflective of a broader societal malaise.

On the other hand, what happens if you are aware that you are alienated from your Centre? One response is to look for meanings in the life of others. This is where MacCannell's thesis comes into play. The quest for meaning and the quest for a Centre begins with a search, which in this case is the modern tourism experience. Experiential tourists want authentic experiences, and therefore the opportunity to reconnect, which are available to those people who try to break the bonds of their everyday existence and begin to 'live.' Tourism, thus absorbs the social function of religion for experiential travelers, where they try to experience other cultures through their touristic travels.

But, this is where the conundrum kicks in. The tourism industry stages authenticity that leads these types of tourists to believe that they have broken through the contrived front of the unauthentic. Still with me? But, they haven't really done that. Unfortunately, these types of tourists can only experience these things in a vicarious and ultimately shallow way. As such, the inauthentic may become authentic.

On the other hand, Experimental tourists go that next step by actively going on a quest for Centres in other societies. These people seek authentic life and lifestyle in other cultures, but do not fully commit to them. Instead, they sample a variety of things, but ultimately return to their old life. This type of traveler was typified by the hippies of the late 1960s and early 1970s who went looking for something different and better. Most of them sampled alternative lifestyles for a while, but most ultimately returned to their middle class lifestyle. Today's backpacker may be seeking similar experiences.

Lastly, Existential Tourists are those people who not only sought different Centres, but who fully committed themselves to it. They may be people who, for example, joined an ashram in India, or who liked visiting Australia so much that they decided to emigrate there. For them they seek and become immersed in the authenticity of the culture they visit.

Clearly, these five types of tourists are very different. The most commonly found types are the recreational and diversionary tourist, with the experiential tourist ranking third. As the level of commitment to seeking and adopting a different Centre increases, the number of tourists diminishes.

Tourism as a liminal experience

As discussed in Chapter 2, tourism involves a journey away, a stay at some distant location and a return journey home. But, it involves far more than the physical act of traveling, for it is also an activity that occurs outside of physical, social and psychological time and space. It thus involves both geographical and symbolic separation from one's home (Beckstead, 2010). As such, tourism is imbued with many meanings, where the act of becoming and then being a tourist, is associated with a temporary psychological and emotional metamorphosis of disassociation from the normal (to a great or lesser extent) (Carr, 2002a, b) and the assumption of a 'tourist' persona with its own roles, and associations. A number of sociologists and anthropologists have likened tourism to a liminal experience, where people experience communitas upon entering a liminoid landscape (Cohen, 1988).

A liminal state refers to a temporary transitional state between more permanent states where the person lacks social status and structure (Turner, 1974). The root of the word can be traced to the Latin term *limen* which refers to a state of being on a threshold or in an intermediate condition (Huang et al., 2018). It can be applied to individuals, groups or people or populations (Gisolf, 2013). As Turner (1974: 59, 60) notes, in liminality, profane social relations may be discontinued, formal rights and obligations are suspended and the social order may have been turned upside down. Instead, an anti-structure sets in with its own rules, rights and duties.

The anthropologist van Gennep's (1906, as cited by Turner, 1974) was the first to coin the phrase when he examined rites of passage of people in traditional agrarian societies as boys moved symbolically from childhood to adulthood. He suggested that such rites involved separation from the ordinary, a period of marginality or liminality and then re-aggregation or reintegration, where the person comes back to the normal society as a changed person (Graburn, 1983: 13). The person thus enters a betwixt and between state where, according to Wels et al. (2011), the person is understood to be 'no longer' and simultaneously 'not yet'. Typical examples include teenage years, where the person is no longer a child and simultaneously not yet an adult.

Turner (1974) was among the first to regard leisure, and by extension tourism, as a form of a liminal state, noting "*'leisure', then, presupposes 'work': it is a non-work, even an anti-work phase in the life of a person who also works*" (Turner 1974: 67). He saw a direct link to tourism for he felt the passage from one social state to another also often involved a parallel passage in space, or a geographical movement from one place to another. Importantly, as well, he noted liminal spaces provide two types of freedom: *freedom from* institutional obligations and *freedom to* play with ideas, with fantasies, with words (Turner, 1974: 68). It is for this reason that Foster and McCabe (2015: 48) observe that the liminal phase is simultaneously destructive and constructive. It is a time when social structures are inverted and through various ritual practices individuals are encouraged to question tacit knowledge. Pritchard and Morgan (2005:764) add such spaces are intangible, elusive, and obscure that lie in a limbo-like space often beyond normal social and cultural constraints.

9

As McKercher and Lui note (2014), the core premises are that travel allows individuals to set aside conventional standards of conduct that would normally be repressed at home (Selanniemi, 2003), that being in a liminal touristic space provides relative anonymity and freedom from scrutiny (Kim & Jamal, 2007). Epstein and Kheimets (2001:42) add "*becoming a tourist, however briefly, means shedding part of one's old identity and normal behavior [and]...adopting a new, temporary identity that necessarily incorporates some elements that are the opposite of the habitual personality and behavior.*"

Typically, entry into and exit from a liminal state involves three stages (Nash & Smith, 1991). Individuals reside in a structured situation. They are then separated from this situation, where they are distanced from their routine social groupings. Then they enter the liminal phase in which the structured requirements of daily life disappear into an undifferentiated state. The third stage involves a reintegration into which the person passes back into his or her structured life.

Where does tourism fit in? Well, tourism can represent the symbolic three stage process where by going away from a familiar place (separation) to a far place (liminality) and then returning to the familiar place (reintegration) one can reconnect to their Centre. The liminal state or far place in tourism represents another Centre that is temporarily visited before coming back to one's Centre (Cohen, 1988). This 'Centre Out There' can play a key role in helping reconnect to who people are. As a result, they can do things that they normally would not do: spontaneity, wholeness, social togetherness and participation in role reversal activities. Traditionally, men have rarely played a significant role in cooking. Cooking has always been (and to a large extent still is) considered to be women's work. But, tourism provides the opportunity for these roles to be reversed. The barbeque is the one place where it is socially acceptable for men to be the cooks. In a similar fashion, men often assume a much greater child care role on vacation than they do at home.

Jafari (1987) popularized the idea of tourism as a liminal experience. He identified a five stage process (Figure 9.1) beginning with the Corporation phase which represents the individual's profane and ordinary life. Some factor or factors push the person to want to escape from his or her ordinary life and the person begins to mentally prepare for an escape.

The Emancipation phase corresponds to the traditional separation phase in the three phased model. He identifies two stages in this phase. The first is the spatial notion where the person physically separates from the home region. The second, equally important stage, is one of declaration, where the individual declares he or she is no longer tied to the home situation. As Jafari (1987: 153) states these two stages work in consort for "separation signifies the *growing distance* between him and the ordinary, while declaration denotes the *shrinking distance* between him and the non-ordinary. Combined into one, the emancipation whole disconnects the cords of the ordinary womb and suddenly the tourist is born into the non-ordinary world." Cohen (1988) adds this stage is vital for the person is taken to an unfamiliar place peripheral to the ordinary place of abode and separated from the individual's ordinary social group. Here the process of leaving one's past behind and entering touristhood begins.

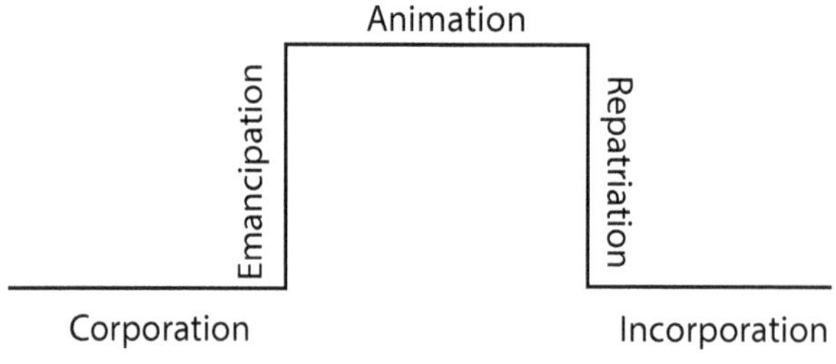

Figure 9.1: Jafari's liminal model of tourism

In the destination, the person enters a liminal state, or as Jafari suggests the Amination Phase, where the tourist enters a tourist culture and leaves behind the residual culture to a greater or lesser extent. Jafari (1987: 152) writes "*as immersion into touristhood deepens, the tourist further internalizes the idea that he is no longer imprisoned by his former self but is transformed into a new person with a new identity (the tourist), that he is playing on a new stage (the magnet), and that he is living up to the rhythm of a new culture (the tourist culture)."* Carr (2002a) suggests this process may be rapid or gradual depending on the nature and duration of the trip, travel party composition, the degree of spatial separation from home and also the degree of psychological separation desired.

Tourism involves entering liminal space, as shown in these pictures of a safari in South Africa. Liminality in tourism involves at least three phases. The emancipation phase occurs when one simultaneously leaves his or her normal world and then enters the world of being a tourist. This step involves both a physical and psychological escape. In the animation or liminal stage, the tourist leaves behind his or her old world and enters a new world as a tourist, sharing that new, but temporary environment with other tourists. But alas, the trip must come to an end and the person returns to his or her ordinary life both physically and psychologically. While in the liminal state, the tourist has no status other than being a tourist and social norms of behaviour can be jettisoned to some extent. This break from social pressures enables the person to release a range of subterranean values and thus can return to the normal life as a much more complete person.

Liminal spaces are typically associated with freedom (Pritchard & Morgan, 2006) where antistructure rules and where the tourist temporarily abandons social status and home influence (Gisolf, 2013). While the person may physically leave the home environment, entry into the destination environment is incomplete (Ryan & Martin, 2001), heightening the sense of social marginality and opening up possibilities. Instead, one enters a netherworld of the tourist environment, populated by other tourists who are temporary residents of the destination and who function under their own sets of implied and semiotically induced norms and rules. Neither the rules of home or of the destination apply. As Jafari (1987) notes, there is a palpable sense of freedom, enabling the tourist to play any role of his or her choice.

This freedom has been labelled by Graburn (1983:21) as the idea of inversion, whereby certain meanings and rules of ordinary behavior are changed, held in abeyance or even reversed. Inversion does not mean revolution and entry into the opposite. Instead, it exists on a continuum, whereby tourists can choose which behaviours to change and the degree of change from their normal style. Inversion could occur at a spatial level (urban tourists going to rural locations); be seasonal (winter tourists becoming 'snowbirds') or be class based (middle class people becoming backpackers).

Trips must come to an end. The reintegration phase has been labeled as the Repatriation stage by Jafari (1987). Again, this stage involves two phases, reversion and submission. Reversion refers to the process of letting go as a tourist. It involves both a physiological process of leaving touristhood behind that culminates in the actual departure from the destination region. Submission is the process of reintegrating into one's profane life. After this stage, the tourist enters Incorporation, where he or she is subsumed into the ordinary life.

The extent to which a person can have a liminal experience, or simply enter a liminoid state depends on a number of factors. Indeed, many people argue that modal types of tourism do not necessarily represent a liminal transition from one stage to another and instead represent a temporary entry into liminoid spaces where the normal codes of everyday life are not valid and place and time lose their meaning (Cohen, 1985; Mustonen, 2006). Cohen (1985) notes the person may return refreshed to his or her mundane life, but does not spiritually change the person's life. As Beckstead (2010) notes liminality in tourism is neither static nor a given. Instead, it is dynamic and negotiated by various social actors. People are far less likely to engage the liminoid during short duration holidays where they travel to culturally similar places and travel with a large group of closely associated companions (Gisolf, 2013). However, the longer the trip lasts, the more culturally distant the destination and the fewer the number of travel companions, the more likely one is able to peel back the onion of one's home culture and leave a nub of their residual culture as they fully enter the touristic culture (Carr, 2002b).

There are many examples of liminal spaces in a tourism context. Pritchard and Morgan (2006) suggest hotels are liminal sites that exist out of time and out of place. They write (Pritchard & Morgan, 2006: 769,770) "*to enter a hotel is to cross an imagined threshold into a liminal place which is strange, yet familiar, which offers freedom for some, but constraint, risk and unease for others. As potentially liminal and transgressive spaces they offer 'the spatial presence and practice outside of the norms of the prevailing (enforced) social specialization*." Underthun and Jordhus-Lier (2018) also see hotels as a liminal space for transient workers who can work with different expectations and commitments than permanent staff. Huang et al. (2018) write about airports as being liminal spaces for passengers for they are between time zones and countries and are metaphorically structurally invisible in a land of nowhere. Cohen (2018) makes a similar observation about backpacker enclaves, where over time such places have become spaces of suspension in which backpackers are suspended between the local and global culture. Here a distinctive hedonistic lifestyle develops marked by one's own ability to do one's own thing. Foster and McCabe (2015) feel residential activity camps satisfy the criteria of liminal spaces, while Mustonnen (2006) writes about volunteer tourism as a liminal activity where volunteers can engage projects to the extent they feel comfortable, but can leave any time they wish.

Hotels and resorts, like this one in Singapore, are liminal places. The guests are temporary residents where their status is unknown. All guests share the same experience, regardless of what they do in the profane world. The social order is turned upside down as all guests enjoy an elite status of being served. But stays are short and they must leave this unusual space and return to the normal world from whence they came.

Indeed, the idea of antistructure, suspension of social norms and freedom are essential elements of the liminal nature of tourism (Brooker & Joppe, 2014). These marginal places are privileged sites not normally afforded us where we can escape the oppressive places (Johnston, 2001). Huang et al. (2018) remind us this freedom equates to freedom from social norms and the existing social hierarchy we find ourselves in and freedom to mix with different people and exist within a different social structure.

The benefits are well-recognized, for the opportunity to play gives access to valuable compensatory experiences (Cohen, 1988). Gisolf (2013), summarizing the literature, identifies four main benefits of entering a liminal space. First, people can analyze their lives and background more clearly than they can at home and decide if the path they are one is the one to maintain. Second, social differences are deemphasized or ignored, resulting in a state of equality and solidarity among a group of tourists sharing the same experience. Third, the liberation of social constraints opens up the possibility of the more authentic self-emerging, at least temporarily. Finally, it may help people re-find themselves.

Matza and Sykes (1961: 716) suggest an even more profound benefit by arguing tourism provides an opportunity for the release of subterranean values, those which are in conflict or competition with other deeply held societal values but which are still recognized and accepted by many, in a socially acceptable space and time. Buzzell (2005:30) added later holidays "*include behaviours that are by design out of the ordinary, expressions of misrule rather than rule. These occasions are moments when the dynamics of social control are altered, shifting from normative to acceptable non-normative acts.*" The ability to release these values in a safe and non-judgmental environment performs a number of important social roles. A holiday, therefore, is as much a psychological escape from the range of duties and obligations one might feel at home as it is a physical escape (Prideaux et al., 2004). Much has been written about the therapeutic benefits of such a break (Cohen, 1972; Urry, 1990a, 1990b), with Zhu (2012:1498) observing the liminal space of tourism enables tourists to participate in temporary forms of transgression that enable their secret selves to be displayed.

Conclusion

Sociologists and anthropologists see tourism as an escape or quest, where the act of being in a different physical, social and psychological place provides a whole range of benefits to the individual that helps her or him to be better adjusted when returning home. The idea that tourism is a quest seems to form the basis for much of the thinking about this topic. It is worthwhile referring to the chapter on motivations as well, to place these works in perspective. The need to escape either because of anomie, disequilibrium or seeking creates the initial impetus to travel. The destination provides the venue where the break can occur, or where the seeking can be satisfied.

Discussion points / Exercises

1. Cohen's typology based on alienation from one's centre and Plog's ideas overlap. Prepare a table showing how Cohen's and Plog's five groups are similar. Discuss.
2. Cohen's model may relate to Crompton's motives. Discuss.
3. Give examples of how tourism can release subterranean values.
4. How can tourism represent the different liminal stages?

Points to consider for revision

- The sociological perspective of tourism has evolved from seeing it as an essentially superficial and negative experience to a positive one to one now that is more neutral in its views.
- The environmental bubble is a core concept in tourism. It explains much of tourist behaviour and further explains the basis for the commercial delivery of tourism products.
- Cohen sees tourists as fitting four roles depending on the level of institutionalisation involved in the travel experience.
- Cohen also identified five types of tourist based on their level of alienation from their home culture.
- Liminality in tourism explains why tourists behave the way they do, for tourism provides a powerful release valve from the ordinary pressures one faces.
- Tourism provides the opportunity to release subterranean values.

Suggested Additional Readings

Cohen, E. (1972). Toward a sociology of international tourism. *Social Research* 39(1), 164

Cohen, E. (1979) A phenomenology of tourist experiences, *Sociology*, 13, 170-201.

Jafari, J. (1987) Tourism models: the sociocultural aspects. *Tourism Management* 8(2), 151-159

References

Andreu, L., Kozak, M., Avci, N., & Cifter, N. (2005). Market segmentation by motivations to travel. *Journal of Travel and Tourism Marketing,* 19 (1), 1-14.

Beckstead, Z. (2010) Liminality in acculturation and pilgrimage: when movement becomes meaningful. *Culture & Psychology,* 16(3), 383-393.

Boorstin, D. (1964) *The Image: A guide to pseudo-events in America.* New York: Harper and Row.

Brooker, E. & Joppe, M. (2014) Developing a tourism innovation typology: Leveraging liminal insights. *Journal of Travel Research* 53(4), 500-508

Buzzell, T. (2005) Holiday revelry and legal control of fireworks: a study of neutralization in two normative contexts. *Western Criminology Review* 6, 30-42.

Carr, N. (2002a). The tourism-leisure behavioral continuum. *Annals of Tourism Research,* 29(4), 972-986.

Carr, N. (2002b). A comparative analysis of the behavior of domestic and international young tourists. *Tourism Management, 23*(3), 321-325.

Chang, S. Y. (2011). The influence of novelty-seeking and risk-perception behaviour on holiday decisions and food preferences. *International Journal of Hospitality & Tourism Administration, 12,* 305-329.

Cohen, E. (1972). Toward a sociology of international tourism. *Social Research* 39(1), 164

Cohen, E. (1979) A phenomenology of tourist experiences, *Sociology,* 13, 170-201.

Cohen, E. (1985) Tourism as play, *Religion,* 15(3), 291-304.

Cohen, E. (1988) Traditions in the qualitative sociology of tourism. *Annals of Tourism Research* 15, 29-46.

Cohen, E. (2018). Backpacker enclaves research: Achievements, critique and alternative approaches. *Tourism Recreation Research,* 43(1), 105-116. doi: 10.1080/02508281.2017.1388572

Craik, J. (1997) The culture of tourism sites in Rojek, C., Urry, J. (eds) *Touring Cultures: Transformations of Travel and Theory* Routledge, London, pp 113 – 136

Epstein, A., & N. Kheimets (2001) Looking for Pontius Pilate's Footprints near the Western Wall: Russian Jewish tourists in Jerusalem. *Tourism, Culture & Communication,* 3: 37–56.

Foster, C., & McCabe, S. (2015). The role of liminality in residential activity camps. *Tourist Studies, 15*(1), 46–64.

Gisolf, M (2013) Liminality in tourism. *Tourism Theories* www.tourismtheories.org/?p=700 .

Graburn, N. (1983) The anthropology of tourism. *Annals of Tourism Research* 10, 9-33.

Huang, W., Xiao, H. & Wang, S. (2018) Airports as liminal space. *Annals of Tourism Research* 70, 1-13.

Hughes, H. L. (1998) Theatre in London and the Interrelationship with Tourism. *Tourism Management* 19(5), 445-452.

Jaakson, R. (2004). Beyond the tourist bubble? Cruiseship passengers in port. *Annals of Tourism Research*, 31(1), 44-60.

Jafari, J. (1987) Tourism models: the sociocultural aspects. *Tourism Management* 8(2), 151-159

Johnston, L. (2001) (Other) bodies and tourism studies. *Annals of Tourism Research,* 28(1), 180-201

Karl, M. (2018). Risk and Uncertainty in travel decision-making: tourist and destination perspective. *Journal of Travel Research*, 57(1), 129-146.

Kim, H., & Jamal, T (2007) Touristic quest for existential authenticity. *Annals of Tourism Research,* 34(1): 181-201.

MacCannell, D. (1973) Staged Authenticity: Arrangements of Social Space in Tourist Settings. *American Journal of Sociology* 79(3):589-603

MacCannell, D. (2018) The Making of *The Tourist,Via* 13. http://journals.openedition.org/viatourism/2239

Matza, D. & Sykes, S. (1961) Juvenile delinquency and subterranean values. *American Sociological Review* 26, 712-719

McKercher, B. & Lui, S. (2014) Becoming a tourist. *International Journal of Tourism Research* 16, 399-406.

Mustonen, P. (2006) Volunteer tourism: Postmodern pilgrimage? *Journal of Tourism and Cultural Change*, 3(3), 160-177.

Nash, D., & Smith, V. (1991) Anthropology and tourism. *Annals of Tourism Research.* 18, 12-25.

Newman, H. (2002) Race and the tourist bubble in Downtown Atlanta. *Urban Affairs Review* 37(3), 301-321

Pratt, S., (2013). Same, same but different: Perceptions of South Pacific destinations among Australian travelers. *Journal of Travel & Tourism Marketing*, 30(6), 595-609.

Prideaux, B. Agrusa, J., Donlon, J. & Curran, C. (2004) Exotic or erotic - Contrasting images for defining destinations, *Asia Pacific Journal of Tourism Research*, 9(1), 5-18.

Pritchard, A., & Morgan, N. (2006) Hotel Babylon? Exploring hotels as liminal sites of transition and transgression. *Tourism Management* 27(5), 762-772.

Reisinger, Y., & F. Movando (2005) Travel anxiety and intentions to travel internationally: implications of travel risk perception. *Journal of Travel Research* 43(3): 212-225

Rittichainuwat, B., & Chakraborty, G. (2009). Perceived travel risks regarding terrorism and disease: The case of Thailand. *Tourism Management,* 30(3), 410-418.

Ryan, C & Martin A. (2001) Tourists and strippers: Liminal theater. *Annals of Tourism Research* 28(1), 140-163.

Scott, K. & Mowen, J. C. (2007). Travelers and their traits: a hierarchical model approach. *Journal of Consumer Behaviour*, 6, 146-157.

Selanniemi, T. (2003). On holiday in the liminoid playground: Place, time and self in tourism. In T. Bauer and B McKercher (eds.), *Sex and Tourism: Journeys of romance, love and lust* (pp. 19-34). Binghamton: Haworth Press.

Thapa, B., Cahyanto, I., Holland, S. M., & Absher, J. D. (2013). Wildfires and tourist behaviors in Florida. *Tourism Management*, 36, 284-292.

Turner, V. (1974) Liminal to liminoid in play, flow, and ritual: an essay in comparative symbology, *Rice University Studies* 60(3), 53–92

Underthun, A. & Jordhus-Lier, D. (2018) Liminality at work in Norwegian hotels. *Tourism Geographies*, 20(1), 11-28, DOI: 10.1080/14616688.2017.1314546

Urry, J. (1990a). The consumption of tourism. *Sociology* 24, 23-35

Urry, J. (1990b) *The Tourist Gaze*. London: SAGE Publications

Wels, H., van der Waal, K., Spiegel, A. & Kamsteeg, F. (2011) Victor Turner and liminality: An introduction. *Anthropology Southern Africa*, 34(1-2), 1-4, DOI: 10.1080/23323256.2011.11500002

Wickens, E. (2002) The sacred and the profane: A tourist typology. *Annals of Tourism Research* 29(3), 834-851.

Zhu, Y. (2012) Performing heritage: rethinking authenticity in tourism. *Annals of Tourism Research* 39

10 Tourism and the Individual

By the end of the chapter, the reader will be able to

- Define propensity and intensity
- Analyse barriers to travel and know how to disaggregate interest as a discrete barrier
- Understand the role of the leisure paradox in influencing travel over one's life span
- Analyse whether generational cohort theory or age-period cohort theory better explains changes in travel behaviour over time

Introduction

Tourism is a socially selective activity. While we may think that everybody has an opportunity to travel this is clearly not the case. Further, even among those who do travel, social, personal and demographic factors exert a significant effect on both propensity to and intensity of travel. This Chapter examines a range of factors that influence tourism propensity and intensity among individuals. It begins with a definition of these terms and them explores barriers to participation, the traditional leisure paradox and how it affects tourism and finally the role of generational cohort theory and tourism activity.

Propensity and intensity

Propensity relates to the participation rate, or the percentage of individuals within a population who travel. It is defined as the share of the population that makes at least one (holiday) trip of a specified number of overnight stays during a given reference period (Eurostat, 2010). Like all things in tourism, though, this simple definition is not absolute, for propensity can be measured in a number of ways. Usually, the reference period is one calendar year or 12 months, depending on when the survey was completed. The length of stay can vary, with some European statistics looking at trips of four nights or longer (Eurostat, 2010), while others consider any overnight pleasure trip (Eurostat, 2017). The UNWTO, being an organisation that is interested in international travel, calculates propensity on the basis of the number (or percentage) of international travelers as a proportion of a market's total population.

Intensity or frequency is a simpler concept for it relates to the average number of trips taken over a given period of time. Again, though, the denominator can vary using either the entire population, and, therefore, calculating non-travellers as part of the intensity figure, or by focusing only on those who travelled. It is interesting to note that during COVID-19, a relationship was found between past and current intensity, with those who travelled frequently prior to the pandemic still likely to continue to travel, albeit at a less intense level, while those who travelled infrequently were most likely to forsake this activity (Boto-García & Baños-Pino, 2023; Song et al., 2023). This pattern has been attributed to pre-existing risk aversion attitudes (Lin et al., 2023).

Non-travel or limited interest in travel

One of the myths of tourism, especially in developed economies, is that everybody travels or at least has the potential to travel. Yet, one does not need to look very far to see the inherent flaw in this assumption. Some people may not be physically robust enough to travel. Some people may simply not want to travel. Others may not have any travel companions. Others still may not be able to afford to travel. In Europe, about 56% of residents took either a domestic or international pleasure trip during COVID-affected 2021 (Eurostat, 2023), down from the 62% reported in 2016 (Eurostat, 2017). Propensity varied significantly by country of origin, though, ranging from a low of a 24% in Bulgaria to a high of 81% in the Netherlands. Again, these figures are for COVID-impacted 2021 and are lower than in pre-COVID years. A study of Hong Kong (McKercher, 2009) showed propensity varied from the mid 60 percentile to the low 70 percentile depending on the state of the economy. Smith and Carmichael (2005) concluded that one in three Canadians do not travel. Overall, it appears that between one-quarter and one-third of residents of developed countries do not take an overnight pleasure trip in any given year. Obviously, this figure collapsed during COVID-19, while some say propensity may spike in the aftermath of the pandemic due to revenge travel (Tiwari & Chowdhary, 2021), although evidence supporting this assertion is mixed (Liu & Wang, 2023).

Our conceptualization of why people do not travel grew out of leisure constraints theory. This theory postulates that everyone wants leisure but due to a variety of reasons not everyone can have it. It argues further that if constraints can be removed, then participation in leisure activity will follow. Crawford and Godbey (1987) and Crawford, Jackson and Godbey (1991) proposed that leisure constraints could be classified into one of three categories of intrapersonal, interpersonal, and structural barriers, and saw leisure participation as a function of the ability to negotiate each of the constraints.

Intrapersonal constraints represent psychological states or attributes that reflect preferences. They include such items as lack of interest, religion, reference group attitudes and perceived skill. Interpersonal constraints relate primarily to the lack of travel partners, while structural constraints lie somewhat beyond the individual's own control and can include time, cost, lack of opportunity, family commitments and the like. Leisure constraints theory argues that these three classes of constraints are hierarchical in nature, with intrapersonal issues needing to be resolved before interpersonal constraints can be considered. Interpersonal constraints must also be resolved before structural barriers can be addressed. However, in a more recent retrospective piece, Godbey et al. (2010) identified an intertwining relationship between the three dimensions, which challenges the linear hierarchical proposition associated with the model.

The model has been tested a number of times in a leisure context and found to be valid (Godbey et al., 2010). Jackson et al. (1993) evaluated this model from a tourism context, while it has also been used by a number of other researchers (Nyaupane & Andereck, 2008; Pennington-Gray & Kerstetter, 2002) with modest success. Intuitively it makes sense and some empirical evidence seems to support the existence of such barriers. For example, when considering intrapersonal constraints, Lee and Tideswell (2005) and McGuire (1984) found that societal pressures and, in particular, reference group approval from family and friends precluded many seniors from travelling. March (2000) came to similar conclusions examining propensity among Japanese women, where they are expected to be travel active in their 20s, but once they marry are expected not to travel. Lack of travel partners has long been identified as a core intrapersonal constraint, especially among people with disabilities (Smith, 1987).

The impact of structural constraints has attracted most attention in the literature. Tourism Research Australia (2011a, 2011b) identifies income as the primary driver for both domestic and inbound tourism while Eurostat (2010) found a strong correlation between household income and overall travel propensity. The inability to speak foreign languages precludes some people, especially seniors from travelling abroad (Chen and Wu 2009). Many others have identified work commitments as a factor that inhibits travel (Dickinson, 2014; Losada et al., 2016; Nickerson, 2000; Smith et al., 2011).

Reported constraints vary by age and life stage (Kattiyapornpong & Miller, 2009; Smith et al., 2011). Older less affluent working families, in their 50s with children still living at home perhaps face real structural constraints of family obligations and cost factors as inhibiting their ability to participate in tourism, while middle aged affluent couples and families cited work commitments as their main reason to not travel. Interestingly, Smith et al. (2011) also conducted a study of non-tourism and concluded that many people identified lack of money and time as factors that precluded travel, yet, their research found that most non-travellers can in fact afford to travel and have the time available to do so. But, as they note, something gets in the way as they make their purchase decisions.

Herein lies one of the dilemmas of importing theory from one discipline and applying it to tourism without necessarily considering the unique nature of tourism. If one frames research within the confines of leisure constraints, then it is easy to identify leisure constraints as factors that inhibit travel. But, is the use of this framework really unveiling the truth? Kattiyanornpong and Miller (2009) argue that travel constraints are quite different from general leisure constraints. Others note traditional leisure constraints theory may not be particularly relevant within a tourism context (Chen & Wu, 2009).

To begin constraints are not absolute (McGuire et al., 1986) and instead their impact varies depending on how people react to them. The same situation may act as a prohibitor that stops some people from travelling, a limitor that reduces participation or requires a modification of activities among others or may be a non-issue for others still. For example, reference group pressures may stop some seniors from travelling, may modify their travel experiences by limiting them to travelling in seniors groups or with other family members or may have no impact at all if the person is a keen tourist. The same can be said for interpersonal constraints where the growing body of solo backpacking literature, especially among women suggests it can be a liberating and life-changing event and not a prohibitor of travel (Bianchi, 2016; McNamara & Prideaux, 2010; Wilson & Little, 2008). Likewise, income as a structural constraint may indeed stop some people from travelling, but is more likely to limit others to taking a lower cost, short break or short duration holidays, or it may be a non-issue for others. For example, Zhang et al. (2020) indicated lower income individuals will substitute car travel instead of flying to their destination.

More than that, though, lack of interest may play a far more important role than packaging it in a bundle of intrapersonal constraints suggests. Haukeland (1990) categorized travel along a 2 x 2 matrix. He identified four types of people who did not travel, including three groups who felt various constraints. His Type A individual, though, did not face economic or social barriers, but simply preferred to stay at home. Litvin, Smith and Pitts (2012) found the stated reasons for non-participation, such as lack of money, contradicted the demographic characteristics of respondents. They concluded that some element of social desirability bias prevented respondents from identifying the real root cause of their non-participation.

In a similar manner Li, Meng and Zheng (2016) found that Chinese non-tourists had significantly lower internal motivations and attitudes towards tourism, while Popp et al. (2021) divided non-travellers into two groups: those with medium to high levels of travel motivation those with no travel motivation. Dolnicar, Lazarevski and Yanamandram (2013) and Dolnicar, Yanamandram and Cliff (2012) recognise that not all people want to go on vacation.

Lack of interest, more so than any other barrier may be the main reason why some people don't travel, especially if suitable substitute activities are available close to home that provide the same benefits. Since these skiers can find high quality experiences within an hour's drive of their homes, they may not be inclined to travel longer distances for essentially the same type of experience. Simply stated, for some people, travel is not a life priority or a high enough life priority to engender action. While we think travel is available to all, some people simply do not have the travel bug.

McKercher and Chan (2014) found a strong relationship between interest or the importance one places on travel and tourism propensity. Their study asked people to rate whether tourism was more, less or as important as a variety of other activities such as spending time with family and friends, saving money for long-term goals or pursuing recreational interests to improve their skills. People who identified tourism as being very important, felt it was equally or more important than the other items measured, while those who felt travel was unimportant or very important to them ranked every other item as being of equal or greater importance than travel. The study concluded that intensity of and propensity to travel are related to the personal importance respondents place on tourism. They argued people who have the travel bug place the opportunity to travel as a high life priority, while those who do not see many other things in life as being of greater importance.

In a follow-up study, McKercher et al. (2022) wanted to examine disinterest in travel in the absence of objective constraints. Their study asked people to rate their interest in travel on a five point Likert scale. About 25% of respondents had no to very limited interest in pleasure travel. Disinterest does not necessarily equate with non-travel, though, but is closely associated with lower overall propensity and intensity, compared to those who valued travel moderately to highly. Respondents were asked to explain their lack of interest in an open-ended, non-judgmental fashion free from social desirability bias. A total of 11 different themes emerged that could be grouped into four thematic domains and ordered along a continuum from those who were totally disinterested to those who were moderately disinterested, but still did not travel (Figure 10.1). The lack of interest cohort, encompassing people who previously had interest in travel but not now through to those people who described themselves as homebodies represented the largest group. A smaller group who had concerns about travel felt the personal and financial cost of travel outweighed the limited benefits it provides. Another smaller subset were also concerned about the lingering effects of COVID-19. The third group saw some benefits of travel, but still did not value it highly, or as highly as other activities, while the fourth group recognized the personal growth opportunities of travel, but again expressed little to limited interest in this activity.

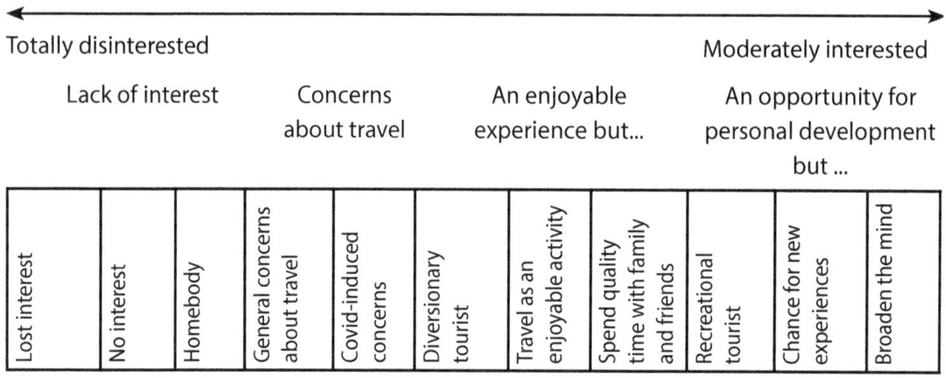

Figure 10.1: A continuum of disinterested tourists

These findings along with aforementioned work by Haukeland, Smith and his colleagues and Dolnicar and her team suggest we need to reconsider the leisure constraints model as it relates to tourism. In particular, the issue of lack of interest emerges as being the most critical early factor that influences travel participation. Again, leisure theory suggests lack of interest is just one of a bundle of constraints that can be easily overcome by removing barriers. Within a tourism context, though, it is difficult to overcome a deep-seated lack of interest, especially when other opportunities compete effectively for discretionary leisure time and expenditure. Instead, lack of interest is of such profound importance that it deserves to be identified as a discrete barrier category in its own right. As such, it is proposed that the leisure constraints model be modified by adding a fourth foundation tier of lack of interest as shown in Figure 10.2. Lack of interest, therefore, represents the primary barrier and unless, and until, it can be overcome many people will not travel or will travel infrequently.

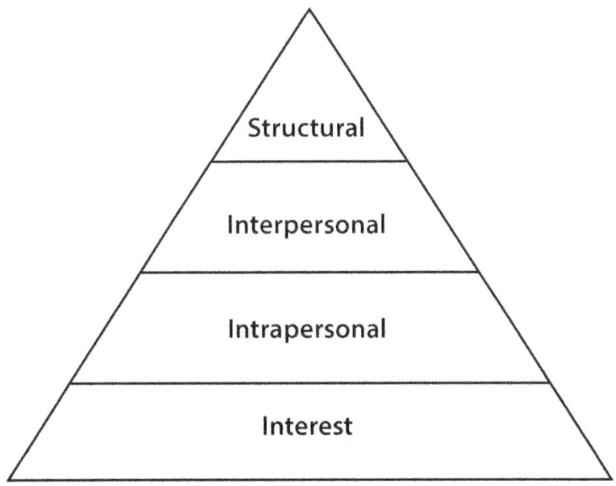

Figure 10.2: Tourism constraints model. Source: from McKercher and Chan, 2015

The leisure paradox

Constraints theory has its merits, but life is rarely as simple as theory suggests, for a variety of intervening factors also affect one's ability to travel and the type of tourism activity pursued throughout one's life. Cooper et al. (1998) introduced the idea of the leisure paradox in tourism (Figure 10.3) where there is an inverse relationship between income and time availability. As Mill (2008) notes the paradox suggests that when people have money to engage in various leisure pursuits they lack time and when they have the time availability, discretionary income is at a premium. Added to this are different health, fitness and physical issues that mean both intensity and the type of travel will vary with age (Hong et al., 2005).

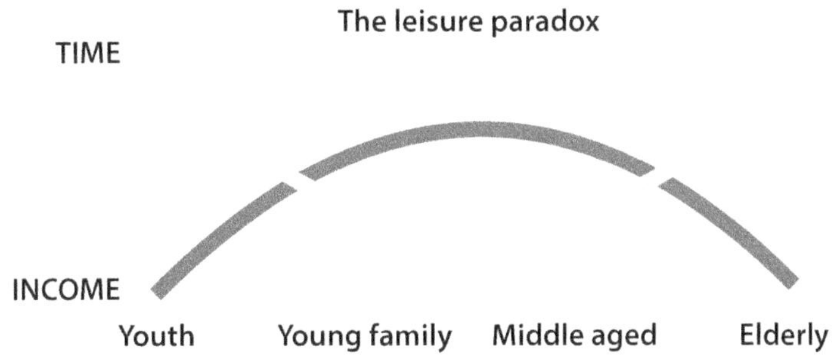

Figure 10.3: The Leisure Paradox. Source: Cooper et al. 1998

Ironically, the leisure paradox is becoming an increasingly challenging issue for digital nomads. Many of them feel the appeal of being a digital nomad in a tropical paradise enables them to have the best of both words. But Cook (2020) highlights three challenges they face. First, the daily realities of the need to get things done quickly shatters the utopian dream of working in paradise. Second, the novelty of working remotely becomes normalized and loses much of its appeal. Third and most importantly, work tends to bleed into every aspect of daily life, making it hard to turn off the work mind. Thus, while being a digital nomad may hold romantic appeal, the reality is that many have less leisure time than they would in a standard work environment.

Travel activities vary according to one's lifecycle stage. Singles, for example, are usually able to participate in a wider variety of travel experiences than are those people who have families, simply because they have fewer responsibilities. More than just simply the availability of time, life stage also shapes the motivation for and needs of a holiday. Prior to the onset of children, singles and couples are motivated to participate in an activity primarily for the intrinsic satisfaction that participation brings. Once children arrive, however, the primary motivation becomes one of providing benefits to the children and to the family, more so than for the individual. By the same token, after children arrive, the social and financial realities of the couple or individual change. Travel is often curtailed as the family spends far more time at home.

Hong et al. (2005) found married people without children, full nesters with older children, and empty nesters spend significantly more than singles, whereas solitary survivors spend significantly less on leisure travel. Families with a higher number of members are more inclined to make a domestic trip than one abroad. However, the oldest cohorts are more inclined to participate in tourism than the youngest one, but they are less inclined to spend on tourism (Bernini & Cracolici, 2015). Backer and King (2017) saw an increase in visiting friends and relatives travel among the over 60s, while Huang and Petrick (2010) found older people travelled for longer periods of time and travelled greater distances. Boksberger et al. (2009) saw a greater tendency to seek more comfort as people age.

Fleischer and Pizam (2002) and Nickerson (2000) found that travel participation often peaks between the age of 55 and 65 when leisure time and income are at their peak and one's health is still strong. Children have grown, mortgages are paid and those in long term employment have plenty of leave time available. Demand remains strong until the early 70s when propensity declines. But the frequency with which seniors travel is lower than other age groups (Hung et al,. 2007; Lung, 2017). Lung (2017) determined that age alone is not a factor, but that post retirement income is the key determinant. Fleischer and Pizam (2002) add, somewhat axiomatically, that income and health are the major constraints that affect senior's travel propensity. Zimmer et al. (1996) add education, mobility problems, self-assessed health status, rather than absolute health condition and interest in spending money on recreation are other factors that influence participation among seniors.

The changing profile of families, split families and the emergence of three generation families is changing travel patterns (Glover & Prideaux, 2008). Bernini and Cracolici (2015) cite a higher life expectancy, a healthier ageing population, smaller family size, a higher share of the population with a high level of education and population migration as affecting tourism. Schänzel and Yeoman (2015), for example noted families are becoming more vertical rather than statically horizontal, with grandparents enjoying more time with their grandchildren as they live longer. As a result, we are seeing an increase in multi-generational travel.

Education levels are closely related to travel propensity and the type of activity pursued (Bernini & Cracolici, 2015; Zimmer et al., 1996), in part because the better one's education the greater the likelihood of having a higher income and in part because people who are better educated are more interested in what is going on in the world (Plog, 1974). This in turn, often stimulates a desire to see different areas and to experience different cultures. A well educated person may seek status and learning opportunities from their travel, while a less highly educated person may seek simply escape.

Education also relates to the ability of the individual to want to break free of his or her own cultural confines. Better educated tourists are generally more willing to seek holidays in unfamiliar and culturally different places. Less well educated tourists, who have less of a world view, feel intimidated when travelling in strange cultures and are more likely to confine their activities to culturally proximate destinations. Ecotourism, adventure travel, alternative tourism, nature based tourism, cultural tourism and much special interest tourism (festivals and the like) all appeal to highly educated individuals.

What influences changes behaviour over time – generation or age?

There is no doubt that members of different generations exhibit substantially different tourist behaviours, to the extent some feel the very nature of tourism will change as subsequent generations mature and become the dominant consumer group (Benkendorff & Moscardo, 2013; Gardiner et al., 2014). But what is the cause of these differences? Two competing theories have been developed. One is generational cohort theory that basically argues that each generation experiences life differently (Alwin & McCammon, 2003) and that important historical events and social changes encountered by a group of people who were born at approximately the same time and experience these events during their critical development stages shape their attitudes, values, beliefs and inclinations (Moss, 2016; Tang et al., 2017). Generational Cohort theory further argues that consumption patterns are driven by a person's system of value priorities, and, therefore, each generation tends to display patterns that are internally consistent but differ across generations (Li et al., 2013). It argues that differences in the behaviours of Baby Boomers and Millennials are due to fundamental differences in generation that are enduring over time. Thus, as they age, Millennials will continue to exhibit travel behaviours that are uniquely different from those of other generational cohorts.

The other theory, labelled age-period-cohort theory, posits that development is not completed at adulthood but continues across the entire life course of the person (Baltes et al., 1999). As such, our values, attitudes and beliefs are dynamic and malleable. They will change due to three interrelated functions of: changes in age; period changes or changes due to the events in particular years; and generational changes as a younger generation supplants an older one (Fosse & Winship, 2019). In essence, the age-period-cohort model argues that people of the same age may display similar behaviours, regardless of generation, notwithstanding period effects. As a result, this theory argues, for example, that few real differences existed between Baby Boomers and Millennials when they were the same age, and more importantly today's younger generation will exhibit similar travel patterns to those of older generations when they reach the same age, notwithstanding period effects.

■ Generational cohort theory

The concept of generational cohort theory argues that that generations share much in common for they are similarly located, by being members of the same stage as the collective process, experiencing the same events and by being born within the same historical and cultural region (Mannheim in 1923, as cited in Mannheim, 1952). They are also substantially and fundamentally different from other generations that come of age before or after them. Members of a birth cohort share a social history, experience their life cycle at the same time, and importantly, share the experience of the cohort itself (Alwin & McCammon, 2003: 25), and because of that each generation experiences life differently (Pendergast, 2009). As such, how people think about the social world around them may depend as much on what was happening in the world at the time they were growing up as it does on what is happening in the present. Ryder (1965: 843) notes each birth cohort acquires coherence and continuity from the distinctive development of its constituents and from its own persistent macro analytic features. Successive cohorts are differentiated by the changing content of formal education, by peer-group socialization, and by idiosyncratic historical experience

A few features are critical when considering the impact of generational cohorts on behaviour. Proponents of this theory argue that attitudes and values crystalise during one's formative years from adolescence to the early 20s (Benkendorff & Moscardo, 2013; Cheng & Foley, 2017) and remain largely unchanged thereafter (Hung et al., 2007; Inglehart, 2008). Thus how people think about the world may depend more on what was happening in the world when they were growing up than what is occurring in the present (Alwin & McCammon, 2003: 25) as individuals born during the same generational cohort may share the same inclinations and cognitive styles (Moss, 2016).

Generational cohort theory suggests your travel patterns are defined more by people born in your generation than your family members, while age-period cohort theory suggests different generations behave in a largely similar manner at the same age, notwithstanding period effects, such as technology changes and changing geo-political realities. Proponents of generational cohort theory point to the fact that many elderly people grew up during time of great hardship, when a culture of pleasure travel had not been established. As such, they tend to travel less. People born later grew up during a time of relative affluence, when a culture of pleasure travel started to emerge. Young people today are growing up in unique circumstances that will translate into travel patterns that are unique to them. Proponents of age-period cohort theory, on the other hand suggest that few real changes in travel behaviour exist across generations in developed economies. Young people have always seen and will continue to see travel as a rite of passage, while family considerations will influence travel decisions of people in their 30s and 40s, regardless of which generation they belong to.

Each generation has two essential components, a common location in historical time and distinct consciousness of that historical position (Tang et al., 2017). Generations, thus are as much social creations as they are biological ones (Sessa et al., 2007), with the length of the generation depending on the duration of events that shaped it during the person's formative years. A wide array of factors can influence the values that are formed in a generation. Sessa et al. (2007) identified such issues as a traumatic or formative events such as war, a dramatic shift in demographics, an interval that connects a generation to great success or failure, the creation of sacred space that creates a collective memory or the presence of mentors and heroes that give a voice to a generation. Coomes (2004) adds that history and popular culture can play a vital role in defining a generation, while the futurist O'Hara-Deveraux (2013) suggests technological changes, the emergence of an industrial revolution and transition from a largely rural agrarian economy to an urban society can also shape generations.

Generation cohort theory has received some attention in the tourism literature. Benkendorff and Moscardo (2013) suggest that changing values of Baby Boomers and members of Generation X and Y may result in the potential decline of eco-tourism and the need to change how products are delivered. Ishida et al. (2017) identified markedly different attitudes toward travel in general and international in particular by successive generations of Japanese tourists. Li, Li & Hudson (2013) found members of younger generations were more experienced tourists than members of older generations at the same age, but that older consumers had more accumulated travel experience. They found that Baby Boomers and Generation X were similar in their past destination experiences but different than Generation Y. Gardiner et al. (2014) concluded that generational cohort membership influences the travel decision making process of Australian domestic tourists, resulting in a generational behaviour gap.

Age-period-cohort model

The age-period cohort model, on the other hand, acknowledges the impact of generations on some core values and attitudes, but also argues that both the period when something occurs as well as the age of the individual at a given point in time play a key role. It is predicated on the lifespan development theory (Baltes, 1987) that argues development is not completed at adulthood but continues across the entire life course of the person (Baltes et al., 1999). As an example, it has been observed that a range of psychosocial changes occur over time, including changes to attitudes, personality and needs (Faranda & Schmidt, 2000). As such, development is continuous and malleable (Baltes, 1987), caused by a series of gains and losses that may relate to accumulated experience, advancement in careers, greater wealth or a quest for greater safety and security as we age (Staudinger, 2001). Whereas values were once thought to be stable, now many people argue that they change over time as we age (Fitzenberger et al., 2022). Within a tourism context, for example, the European Tourism Commission (ETC, 2020) argues that young people are still exploring, discovering and solidifying their consumption preferences.

It has received much less interest in the tourism literature, primarily because of methodological challenges in assessing how values may change over time, and their impact on travel choices. Indeed, it appears that only one study has attempted to examine tourist behaviour of different generations when they were at the same age (McKercher, 2023). This study examined four generations of 20 somethings by conducting a meta-analysis of published literature spanning from the late 1960s / early 1970s to the present.

The research identified virtually no differences in attitudes to, values about and experiences sought when travelling, other than some period changes (technology, opening or closing of boarders, etc.). Tourism has always been and remains a rite of passage for young people (Hardington-Gill, 2020). Young people demonstrate a lack of loyalty, regardless of the generation studied, in part because as new travellers, they have experienced few options of repeating to the same destinations (Almeida-Santana & Moreno-Gil, 2018). Hedonism, immediate gratification, spontaneity and selfishness are common characteristics of travel by each generation of 20 somethings, along a quest for stimulation, exoticism and novel experiences. Likewise, some exploratory research on travel by seniors reveals no real differences across almost 40 years of research.

■ Which model is most appropriate?

Which model uncovers what Tribe (2006) calls the 'truth about tourism'? Which 'truth' you believe depends on the method adopted and the potential in-built bias imposed by that method. Both generational cohort theory and age-period cohort theory acknowledge that travel preferences differ across generations. Where they disagree is the cause of these differences. Generational cohort theory argues that these differences are locked in at a young age while age-period cohort theory says people's values change over time, and as such, so will their travel preferences. Thus, one theory argues travel patterns of generations are unique, while the other argues they are largely similar. Both cannot be right. Generational cohort theory comes to its conclusions using cross sectional studies that capture differences in behaviour by generations at a single point in time. Much of this research begins with the assumption that generational differences exist, with cross sectional studies confirming the existence of such differences. The circle is then closed when a cause-effect relationship is assumed to be attributed to generational cohort theory. This simple but flawed step assumed that the obverse of an effect and cause relationship exists, whereby an observed effect must be attributable to a single case. Yet, when a different lens is applied, as with the study on age-period cohort theory an alternative explanation of differences in behaviour with age emerges.

Conclusion

A number of factors influence travel propensity and intensity. This chapter began by examining barriers to travel and illustrated how the traditional leisure constraints theory is not applicable to tourism for it does not consider the profound role that lack of interest plays as an inhibitor of travel. People will not travel if it is not a life priority in the first place, or not a high enough priority to engender regular action. No amount of barrier removal will overcome such resistance. As such, interest needs to be disaggregated from the bundle of intrapersonal constraints and identified as its own discrete barrier. However, even if interest exists, the presence of the leisure paradox may have a significant effect on travel by individuals, for an inverse relationship exists between time availability and affordability. Those with the most time may be able to afford to travel least and therefore could be constrained in their behaviour. Those with the most income, on the other hand, may have to overcome the most obstacles to enable travel. Together, they reflect the trend toward short break trips, rather than taking one long duration holiday. Finally, the chapter discussed whether differences in behaviour can be attributed to lock-in features of generations or by changing attitudes attributed to age. Both models observe the same features, but argue diametrically opposite causes of these differences.

■ Discussion points / Exercises

1. Barriers to travel affect people differently. The same barrier may be a prohibitor, a limitor or have no impact. How can lack of money affect people in different ways?
2. Lack of interest is a critical feature stopping some people from travelling. Can this barrier be overcome?
3. What impact do children have on family travel?
4. Generational cohort theory and age-period cohort theory suggest travel propensity, intensity and activity change with age, but attribute these changes to different causes. Interview people from three different generations to determine when they first travelled for pleasure, flew and importantly whether they felt long haul travel was something that was available to them.

Points to consider for revision

- Barriers are dynamic and need to be negotiated. Depending on the person, the same barrier may represent a prohibitor to travel, may necessitate moderation of travel or may have no impact on travel.
- Leisure constraints theory says barriers are hierarchical, as one must overcome intrapersonal, interpersonal and structural barriers in order to participate. However, tourism constraints theory identifies a fourth more fundamental level of lack of interest. Lack of interest represents the single greatest barrier to travel for many people. Many people who have little or no interest in travel need to voice socially acceptable answers (such as lack of time and money) to mask their lack of interest.

- The leisure paradox suggests an inverse relationship exists between time availability and income.
- Generational cohort theory suggests people resemble their generations more so that their families. In a tourism context, it relates primarily to whether a culture of pleasure travel was created for each generation.
- Age-period-cohort theory, on the other hand, argues that people of the same age, regardless of generation display similar travel patterns.

Suggested additional readings

Dolnicar, S., Lazarevski, K, Yanamandram, V. (2013) Quality of life and tourism: A conceptual framework and novel segmentation base. *Journal of Business Research*. 66(6) 724-729.

Litvin, S. W., Smith, W. W., & Pitts, R. E. (2012). Sedentary behavior of the non-travel segment: a research Note. *Journal of Travel Research*, 52(1), 131-136

McKercher, B. (2023) Age or Generation. *Annals of Tourism Research*. 103, 103656. https://doi.org/10.1016/j.annals.2023.103656.

References

Almeida-Santana, A., & Moreno-Gil, S. (2018) Understanding tourism loyalty: Horizontal vs. destination loyalty, *Tourism Management*, 65: 245-255.

Alwin, D. F. & McCammon, R. J. (2003) Generations, cohorts, and social change, in J. T. Mortimer & M .J. Shanahan (Eds.) *Handbook of the Life Course. Handbooks of Sociology and Social Research*, Springer, Boston, MA, 23-49 https://digitalcommons.kennesaw.edu/cgi/viewcontent.cgi?article=1529&context=ajis

Backer, E., & King, B. (2017). VFR traveller demographics: The social tourism dimension. *Journal of Vacation Marketing*, *23*(3), 191-204. https://doi.org/10.1177/1356766716665439

Baltes, P. B. (1987). Theoretical propositions of life-span developmental psychology: On the dynamics between growth and decline. *Developmental Psychology*, *23*(5), 611-626. https://doi.org/10.1037/0012-1649.23.5.611

Baltes, P., Staudinger, U. & Lindenberger, U. (1999) Lifespan psychology: Theory and application to intellectual functioning. *Annual Review of Psychology* 50(1), 471-507.

Benkendorff, P. & Moscardo, G. (2013) Generational cohorts and ecotourism. In Ballantyne, R. & Packer, J. (Eds) *International Handbook on Ecotourism*. Cheltenham: Edward Elgar publishing. pp 135-154.

Bernini, C., & Cracolici, M. (2015) Demographic change, tourism expenditure and life cycle behaviour. *Tourism Management* 41, 191-205.

Bianchi, C. (2016) Solo holiday travellers: Motivators and drivers of satisfaction and dissatisfaction. *International Journal of Tourism Research*, 18, 197-208

Boksberger P.E., Sund K.J., Schuckert M.R. (2009) Between Past, Present and Future – Implications of Socio-demographic Changes in Tourism. In: Conrady, R., & Buck, M. (eds) *Trends and Issues in Global Tourism 2009*. Springer, Berlin, Heidelberg

Boto-García, D. & Baños-Pino, J.F., (2023). Deep habits and travel resilience after COVID-19. *Tourism Economics*, 29(2), 543-550.

Chen, C. & Wu, C. (2009) How motivations, constraints, and demographic factors predict seniors' overseas travel propensity. *Asia Pacific Management Review* 14(3), 301-312.

Cheng, M. & Foley, C. (2017) Understanding the distinctiveness of Chinese Post-80s tourists through an exploration of their formative experiences. *Current Issues in Tourism*, 21(11), 1312-1328. DOI: 10.1080/13683500.2017.1406462.

Cook, D., (2020). The freedom trap: digital nomads and the use of disciplining practices to manage work/leisure boundaries. *Information Technology & Tourism*, 22(3), 355-390.

Coomes, M. (2004) Understanding the historical and cultural influences that shape generations. *New Directions for Student Services*. 106, 17-31.

Cooper, C., Fletcher, J., Gilbert, D., Shepherd, R., & Wanhill, S. (1998), *Tourism Principles and Practices*, (2nd ed.), England: Addison-Wesley, Longman.

Crawford, D. & G. Godbey (1987) Reconceptualizing barriers to family leisure. *Leisure Sciences*, 9(2), 119-127.

Crawford, D., Jackson, E., & G. Godbey (1991) A hierarchical model of leisure constraints. *Leisure Sciences* 13(4), 309-320.

Dickinson, J. (2014) Time, tourism consumption and sustainable development. *International Journal of Tourism Research* 16(1), 11-21.

Dolnicar, S., Lazarevski, K, Yanamandram, V. (2013) Quality of life and tourism: A conceptual framework and novel segmentation base. *Journal of Business Research*. 66(6), 724-729.

Dolnicar, S., Yanamandram, V., & Cliff, K. (2012) The contribution of vacations to quality of life. *Annals of Tourism Research*, 39(1), 59-83.

ETC (2020) *Study on Generation Z Travellers*. European Travel Commission. https://etc-corporate.org/uploads/2020/07/2020_ETC-Study-Generation-Z-Travellers.pdf.

Eurostat (2010) *Tourism Statistics in the European Statistical System 2008 data*. https://ec.europa.eu/eurostat/documents/3888793/5848705/KS-RA-10-010-EN.PDF.

Eurostat (2017) *Tourism statistics - participation in tourism*. https://ec.europa.eu/eurostat/statistics-explained/index.php?title=File:Participation_in_tourism_of_EU_residents_(aged_15_and_over),_2016.png.

Eurostat (2023) *Tourism Statistics*. Eurostat. https://ec.europa.eu/eurostat/statistics-explained/index.php?title=Tourism_statistics.

Faranda, W. and Schmidt, S., (2000). Segmentation and the senior traveler: Implications for today's and tomorrow's aging consumer. *Journal of Travel & Tourism Marketing*, 8(2): 3-27.

Fitzenberger, B., Mena, G., Nimczik, J., & Sunde, U. (2022) Personality traits across the life cycle: disentangling age, period and cohort effects. *The Economic Journal*, 132(646): 2141–2172. https://doi.org/10.1093/ej/ueab093

Fleischer, A., Pizam, A. (2002) Tourism constraints among Israeli seniors. *Annals of Tourism Research* 29(1), 106-123.

Fosse, E. & Winship, C. (2019) Analyzing age-period-cohort data: a review and critique. *Annual Review of Sociology* 45(1)

Gardiner, S., King, C., & Grace, D. (2012) Travel decision making: an empirical examination of generational values, attitudes, and intentions. *Journal of Travel Research*. 52(3), 310-324

Glover, P., & Prideaux, B. (2008), Using population projections to identify aspects of future tourism demand, in Joseph S. Chen (ed.) *Advances in Hospitality and Leisure* Emerald Group Publishing, pp.185 – 209.

Godbey, G., Crawford, D., & Shen, X. (2010) Assessing hierarchical leisure constraints theory after two decades, *Journal of Leisure Research*, 42(1), 111-134, DOI: 10.1080/00222216.2010.11950197

Hardington-Gill, T. (2020) Backpacking is a rite of passage for many travelers. Covid could end it forever. CNN Travel. https://edition.cnn.com/travel/article/will-covid-end-backpacking-travel/index.html

Haukeland, J. V. (1990). Non-travelers: The flip side of motivation. *Annals of Tourism Research*, 17(2), 172-184.

Hong, G., Fan, J. Palmer, L. & Bhargava, V. (2005): Leisure travel expenditure patterns by family life cycle stages. *Journal of Travel & Tourism Marketing*, 18(2), 15-30.

Huang, Y. & Petrick, J. (2010) Generation Y's Travel Behaviours: A comparison with Baby Boomers and Generation X. In Benckendorff, P., Moscardo, G. and Pendergast D (Eds) *Tourism and Generation Y.* CABI pp 27 – 37

Hung, K., Gu, F. & Yim, C. (2007) A social institutional approach to identifying generation cohorts in China with a comparison with American consumers. *Journal of International Business Studies* 38, 836-852. https://doi.org/10.1057/palgrave.jibs.8400288

Inglehart, R. (2008) Changing values among western publics from 1970 to 2006. *West European Politics*, 31(1-2), 130-146, DOI: 10.1080/01402380701834747

Ishida, Y., Miyaki, M., Fujisawa, Y., & Iwasaki, K. (2017) How does tourism differ among generations? Tourists from the United States and their willingness to visit Japan, *International Journal of Tourism Sciences*, 17(1), 49-60, DOI: 10.1080/15980634.2016.1274526.

Jackson, E. L., Crawford, D. W., & Godbey, G. (1993). Negotiation of leisure constraints. *Leisure Sciences*, 15(1), 1-11.

Kattiyapornpong, U. & Miller, K. (2009), Socio-demographic constraints to travel behavior, *International Journal of Culture, Tourism and Hospitality Research*, 3(1), 81-94. DOI:10.1108/17506180910980546

Lee, S. & Tideswell, C. (2005), Understanding attitudes towards leisure travel and the constraints faced by senior Koreans, *Journal of Vacation Marketing*, 11(3): 249-263.

Li, X., Li, R., & Hudson, S (2013) The application of generational theory to tourism consumer behavior: An American perspective. *Tourism Management* 23, 147-164.

Li, H., Meng, F., & Zhang, Z. (2016) Non-participation of domestic tourism: analyzing the influence of discouraging factors. *International Journal of Tourism Research.* 18(6), 567-578.

Lin, V.S., Jiang, F., Li, G. & Qin, Y., (2023). Impacts of risk aversion on tourism consumption: A hierarchical age-period cohort analysis. *Annals of tourism research*, 101, 103607.

Litvin, S. W., Smith, W. W., & Pitts, R. E. (2012). Sedentary behavior of the non-travel segment: a research Note. *Journal of Travel Research,* 52(1), 131-136

Liu, X. & Wang, W., (2023). REVENGE travel: fact or myth?. *Current Issues in Tourism*,27(4), 517-523, DOI: 10.1080/13683500.2023.2220956

Losada, N., Alen, E., Dominguez, T & Nicalou, J (2016) Travel frequency of seniors tourists. *Tourism Management* 53, 88-95.

Lung, R. (2017) Travel wanderlust: Age is not a constraint, but income might be. *Visa.* https://usa.visa.com/partner-with-us/visa-consulting-analytics/senior-travel-wanderlust.html.

Mannheim, K. (1952) The problem of generations, in P. Kecskemeti (Ed.) *Essays on the Sociology of Knowledge: Collected Works*, Routledge, New York, pp. 276-322

March, R. (2000) The Japanese travel life cycle. *Journal of Travel & Tourism Marketing,* 9(1-2), 185-200.

McGuire, F. (1984) A factor analytic study of leisure constraints in advanced adulthood. *Leisure Sciences*, 6, 313- 336.

McGuire, F. A., Dottavio, D., & O'Leary, J. T. (1986). Constraints to participation in outdoor recreation across the life span: A nationwide study of limitors and prohibitors. *The Gerontologist*, 26(5), 538-544.

McKercher B. & Chan, A. (2005) How special is special interest tourism? *Journal of Travel Research* 44(1), 21-31.

McKercher, B. (2009). Non-travel by Hong Kong residents. *International Journal of Tourism Research*, 11(6), 507-519

McKercher, B. (2023) Age or generation? *Annals of Tourism Research.* 103, 103656. https://doi.org/10.1016/j.annals.2023.103656.

McKercher, B., Hughes, K. & Mkono, M., (2022). Unpacking the disinterested tourist. *Tourism Management Perspectives*, 44, 101021.

McNamara, K. E. & Prideaux, B. (2010). A typology of solo independent women travellers. *International Journal of Tourism Research,* 12(3), 253-264.

Mill, R. (2008) The Inter-Relationships between Leisure, Recreation, Tourism, and Hospitality in Brotherton, B. and Wood, R. (eds) *The SAGE Handbook of Hospitality Management* London: Sage.

Moss, S. (2016) Generational cohort theory. *SICO-Tests* https://www.sicotests.com/newpsyarticle/Generational-cohort-theory

Nickerson, N. (2000) Travel and recreation outlook 2000: Focusing on demographics. *Montana Business Quarterly* 38(1), 23-26.

Nyaupane, G. P., & Andereck, K. L. (2008). Understanding travel constraints: Application and extension of a leisure constraints model. *Journal of Travel Research*, 46(4), 433-439.

O'Hara-Deveraux, M. (2013) China's 5 generations: Diverging lifestyles. *Global Foresight.* http://www.globalforesight.dreamhosters.com/future-of-china-emerging-markets/chinas-5-generations-diverging-lifestyles.

Pendergast, D. (2009) Getting to know the Y generation. In Benkendorff, P., Moscardo, G. & Pendergast, D (Eds) *Tourism and Generation Y* CABI. pp. 1-15

Pennington-Gray, L., & Kerstetter, D. (2002). Testing a constraints model within the context of nature-based tourism. *Journal of Travel Research* 40(4), 416–423.

Plog, S. (1974) Why destination areas rise and fall in popularity *Cornell HRA Quarterly* 14(4): 55 – 58.

Popp, M., Schmude, J., Passauer, M., Karl, M., & Bauer, A. (2021). Why don't they travel? The role of constraints and motivation for non-participation in tourism. *Leisure Sciences,* 1-26.

Ryder, N. (1965) The cohort as a concept in the study of social change. *American Sociological Review,* 30(6): 843-861. http://www.jstor.org/stable/2090964

Schänzel, H.A. & Yeoman, I. (2015) Trends in family tourism. *Tourism Futures.* 1(2), 141-147.

Sessa, V., Kabacoff, R., Deal, J. & Brown, H. (2007). Generational differences in leader values and leadership behaviors. *The Psychologist-Manager Journal,* 10(1), 47–74

Smith, R. (1987). Leisure of disabled tourists: barriers to participation. *Annals of Tourism Research,* 14(3), 376-389.

Smith, W. W., & Carmichael, B. A. (2005). A geographical analysis of the rates of non-travel across the regions of Canada. *Tourism Geographies,* 7(3), 257-271.

Smith, W., Fralinger, E & Litvin, S. (2011) Segmenting the U.S.A. non-travel market. *Enlightening Tourism. A Pathmaking Journal,* 1(2), 137-151.

Song, H., Qiu, R.T. & Park, J. (2023). Progress in tourism demand research: Theory and empirics. *Tourism Management,* 94, 104655.

Staudinger, U., (2001) Theory of lifespan development. In N.J. Smelser & P.B. Baltes (Eds), *International Encyclopedia of the Social & Behavioral Sciences,:* Pergamon, pp. 8844-8848.

Tang, N., Wang, Y., & Zhang, K. (2017) Values of Chinese generation cohorts: Do they matter in the workplace? *Organizational Behavior and Human decision Processes.* 143, 8-22.

Tiwari, P., & Chowdhary, N. (2021). Has COVID-19 brought a temporary halt to overtourism? *Turyzm/Tourism,* 31(1), 89–93.

Tribe, J. (2006) The truth about tourism. *Annals of Tourism Research,* 33(2), 360-381

Wilson, E & Little, D. E. (2005) 'A 'relative escape'? The impact of constraints on women who travel solo', *Tourism Review International,* 9, 155-175.

Zhang, J., Wang, F., Shao, C. & Mi, X. (2020). Structural equation modelling of household long-distance flexible travel behavior. *International Journal of Sustainable Development and Planning,* 15(4), 461-467.

Zimmer, Z., Brayley, R. and Searle. M. (1996) Whether to go and where to go: identification of important influences on seniors' decision to travel. *Journal of Travel Research,* 33(3):3-10.

11 How Tourists Consume a Destination

By the end of this chapter, the reader will:

- Be able to define a destination and its morphology
- Analyse the impact of seasons on tourist behaviour
- Discuss and critique the different roles a destination plays and how that affects behaviour
- Understand how being a first time or repeat visitor influences consumption
- Explain how motive affects in-destination behaviour
- Discuss the impact of accommodation choice on consumption patterns.

Introduction

While tourism activity occurs in a local destination, relatively little research has been undertaken understanding the factors that influence how tourists consume destinations. This chapter examines that issue. It begins by defining a destination and its geomorphology. A range of issues explaining how and why tourists behave the way they do follows. This examination explores such issues as the impact of seasons on product and experience availability, the role of destinations as the main or secondary goal of a trip and how that affects behaviour, the impact of prior visitation history and some insights into the role of accommodation choice as a core factor that influences behaviour. The chapter concludes with a brief overview of how emergent technologies may fundamentally transform behaviour.

Defining a local destination

The UN World Tourism Organization identifies local destinations as the key to tourism, for they are the locus of where tourism activity occurs and where impacts are felt directly. A local destination is defined as:

> *"a physical space with or without administrative and/or analytical boundaries in which a visitor can spend an overnight. It is the cluster (co-location) of products and services, and of activities and experiences along the tourism value chain and a basic unit of analysis of tourism. A destination incorporates various stakeholders and can network to form larger destination areas. It is also intangible with its image and identity which may influence its market competitiveness"* (UNWTO n.d.).

This definition clarifies the scope of a destination by setting its upper and lower limits. A 'destination', like much in tourism, is a conflated term that is used by different stakeholders to suit their own needs. National or state tourism organisations refer to countries, states or regions as 'destinations', while individual attractions including airports also use the term. Such imprecision is understandable because the mandate of marketing bodies varies by the geographic scale. But it does little to help us understand how and where tourism is consumed.

Defining a destination as a local area with or without administrative boundaries clarifies this term and sets a framework for further analysis. In doing so, it excludes most individual attractions as destinations in their own right, and instead recognises them as critical components of a local destination. Likewise, states, provinces and countries are not discrete destinations, but consist of many individual destinations.

Two other key issues must be recognised. The first is that a destination's scope may extend beyond the administrative boundaries of a community and and potentially include multiple communities. This feature adopts a pragmatic approach to destinations by recognising tourist behaviour and/or tourism development are not necessarily constrained by artificial administrative boundaries. The second is the ability of local destinations to network with other local destinations to create larger destination regions, further recognising the need for collaboration to create appealing experiences for mobile visitors.

Geomorphology of destinations

Destinations are complex socio-geomorphic entities that consist of three types of place: tourism place, shared place and non-tourism place (McKercher et al., 2015). Their structure is shown in Figure 11.1.

Tourism place is designated for the primary use of tourists. It is something that is either constructed or signified for and by tourists as a medium through which the tourist experience is negotiated (Crouch et al., 2001). Wall (1996), suggests tourism place can exist as points, lines and areas. Points represents individual, isolated attractions, lines represent the linear routes that join sets of attractions, and areas represent concentrations of attractions that create nodes. Hotels, major attractions, tourist markets, shopping and dining districts catering primarily for tourists define tourism place.

Tourism place also involves more than simply built superstructure, however, for it is also defined by both mental and social place. Mental place is signified semiotically by a series of markers that create an appealing destination image (Timothy & Boyd, 2006), that says this place is worth visiting (MacCannell, 2001). Social place is signified by the behaviour of fellow tourists. We look for signs or signals that guide our behavior unconsciously (Donaire & Gali, 2008), to such an extent that Andreu, Kozak, Avci & Cifter (2005) call this pattern the 'herding effect'.

Shared place is hybrid place used by tourists and locals. Tourists are welcome as long as they behave as locals would. It is part of the everyday fabric of the destination, such as transport hubs, shopping and dining districts, where both tourists and locals strive to have the same goals met (Hayllar et al. 2008). Snepenger, Murphy, O'Connell and Gregg (2003), for example, have written about how tourists and locals use and share shopping space, even if they purchase different items. Much of the downtown core of urban destinations, as well as trendy shopping and dining districts reflect shared placea.

Non-tourism place represents areas of a destination that have no relevance to tourists and further they have no reason to visit. In many ways it is terra incognita, unknown, unexplored and for all intents and purposes unexplorable, for it has no intrinsic or extrinsic worth. Non-tourism place represents the largest spatial component of most destinations and includes residential, industrial and commercial areas along with their associated retail and recreational precincts. Tourists may may transit through such places on their way to tourism place or shared place, but otherwise feel unwelcome there, and therefore, have no reason to stop.

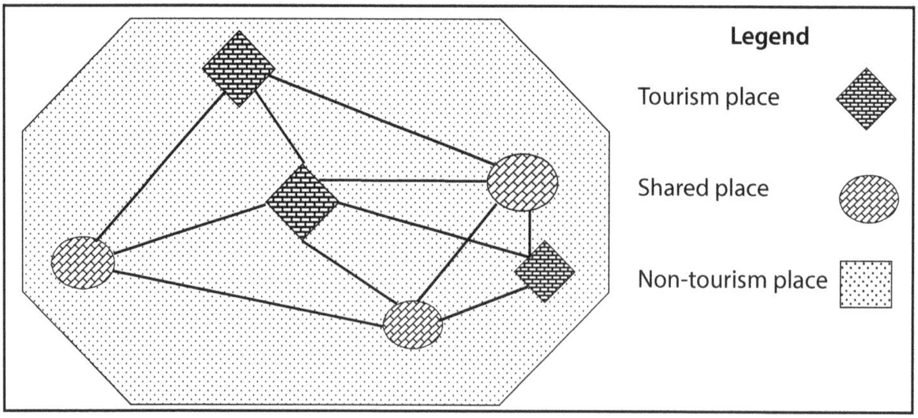

Figure 11.1: Conceptualization of the destination as consisting of three types of place

Destinations, strive to concentrate tourism activities in tourism or shared place or to create new tourism place through the development of major attractions, zoning areas for hotel and other commercial development, building convention centres and creating markers to designate certain shopping/dining nodes as being tourist friendly. Lower order attractions and most tourism retail outlets tend to concentrate in these areas. Conversely, attractions located in non-tourism place will struggle to find an audience. Unfortunately, this is the challenge faced by many heritage sites located outside of recognized tourism nodes.

Pearce's (1998, 1999) study of Paris's tourism districts highlights the structure of tourism place. For example, Ile de la Cite in the historic core of Paris has a long-standing architectural and historical legacy which makes it a leading tourism node. It is anchored by Notre Dame Cathedral which has also induced the establishment of arts, markets, restaurants and the like. He contrasts it with Montmartre, whose boundaries are less well defined. None the less it is characterized by a distinctive and colourful history, a place of religious significance, and an artists' colony. In both cases, the precinct is anchored by a major attraction with a high degree of synergy among other activities, experiences and businesses.

Local destinations are comprised of three types of place. Tourism place is signed and signaled for tourist use. It is often anchored by a major attraction, as shown in the first image of the Sovereign Hill historic theme park in regional Victoria, Australia. This attraction is the catalyst for the development of complementary accommodation, dining and retail outlets. Tourist place can be anchored by a single attraction, by multiple, complementary attractions such as theatre districts, or by nodal attractions including hotel, shopping and dining precincts.

Shared place is used by both local residents and tourists, as the name implies. Here tourists are welcome as long as they behave like locals, as is the case of this pre-Christmas open-air food court in Adelaide, Australia.

Non-tourism place is signaled and signed as no-go places where tourists are not welcome, or that are irrelevant to them. Non-tourism place is the most common feature of destinations, for it includes residential, commercial and industrial areas for the exclusive use of residents. Attractions located in non-tourism place will struggle to find a market, as is the case of this historic Chinese temple in Bendigo, Australia. Tourists are not aware of attractions and activities in non-tourism place, and if aware, feel the cost of visiting outweighs the benefit, especially if other more appealing activities can be accessed easily.

Most of the adverse impacts of tourism are caused when place changes. Non-tourism place can be invaded by tourists and transformed into shared or exclusive tourism place, to the detriment of the livelihood of local residents. Shared place can be so overwhelmed by tourists that it becomes tourist place. Most of the time, these three types of place are in equilibrium, especially when tourism becomes an accepted part of the social fabric of the community. This issue is discussed in detail in Chapter 13 that examines social impacts of tourism.

Tourists function within their own unique system

Each of us functions within our own unique tourism system that overlaps with other people's tourism systems (Leiper, 1990). We tend to talk about a tourism system as a collective whole. But, since the system begins and ends with the tourist, and each tourist behaves in a unique manner, we cannot talk about one tourism system. Instead, there are somewhere in the neighbourhood of 15 billion separate tourism systems when all domestic and international travel is considered. Thinking of tourists in the aggregate oversimplifies and under-reports the diverse nature of tourism consumption. As McKercher et al. (2021) note each tourist displays idiosyncratic behavioural patterns. It may be possible to map individual behaviours, but collectively behaviour of all tourists appears to be stochastic and unpredictable. Yet, there is a level of underlying predictability, as tourists tend to follow similar paths, visit similar attractions and participate in similar activities, although not in the same order or to the same intensity.

Each tourist operates within his or her own discrete tourism system, although it may not seem so when crowding is an issue as shown here in the Temple of Horus in Edfu, Egypt. In reality, each independent tourist, or small group of independent tourists behaves uniquely. They follow their own tourism system either at a macro level (from their home community the use of travel intermediaries, transport providers and in destination accommodation, services and facilities), or at a micro level as noted in their movements through a destination. Leiper (1989) reminds us that each system overlaps at multiple points, creating an impression of a single tourism system.

The extent of individual tourism systems was demonstrated in a study of tourist behaviour in Bali, Indonesia (McKercher et al., 2021) as shown in Table 11.1. For example, of the 563 people who participated in beach activities, a total of 553 discrete combinations of participation in other activities along with going to the beach was noted. In other words, 98.2% of people who went to the beach operated in their own tourism system. Indeed, only three people shared the same set of activities. The same pattern was noted in virtually all other activity sets undertaken.

Table 11.1: Multiple tourism systems

Activity / Attraction	n	Number of discrete combinations	% of discrete combinations	Largest number of people displaying the same activity set
Beach activities	563	553	98.2	3
Eating different types of food	505	495	98.0	3
General sightseeing	471	464	98.5	3
Visiting temples	446	445	99.8	2
Shopping	418	410	98.1	3
Visit rice terraces	387	383	99.0	3
Spa or massage	379	375	98.9	3
Visit waterfalls	332	329	99.1	3
Visit Sacred Monkey reserve	325	321	98.8	3
Visit Ubud	279	277	99.3	2
Bars and nightclubs	277	271	97.8	3
Visit palaces	268	267	99.6	2
Scuba or snorkeling	237	235	99.2	3
Attend dance performance	212	212	100.0	0
Bali coffee plantation	203	202	99.5	2

Leiper's ideas, supported by empirical research highlight the risk associated with ascribing discrete labels to tourists, unless the reasons for visiting a destination are analysed. Just because a person participates in a nature based tourism activity does not make that person a nature based tourist. People participate in a wide array of activities, some of which may relate to their reasons to travel in whole or in part, some of which may be convenience based and others which may be pursued for no other reason than to kill time. It is, therefore, vital to distinguish the reason to visit the destination from the activities undertaken, as shown in Table 11.2 which documents the gap between the participation rate in activities in Hong Kong with the motivation rate for these activities. Participants in this study were asked to identify what they did and then asked to identify which of these activities played a role in whole or in part in their visit decision. The gap between motivation rate and participation rate varied tremendously, with no mean or average inflation between the two. While the relationship between participation and motivation for business was close to 1:1, it blew out to almost 20:1 among the group of people who participated in ecotourism activities.

Table 11.2: Gap between participation rate and motivation rate

Activity	Attending business meetings / exhibitions / conferences	Sightseeing	Shopping	Visiting friends	Eating different types food than I normally eat at home	Visiting theme parks and other tourist attractions	Visiting museums and/or art galleries	Visiting festivals or events, including traditional opera	Visiting natural areas and country parks	Going to beaches	Riding public transport	Participating in ecotourism
Participation rate	44.6	65.7	84.2	33.6	70.8	29.0	18.6	10.0	20.5	16.7	70.6	2.8
Motivation rate	40.6	45.3	55.8	20.8	18.6	6.1	3.6	1.2	2.5	1.2	4.4	0.2
Ratio of participation to motivation	1.1	1.5	1.5	1.6	3.8	4.7	5.2	8.1	8.3	14.5	15.9	18.0

Source: McKercher & Chan (2005)

Seasons

The suite of products, activities and experiences available in a destination is seasonally dependent, which in turn influences who is likely to come and what they are likely to do. A process model to illustrate how seasons affect all aspects of tourism is shown in Figure 11.2. The model illustrates how different tourist seasons determine the activity sets available, which in turn, define destination appeal or the pull features of a destination. Destination appeal then affects who is likely to visit and the push factors that drive visitation, while the combination of what is on offer and who comes determines in-destination behaviour.

The model is predicated on the realization that a tourist season dictates the set of actual or perceived products, activities and experiences available. A 'tourist season' is a temporal period dictated by natural, institutional and destination management factors or a combination of some or all elements. They are destination and context specific. Seasons may be defined in one of three ways. The first is by natural factors associated with seasonal weather changes (Baum & Lundtorp, 2001). The second is institutional or socio-cultural factors such as school holidays or religious, cultural and social events (Almeida & Kastenholz, 2019; Chung, 2009; Croce & Woeber, 2010). A third way is for destinations to create artificial tourist seasons, to encourage visitation during otherwise quiet times of the year. Festivals and events are the most common tactics developed.

Seasons, in turn, dictate both the actual activity set available as well as the perceived activity set. Actual and perceived activity sets are quite different. Clearly, some activities are available only at certain times of the year, such as the winter ski season, golf season, the autumn colour season, the summer beach season and even consumptive activities like fishing and hunting seasons (Spencer & Holecek, 2007; Tkaczynski et al., 2015). Other activities, though, are available all year round, but enjoy seasonal peaks in popularity (Mitchell & Hall, 2003). Grey nomads are drawn to sunny and warm climates to escape winter (Hillman, 2013; Vaillon, 2012). Yet, these destinations hold limited appeal at other times of the year, even though the climate is still warm and pleasant.

Combined, the actual and perceived activity sets dictate the pull features of a destination. They, in turn, influence who is likely to be pushed to travel (Dann, 1977). Who is likely to visit and what drives them to visit are interrelated, for one influences the other. Who is likely to travel to a specific destination is a function of why they will travel or what needs they want satisfied, while needs will play a key role in influence who goes, for it will limit the destination choice set.

It is well-recognised that seasonal differences exist in markets, in part due to institutional and demographics factors. School holidays provide one of the few opportunities for families to spend quality time together, while retirees and empty nesters can travel all year round (Almeida & Kastenholz, 2019; Coshall et al., 2015). Other research suggests that the motives for visits vary by season, even among the same demographic or psychographic group. For example, Hodeck and Hovemann (2016) and Giachino et al. (2020) found summer tourists to alpine areas in Germany were motivated by the chance to relax and escape, while winter tourists were looking for more excitement and adventure.

The combination of real and perceived activity sets and the who and why people visit, ultimately determines in-destination behaviour. In some cases, the consumption decision is self-evident as in the case of skiing, with its associated après ski activities, or fall colour viewing with associated autumn harvest visits to farms. In other cases, though, the relationship is less well defined, especially where the perceived popularity of activities and experiences available all year round varies by season. Here a high degree of substitutability is observed. A study of seasonal variations in tourist consumption patterns in the tropical city of Cairns, Australia, for example, revealed that winter tourists were far more likely to engage in land based activities, including visiting the wet tropics rainforest, than going to the Great Barrier Reef, even though the climate was temperate. By contrast, summer visitors travelled to the the Reef but avoided the hot, humid and rainy wet tropic areas. The Autumn visitor spent time with family and friends and engaged in child friendly activities, while the spring visitor preferred to go hiking, view the scenery, go birdwatching and go swimming (McKercher et al., 2024).

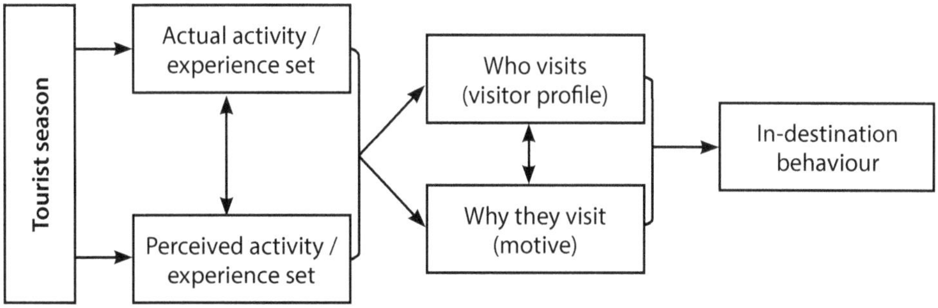

Figure 11.2: A process model of the impact of seasons on in-destination behaviour. Source: McKercher et al. (2024)

Role of destination and visitation history

How tourists consume a destination depends on both the role the destination plays in their itinerary and their past visitation history. A place may function as either the main destination or as a necessary or appealing stopover destination. The tourist may be a first time visitor or a repeat visitor. Combining the two dimensions creates a 2 x 2 matrix showing the relationship between destination role and visitation history, as shown in Figure 11.3. Each dimension will be discussed separately before the combined effect of destination role and visitation history are shown.

Visitation history Role of destination	First time visitor (F)	Repeat visitor (R)
Main destination (M)	FM	RM
Stopover destination (S)	FS	RS

Figure 11.3: Relationship between destination role and visitation history. *Source*: McKercher & Wong (2004)

■ Main vs stopover destinations

It should be axiomatic that destinations can attract two types of visitors: those choosing it as the main or only destination; and those choosing it as a secondary or through destination (Leiper, 1989). The main destination traveller is the focus of most destination management organisation marketing activities, while the stopover visitor is largely ignored. Yet, the stopover market may actually provide a greater share of total visitors at either domestic or international destinations and generate a substantial amount of revenue (Kotsi et al., 2023; Maseiro et al., 2020).

These two groups are so fundamentally different that they can be considered as discrete market segments (Kotsi et al, 2023; McKercher, 2001). To begin, main destination visitors stay longer than stopover visitors, and as such, have more time to explore the destination in depth. Serendipity also plays a key role, as visitors who explore destinations are open to new experiences (Huang et al,. 2014).

Substantial differences are also noted in the degree of psychological investment, for main destination visitors have invested much more time, effort and energy into planning the visit and trying to ensure a satisfactory experience than stopover visitors (McKercher et al., 2006). The level of investment may extend to creating a detailed itinerary outlining what they are going to do and when (Shoval & Raveh, 2004), conducting much more pre-departure research (Litvin & Mouri, 2009) and engaging in various forms of risk reduction (Fuchs & Reichel, 2011).

Stopover tourists, on the other hand are much less invested in the destination, for it will play a minor role in the overall trip satisfaction. As a result, their involvement will be much more superficial. Indeed, the activities pursued will depend on whether the stopover occurs at the beginning or end of the trip. Tourists on the outbound leg of the journey are excited to explore different places on their way to their main destination, but on the inbound leg, they may be emotionally tired and just want to get home (McKercher et al., 2006). In such instances, their activities will be curtailed, and if the weather is inclement, may do little or nothing.

Importantly, as well, tourists have a great deal of discretion as to where they make their stopovers. Drive tourists in particular, have a largely unlimited choice of where they can stop, depending only on how far they wish to drive each day. Inclement weather may convince some people to shorten their stay at stopover destinations, move to the next destination earlier than expected (Becken & Wilson, 2013) or simply keep driving in the hopes of driving through the weather. Long haul international tourists have many options, as well, given the necessity of transiting through aviation hubs on their way to or from their their ultimate destination (Kotsi et al., 2023; Pike & Kotsi, 2020; Poon & Ho, 2021). Here their choices are influence by visa and entry requirements, destination perceptions, the opportunity for unique experiences, cost considerations, and of course the choice or airline.

■ First vs repeat visitation

Differences between first time and repeat visitors have been examined for at least 40 years since Gitelson and Crompton (1984) first began to explore this issue. It continues to be a popular research topic to the present (Hardy, 2020; Rather, 2020). First time visitors, being destination naive tend to explore destinations widely (Sariego-Lopes & Moreno-Melgarejo, 2020) and show preferences for visiting icon attractions (Anwar & Sohail, 2004; Paulino et al., 2022). Many will buy day tours, while others who travel independently will follow a similar itinerary pattern as found in day tours (Bierenboim et al., 2021; Leung, 2016). It is typical for many first time visitors to wander around a destination as they explore its features (Toger et al., 2023).

Destinations must strive for a balance of first time and repeat visitors, for their consumption patterns are markedly different. First time visitors are more likely to visit icon attractions, such as scenic lookouts as shown here in Hong Kong. They are also more likely to explore a destination widely. Repeat visitors, on the other hand, tend to do less, and engage in more hedonic activities, such a shopping and dining at places like this popular seafood restaurant district in the city.

Repeat visitors, on the other hand, are destination aware and have explored it on previous visits. They tend to do less and instead, engage in more hedonic activities, such as shopping and dining (Hardy et al., 2020; Lau & McKercher, 2004), although Yoo and Katsumata (2023) found repeat visitors to Japan seek out moderately unpopular sightseeing spots. They also tend to spend less per visit than first timers (Taveres et al, 2019). Repeaters are less in awe of a destination and, therefore, tend to have a more realistic understanding of its appeal as well as its shortcomings (Su et al., 2021). Nonetheless, both first timers and repeaters are equally satisfied with the destination and and display similar intentions to return (Čaušević et al., 2020).

It is essential that destinations maintain a balance between first time and repeat visitors for certain attractions and activities are more reliant on on one group than the other. For example, first timers want to visit icon attractions, including scenic lookouts and may explore a destination's nooks and crannies if their stay is sufficiently long. Souvenirs are bought disproportionately by new visitors. These places hold little appeal to repeat visitors who have already visited them on earlier visits. As such, attractions and activities that rely on first timers will be at risk if the balance becomes skewed towards repeat visitors.

■ Combined impact of destination role and prior visitation history

Most research has examined the impact of either destination role or past visitation behaviour, but has not combined the two to determine which dimension plays the most vital role in influencing tourists' consumption patterns. To date, though, to the best of our knowledge only one study has examined which dimension explains differences in attitudes and behaviour better. This study focussed on Hong Kong (McKercher & Wong, 2004). The first versus repeat dimension better explains propensity to buy a group tour, motives to discover a place, expenditure per night, movement patterns, dislikes about a destination, and not surprisingly satisfaction. By contrast, the main versus stopover destination dimension played a much more significant role in reflecting differences in tourists' demographic profile, length of stay, motives to relax and recreate, participation in cultural tourism, perceptions of what they liked about the place and perceptions of service quality.

Visitor motive

It should be axiomatic that a relationship exists between motive and in-destination behaviour, even though this subject has received remarkably little attention in our literature. Instead, most of the research on motives focusses on destination choice and not in-destination behaviour. Instead, attempts to link in-destination behaviour to motives have been typified by reverse engineering observed behaviour to a single motive (McKercher & Chan, 2005). Analysing what people do in a destination, without considering why they came is fraught with danger as this type of reverse engineering works only if behaviour can be attributed to a single motive, when in reality it may be attributed to different motives (Ryan, 2003). Kantanen & Tikkanen (2006) add that tourists may visit places for a range of reasons varying from reflecting a core reason to visit to simply passing time between visits to other places.

Various studies have demonstrated a clear link between motive and different behavioural characteristics. Some tourists engage in the same types of activities when they travel that they do at home, just in a different location (Smith et al., 2012). In support of this concept, Lee et al. (2015) found that slow food tourists engaged on all slow tourism activities extending well beyond dining preferences, while Chen et al. (2017) found similarities between the behaviour of family-oriented and the stylish tourist at home and on the road. Ritchie et al. (2010) focusing on cycling tourists found significant differences between motives and cycle touring behaviours, while discrete market segments could be defined among surf tourists depending on their motives (Dolnicar & Fluker, 2003). It is for this reason that Zoltan and Maseiro (2012) conclude that tourists will gain more satisfactory experiences at the destination, if they can find activities according to their desires and that motivation has a significant impact on activeness in a destination.

As an example, people who went to Bali looking for novelty and escape travelled widely through the destination, but tended to avoid temples and other religious sites. Not unexpectedly, people who were travelling for more of a cultural experience were extremely active, choosing to visit places that reflected Bali's diverse tangible and intangible cultural heritage, including its natural heritage, while showing a disdain for built cultural attractions. People whose main motive was to spend time with family tended to participate in resort-based leisure activities at a much higher level than others (McKercher et al., 2021).

Accommodation type and location

An interesting relationship between consumption patterns and accommodation choice has also be discovered. Different types of accommodation attract a different type of clientele (Croes, 2008), who have different motives for visiting (Radojevic et al., 2018; Yavas & Babakus, 2005) and display different behaviour patterns (Nicolau et al., 2018). Two factors come into play. First, a strong inverse relationship has been noted between trip duration and accommodation choice (Losada et al., 2017; Tussyadiah & Pesonen, 2016), with Chen et al. (2017) noting lower quality hotels are chosen by longer stay visitors. Second, the nature of the trip plays a role in accommodation choice (Radojevic et al., 2018; Yavas et al., 2005), with people looking to spend time with family or to make a short, escapist trip being more likely to choose mainstream accommodation (Albaladejo & Diaz-Delfa, 2021), while those who want a more in-depth community experience may select alternative types of accommodation. While the relationship exists, it is important to appreciate that choosing a certain style of accommodation does not 'cause' one to engage in different types of behaviour (McKercher et al., 2023). Instead, the two may be attributable to other factors. None the less, knowing that such a relationship does exist creates innovative marketing opportunities for attractions, activities and services.

Hotel location also has an influence on the sequence in which activities are pursued as well as the likelihood of exploring deeply into a destination (Shoval et al., 2011). Tourists spend a disproportionate amount of time within 500m of their accommodation house, especially on the arrival and departure days, severely limiting how deeply they can engage in a destination, especially if their length of stay is short. Thus, there may only be an effective one-day touring window in a typical three day, two night stay where tourists may be amenable to exploring a destination widely. This effective touring window can be doubled if the length of stay can be increased by one night (four days, three nights). They also tend to prioritise visits to proximate attractions earlier in the day and then visit more distant ones later in the day. More importantly, the geographic layout of the destination and the location of the hotel may create artificial barriers that effectively preclude visiting certain places in inconvenient locales.

Emerging issues – smart phones

We are currently witnessing a seismic, technological shift that is transforming how tourists consume a destination, for they can now gather real time information about activities available and make spontaneous decisions. Smartphones are reshaping the physical and social contexts in which consumption decisions are being made. Liu et al (2022) suggest they have transformed destinations from the purely physical to the 'phygital-social' that supports the development of new travel plans and the re-evaluation of pre-existing plans on the fly. Smartphone interaction may completely transform how we conceive of tourist destination consumption by providing a far greater opportunity for spontaneity and serendipitous experiences (Huang et al., 2014) and also for destination marketers to send unique messages to tourists when they are visiting different parts of a destination.

Conclusion

Tourism activity occurs in a local destination. Yet, surprisingly little research has focused explicitly on understanding how tourists consume destinations. This chapter identified a range of factors that influence consumption patterns, beginning with the impact different seasons on product and experience availability and recognizing the different role destinations can play as the main or secondary goal of a visit. Behaviour is influenced by a wide range of other factors, including prior visitation history, the motive driving the visit and even where they stay.

■ Discussion points / Exercises

1. Examine your local destination and identify tourism, non-tourism and shared place and the reasons why each exists.
2. Analyse your destination to identify different activity sets available by season as well as perceptions of different appeal of non-season dependent activity sets.
3. Discuss how smartphones will transform tourist consumption behaviour.

Points to consider for revision

- Destinations are comprised of three types of place, tourism, shared and non-tourism
- Each tourist operates in a unique tourist system that may overlap with others at various points
- Seasons affect all aspects of the tourism experience by determining the actual and perceived activity sets available, which influence who comes, why they come and ultimately in-destination behaviour
- A causal relationship exists between motives to visit a destination and in-destination behaviour
- While accommodation choice does not determine behaviour, a strong relationship exists between the two

Recommended additional readings

Liu, X., Wang, D. & Gretzel, U., (2022). On-site decision-making in smartphone-mediated contexts. *Tourism Management*, 88, 104424.

McKercher, B., Thompson, M. & Prideaux, B. (2024) The impact of seasons on tourist behaviour. *Tourism Review*

McKercher, B., Prideaux, B. & Thompson, M. (2023). The relationship between accommodation type and tourists' in-destination behaviour. *Tourism Recreation Research*, DOI: 10.1080/02508281.2023.2221070

References

Albaladejo I. & Díaz-Delfa M. (2021). The effects of motivations to go to the country on rural accommodation choice: A hybrid discrete choice model. *Tourism Economics, 27*(7), 1484-1507.

Almeida, A. & Kastenholz, E. (2019) Towards a theoretical model of seasonal tourist consumption behaviour, *Tourism Planning & Development*, 16(5), 533-555, DOI: 10.1080/21568316.2018.1528564.

Andreu, L., Kozak, M., Avci, N., & Cifter, N. (2005). Market segmentation by motivations to travel. *Journal of Travel and Tourism Marketing*, 19(1), 1-14.

Anwar, S. A. & Sohail, M. S. (2004). Festival tourism in the United Arab Emirates: First-time versus repeat visitor perceptions. *Journal of Vacation Marketing*, 10(2), 161–170.

Baum, T., & Lundtorp, S. (Eds) (2001) *Seasonality in Tourism.* Oxford: Pergamon.

Becken, S. & Wilson, J. (2013). The impacts of weather on tourist travel. *Tourism Geographies*, 1-20.

Birenboim, A., Anton-Clavé, S., Russo, A.P. & Shoval, N., (2021). Structure versus agency: which best explains tourist activity in a destination? In Ferante, M., Fritz, O. & Oner, O. (Eds), *Regional Science Perspectives on Tourism and Hospitality*, Cham: Springer, pp.355-373.

Čaušević, A., Mirić, R., Drešković, N. & Hrelja, E., (2020). First-time and repeat visitors to Sarajevo. *European Journal of Tourism, Hospitality and Recreation*, 10(1), 14-27.

Chen, C. Tsai, Y. & Chiu, H. (2017). The decision-making process of and the decisive factors in accommodation choice. *Current Issues in Tourism, 20*(2), 111-119. DOI: 10.1080/13683500.2015.1087476

Chung, J. (2009) Seasons in tourism: A review. *e-Review of Tourism Research* (eRTR), 7(5) http://list.rpts.tamu.edu/ertr/

Coshall, J., Charlesworth, R. & Page, S. (2015) Seasons of overseas tourism demand in Scotland: A regional analysis. *Regional Studies*, 49(10), 1603-1620.

Croce V., & Wöeber K. (2010) Seasons in city tourism: Concepts and measurements. In: Mazanec J., Wöeber K. (eds) *Analysing International City Tourism*. Springer, Vienna.

Croes, H. (2008). Accommodation portfolio and market differentiation: The case of Aruba. *Tourizam*, 56(2), 185-197.

Crouch, D., Aronsson, L., & Wahlstrom, L. (2001). Tourist encounters. *Tourist Studies*, 1(3), 253-270.

Dann, G. (1977) Anomie, ego enhancement and tourism. *Annals of Tourism Research*, 4(4), 184-194.

Dolnicar, S. & Fluker, M. (2003) Behavioural market segments among surf tourists: investigating past destination choice, *Journal of Sport Tourism*, 8(3), 186-196.

Donaire, J. & Gali, N. 2008). Modeling tourist itineraries in heritage cities. Routes around the Old District of Girona. *PASOS*, 6(3), 435-449.

Fuchs, G. & Reichel, A., (2011). An exploratory inquiry into destination risk perceptions and risk reduction strategies of first time vs. repeat visitors to a highly volatile destination. *Tourism Management*, 32(2), 266-276.

Giachino, C. Truant, E. & Bonadonna, A. (2020). Mountain tourism and motivation: millennial students' seasonal preferences. *Current Issues in Tourism*, 23(19), 2461-2475.

Gitelson R. G. & Crompton, J. L. (1984). Insights into the repeat vacation phenomenon. *Annals of Tourism Research*, 11(2), 199-217.

Hayllar, B., Griffin, T., & Edwards, D. (2008). Urban tourism precincts: Engaging with the field. In B. Hayllar, T. Griffin, & D. Edwards (Eds.), *City Spaces – Tourist Places: Urban tourism precincts* (pp. 1-19). Sydney: Routledge.

Hardy, A. (2020). *Tracking Tourists: Movement and mobility.* Oxford: Goodfellows.

Hardy, A., Birenboim, A. & Wells, M., (2020). Using geoinformatics to assess tourist dispersal at the state level. *Annals of Tourism Research*, 82, 102903.

Hillman, W. (2013). Grey nomads travelling in Queensland, Australia: Social and health needs. *Ageing and Society*, 33(4), 579-597. doi:10.1017/S0144686X12000116.

Hodeck, A. & Hovemann, G. (2016) Motivation of active sport tourists in a German highland destination – a cross-seasonal comparison, *Journal of Sport & Tourism*, 20(3-4), 335-348, DOI: 10.1080/14775085.2016.1235988.

Huang, W.J., Norman, W.C., Hallo, J.C., Mcgehee, N.G., Mcgee, J. & Goetcheus, C.L., (2014). Serendipity and independent travel. *Tourism recreation research, 39*(2), 169-183.

Kantanen, T., & Tikkanen, I. (2006). Advertising in low and high involvement cultural tourism attractions: Four cases. *Tourism and Hospitality Research*, 6(2), 99-110.

Kotsi, F., Johnston, K. & Pike, S., (2023). Stopovers as valued-based experience: A conceptual model. *Annals of Tourism Research Empirical Insights*, 4(1), 100084.

Lau L. S. & McKercher, B. (2004) Exploration versus Acquisition: A comparison of first-time and repeat visitors *Journal of Travel Research* 42(3), 279-285.

Lee, K., Scott, N., & Packer, J. (2015) Habitus and food lifestyle: In-destination activity participation of Slow Food members. *Annals of Tourism Research* 48, 207-220.

Leiper, N. (1989) *Tourism Systems; An Interdisciplinary perspective.* Department of Management Systems. Palmerston North: Massey University

Leung, X.Y., Wu, B., Xie, F., Xie, Z. & Bai, B., (2016). Overseas tourist movement patterns in Beijing: the impact of the Olympic games. *Travel and Tourism Research Association: Advancing Tourism Research Globally*. 1. https://scholar-works.umass.edu/ttra/2010/Oral/1.

Litvin, S., & Mouri, N. (2009). A comparative study of the use of 'iconic' versus 'generic' advertising images for destination marketing. *Journal of Travel Research*, 48(2), 152-161.

Liu, X., Wang, D. & Gretzel, U., (2022). On-site decision-making in smartphone-mediated contexts. *Tourism Management*, 88, 104424.

Losada, N., Alén, E., Nicolau, J. & Domínguez, T. (2017). Senior tourists' accommodation choices. *International Journal of Hospitality Management*, 66, 24-34.

MacCannell, D. (2001). Tourist agency. *Tourist Studies*, 1(1), 22-37.

Masiero, L., Qiu, R.T. & Zoltan, J., (2020). Long-haul tourist preferences for stopover destination visits. *Journal of Travel Research*, 59(5), 811-827.

McKercher, B. (2001) A comparison of Main-destination visitors and through travelers at a dual-purpose destination. *Journal of Travel Research* 39(4): 433-441

McKercher B. & Chan, A. (2005) How special is special interest tourism? *Journal of Travel Research* 44(1), 21-31.

McKercher, B. & Wong, D.Y., (2004). Understanding tourism behavior: Examining the combined effects of prior visitation history and destination status. *Journal of Travel Research*, 43(2), 171-179.

McKercher B., Wan, S. M. & Tse, T. (2006). Are short duration cultural festivals tourist attractions? *Journal of Sustainable Tourism*, 14(1), 55-66.

McKercher, B. Tolkach, D., Mahadewi E., & Byomatara, D. (2021) The relationship between motive and in-destination behaviour. *Journal of Hospitality and Tourism Management*. 46, 432-439.

McKercher, B., Prideaux, B. & Thompson, M., (2023). The relationship between accommodation type and tourists' in-destination behaviour. *Tourism Recreation Research*, DOI: 10.1080/02508281.2023.2221070

McKercher, B., Thompson, M. & Prideaux, B. (2024) The impact of seasons on tourist behaviour. *Tourism Review*. https://doi.org/10.1108/TR-06-2023-0420

McKercher, B., Wang, D., & Park, E. (2015) Social impacts as a function of place change. *Annals of Tourism Research*. 50, 52-56.

McKercher, B., Wong, C. & Lau, G., (2006). How tourists consume a destination. *Journal of Business Research*, *59*(5), 647-652.

Mitchell, R. & Hall, C. M. (2003) Seasons in New Zealand winery visitation. *Journal of Travel & Tourism Marketing*, 14(3-4), 155-173, DOI: 10.1300/J073v14n03_09

Nicolau, J., Zach, F. & Tussyadiah, I. (2018). Effects of distance and first-time visitation on tourists' length of stay. *Journal of Hospitality & Tourism Research*, 42(7), 1023-1038.

Paulino, I., Domènech, A. & Bassols, N., (2022). Do visitors visit what the tours industry supplies? Sightseeing tours versus first-timers' photos. *Tourism Planning & Development*, DOI: 10.1080/21568316.2021.2023206

Pearce, D. G. (1998). Tourist districts in Paris: structure and functions. *Tourism Management*, 19(1), 49-65.

Pearce, D. G. (1999). Tourism in Paris: Studies at the microscale. *Annals of Tourism Research*, 26(1), 77-97.

Pike, S. & Kotsi, F., (2016). Stopover destination image—Using the Repertory Test to identify salient attributes. *Tourism Management Perspectives*, 18, 68-73.

Poon, P.C. & Ho, G.K., (2021). Opportunities in transit tourism: A case study of Hong Kong as a transit destination. *Journal of Travel & Tourism Marketing*, 38(1), 31-43.

Radojevic, T., Stanisic, N., Stanic, N., & Davidson, R. (2018). The effects of travelling for business on customer satisfaction with hotel services. *Tourism Management*, 67, 326–341.

Rather, R.A., (2020). Customer experience and engagement in tourism destinations: The experiential marketing perspective. *Journal of Travel & Tourism Marketing*, 37(1), 15-32.

Ritchie, B., Tkaczynski, A. & Faulks, P. (2010) Understanding the motivation and travel behavior of cycle tourists using involvement profiles. *Journal of Travel & Tourism Marketing*, 27(4), 409-425, DOI: 10.1080/10548408.2010.481582

Ryan, C. (2003) Behaviour. In Jenkins, J. & Pigram, J (Eds) *Encyclopedia of Leisure and Outdoor Recreation*. New York: Rotuledge pp. 29-30.

Sariego-Lopes, I. & Moreno-Melgarejo. M.(2020). Repeat visitors in Cantabria (Spain): Exploring tourist behavior. *Revista Espacios*, 41(17).

Shoval, N., & Raveh, A. (2004). Categorization of tourist attractions and the modeling of tourist cities: based on the co-plot method of multivariate analysis. *Tourism Management*, 25(6), 741–750.

Shoval, N., McKercher, B., Ng, E. & A. Birenboim (2011) Hotel location and tourist spatial activity in cities. *Annals of Tourism Research*. 39(4), 1594-1612.

Smith, W., Pitts, S & Litvin, S. (2012) Travel and leisure activity participation. *Annals of Tourism Research*. 39(4), 2207-2210.

Snepenger, D., Murphy, L., O'Connell, R & E. Gregg (2003) Tourists' and residents' use of a shopping space. *Annals of Tourism Research* 30(3): 567-580.

Spencer, D., & Holecek, D. (2007) Basic characteristics of the fall tourism market. *Tourism Management* 28(2), 491-504

Su, L., Hsu, M.K. & Huels, B., (2021). First-time versus repeat tourists: resistance to negative information. *Journal of Service Theory and Practice, 32*(2), 258-282.

Tavares, J., Ivanov, S.H. & Cobanoglu, C., (2019). Does the repeat tourist spend more than the first time? A contribution to the literature. *Revista Hospitalidade*, 16(3), 118-133.

Timothy, D., & Boyd, S. (2006). Heritage tourism in the 21st Century: Valued traditions and new perspectives. *Journal of Heritage Tourism*. 1(1), 1-16.

Tkaczynski A, Rundle-Thiele S., & Prebensen N. (2015) Segmenting potential nature-based tourists based on temporal factors: The case of Norway. *Journal of Travel Research*. 54(2), 251-265. doi:10.1177/0047287513514296

Toger, M., Östh, J. & Persson, S.G., (2023). What you see is where you go: Cruise tourists' spatial consumption of destination amenities. *Economic Themes*, 61(1), 63-84.

Tussyadiah, I. & Pesonen, J. (2016) Impacts of peer-to-peer accommodation use on travel patterns. *Journal of Travel Research*, 55(8), 1022-1040.

UNWTO (n.d.) *Definitions Committee on Tourism and Competitiveness (CTC)*. http://cf.cdn.unwto.org/sites/all/files/pdf/unwto_tourism_definitions.pdf .

Vaillon, P. (2012) Retired snowbirds. *Annals of Tourism Research*. 39(4), 2073-2091.

Wall, G. (1996). Perspectives on tourism in selected Balinese villages. *Annals of Tourism Research*, 23(1), 123-137.

Yavas, U., & Babakus, E. (2005). Dimensions of hotel choice criteria: Congruence between business and leisure travellers. *International Journal of Hospitality Management*, 24(3), 359–367.

Yoo, B. & Katsumata, S., (2023). Sightseeing spot satisfaction of inbound tourists: Comparative analysis of first-time visitors and repeat visitors in Japan. *International Journal of Tourism Cities*, 9(1), 111-127.

Zoltan, J., & Masiero, L. (2012) The relation between push motivation and activity consumption at the destination within the framework of a destination card. *Journal of Destination Marketing & Management* 1 (1/2) 84-93.

Part 5 – Tourism Impacts and Sustainability

Tourism has alternately been portrayed as either a benign 'smokestack-free' activity that produces limited adverse impacts or as a major source of social, cultural and ecological impacts. Part 5 explores this issue in greater detail.

Chapter 12 examines the origins of many of the concerns over the social impacts of tourism and how much of these concerns have been placed within a conflict paradigm. It examines the two core origins of this concept, by Doxey (1975) and Budowksi (1976). The chapter concludes by arguing most impacts are, in reality, a function of place change and residents' ability or inability to cope with it.

Chapter 13 develops this theme more by examining the seemingly never ending challenge of trying to achieve sustainable tourism. It discusses the need to rethink how tourism is delivered, especially the challenge of sustainable development.

Chapter 14 focuses more on issues relating to crisis and resilience. It begins by looking at COVID-19 and then shifts its focus to the more pressing long term issue of climate change. A range of issues relating to how to model scenarios, examine options and better anticipate the dynamic world tourism operates in are discussed.

Chapter 15 ties all these issues together by highlighting the need to rethink planning and policy considerations within a contemporary tourism context. It begins by reviewing the evolution of planning models and then discusses their limitations before embarking on a discussion about how a new approach is needed.

12 Social Impacts – Conflict over Place Change

By the end of this chapter, the reader will be able to:

- Describe Doxey's Irridex
- Evaluate Budowksi's relationships between tourism and the environment
- Analyse how impacts are caused by place change
- Understand how conflict theory underlies early attempts to assess social and environmental impacts of tourism.

Introduction

Concerns about unsustainable tourism practices have been a subject of academic inquiry since the earliest days of tourism scholarship (Butler, 2020; Gössling et al., 2020). Indeed, it seems that a majority of the papers published in the first editions of *Annals of Tourism Research* documented adverse social and cultural impacts of tourism. Since then, a range of topics has emerged with the sophistication of the research evolving as our understanding of the causes of impacts and possible mitigation strategies have matured. Yet, to a large extent, many of the critical issues remain unresolved, in spite of the emergence of sustainable tourism as a dominant paradigm. Indeed, many of the issues have re-emerged under the guise of overtourism (Tokarchuk et al., 2022; Wall, 2020) as if it were a new issue, instead of an age-old problem associated with the numbers of tourists overloading a destination and compromising residents' quality of life (Butler, 2018).

Hundreds, if not thousands of academic papers have been published examining the impacts of tourism on host communities (Nunkoo et al., 2013; Hadinejad et al., 2019). Deery, Jago, and Fredline (2012: 65), though, note that much of this work is derivative, leading them to conclude "*research into the social impacts of tourism appears to be in a state of 'arrested development,' [where] there is a sense that the advances in understanding the impacts of tourists on host communities is incremental at best, or potentially circular*." The reasons are manifold. Far too much of this research is descriptive in nature, and is typified by a 'fill in the blank study – Understanding social impacts/ community attitudes of tourism in [name of destination ___________].' Many papers as well use the same metrics to measure attitudes, and therefore, unsurprisingly, come to the same conclusions. A third issue is that much of this research is of the self-fulfilling prophecy kind, where if one structures the study to look for adverse impacts, they can be found. A fourth and more critical issue is the lack of an agreed upon theoretical underpinning for most of the research, something identified by Ap (1990) 30 years ago that still resonates today.

This chapter explores some of the challenges in the conceptual discussion of tourism impacts. It begins with quest for theory and then reviews the foundational models by Doxey (1975) and Budowski (1976) and then explores the idea that the causes of impacts are a function of place change.

Impact research - A quest for a unifying 'theory'

Woosnam and Ribeiro (2023) building Deery et al.'s (2012) work identified five stages in the development of social impact research. The first stage began with the identification of definitions and parameters of the debate. It occurred in the 1970s. Shortly thereafter, incipient models were developed, with Doxey's (1975) and Budowksi's (1976) models discussed here being the two most commonly cited ones. Stage three then looked at ways to operationalise these models, with Ap and Crompton's (1998) work being among the most respected studies, for it developed rigorous questionnaires and study methods. The fourth stage represented instrument refinement. The fifth stage builds on previous work by developing even greater and more sophisticated theoretical model development and testing.

Three issues confront social impact research. The first is that that there is no such thing as a single community, when one measures community attitudes. Instead, geographic spaces are comprised of multiple communities that react to tourism in different ways. Some may feel they benefit from tourism, others are adversely affected, others still feel tourism has no impact and others yet have a complicated relationship, seeing both benefits and costs. This observation leads to the second point in that impacts are not absolute. Instead, they are perceptual and reflected by attitudes or changes in attitudes (Gu & Ryan, 2008). The third issue relates to what to measure, for the list of potential impacts is only limited by one's imagination. For example, Easterling (2004) identified over 40 different impacts that could be summarised into technological, political, socio-cultural and natural. Woosnam and Ribeiro (2023) conducted a detailed literature review of social impact studies and identified three thematic domains and 17 themes under which various measurement variables have been developed. They include: internal issues such as community attachment, economic dependence and personal values; interactive issues such as emotional solidarity with tourists, interaction and social distance from tourists, and; external issues relating to volume of visitors, seasonality, life cycle stage, etc.

Table 12.1: 'Theoretical' approaches to impact studies

Equity Theory	Assesses perceived costs vs. benefits Tourism is seen as beneficial when positive and negative consequences are in balance or when the positive aspects outweigh the negative.
Growth Machine Theory	Certain stakeholders (those who seek to gain) will support growth in order to maximize their personal economic returns. Others may not support growth or may actively oppose it.
Lifecycle Theory	Attitudes change as tourism evolves through its lifecycle.
Power Theory	Personal power, based on property, money, skills, knowledge and competence, affects one's ability to exploit exchanges. No power residents were more favorable toward tourism development than those with power.
Social Exchange Theory	Voluntary actions that are motivated by expected returns. Individuals or groups will engage in an exchange if they value that which is being exchanged, perceive that costs do not exceed benefits and that the exchange will be rewarding. Residents will engage in exchanges with tourists, as long as they 'profit' or as long as benefits exceed costs.
Stakeholder Theory	As key stakeholders in tourism, residents need to be identified, included and satisfied.
Sustainability	Triple bottom line of social, economic and ecological impacts.
Dependency Theory	Applicable mostly to developing countries where dependency originates from and, eventually, reinforces unequal power relationships between wealthy countries (i.e. the center) and poor countries (i.e. the periphery).
Social Impact Theory	The consequences to human populations of any public or private actions that alter the way in which people live, work, relate to one another or cope.

Source: From Andereck et al., 2016; Easterling, 2004; Monterrubio et al., 2018; Nunkoo & Gursoy, 2019

Perhaps the biggest challenge, though, is the ongoing struggle to find some type of universal conceptual framework to explore impacts. Easterling (2004), Nunkoo and Gursoy (2019) and others summarised this dilemma by illustrating that research has been framed in different theories as shown in Table 12.1. Each of these 'theories' would be considered as lower order theories by Smith and his colleagues. While Easterling (2004) notes they help broaden our understanding of tourism systems, none really explains it fully.

It must also be noted that as a result, attitude, perception and impacts are often conflated. As an example, social exchange theory is an excellent tool to assess residents' support for tourism, but does not necessarily prove effective in documenting impacts. Instead, by extension, ecological, social and economic impacts resulting from host-guest exchange affect resident attitudes to tourism (Nunkoo & Gursoy, 2019). The net result is that social impact research still remains rather fractured and descriptive.

Impacts framed within a conflict paradigm

Much of the work on tourism impacts is framed explicitly or implicitly within conflict theory. Doxey's (1975) famous, though rarely read Irridex, positioned conflict as a cause of social impacts. Budowski (1976) asserted tourism had a volatile relationship with the environment, inevitably leading to conflict and Jackson and Wong (1982) felt conflicting land use demands led to clashes between recreational user groups.

Conflict is goal interference attributed to another's behaviour. Goals, in turn are preferred social, psychological or physical outcomes of a behavior that provide incentives for that behavior (Jacob & Schreyer, 1980). As Almeida et al. (2017) note, conflict can arise from specific issues such as a shortage of resources, power imbalances, a flawed decision-making process and, different values (cultural, social and political). Conflict may exist at two levels: direct, where the actions of others affect one's enjoyment; and indirect, where a general and more pervasive feeling of dislike or an unwillingness to appreciate others' views exist (Jackson & Wong, 1982). In fact, social values conflict can occur just by knowing that others who hold different social values are using an area (Morgan et al., 2007). The likelihood of conflict emerging is enhanced when people must fight over a fixed-pie asset, which inevitably produces winners and losers (Katz-Navon & Goldschmidt, 2009). It is also likely to occur when the perceived power balance between stakeholders shifts, empowering one and disempowering the other. Tourism is ideally suited to become a conflict inducer for it often represents a powerful, new stakeholder with different values to existing stakeholders (McKercher, 1993).

Conflict is defined as goal interference attributed to another's behaviour. Often consumers are seeking similar goals, as in this case of this whitewater kayaker and clients of a commercial white water rafting company on the Ottawa River in Canada. One group is looking for a more solitary experience that challenges skill levels and the other group is looking for a pleasurable experience that requires far less skill. The key to this type of conflict is that both types of users must share the same fixed pie resource. Whitewater kayakers discovered this portion of the river and only later did commercial rafting companies start to use this resource. Tourism is often seen as a powerful source of conflict for it represents a new type of user that does not easily fit into either the biocentric or anthropocentric stakeholder group. On the one hand, nature based tourism shares much with biocentrism, for both want a clean environment. But on the other hand, it shares much with anthropocentrics who see the use value of environmental resources being of greater importance than their existence values. As a result, tourism is often a stakeholder that is not really trusted by either stakeholder group.

Conflict is dynamic, evolutionary or revolutionary in style and rarely if ever permanent. Yet, as will be discussed, much of the early work in this field assumed conflict was an end state. It is understandable, for tourism was undergoing rapid development at the time and systems that had been pushed into phase shift have not had the chance to revert to a steady state (Russell & Faulkner 1999). However, as Perdue, Long and Kang (1999) note, social disruption is not permanent. Instead it is transitional where, eventually communities adapt to these changes. For example, perceived quality of life may decline initially but eventually stabilizes and then improves for both individuals and communities are resilient (Greider et al., 1991).

Doxey's Irridex

George Doxey's (1975) work examining the causes of social impacts resonates today, in part because he coined the term 'Irridex' which has become part of the tourism lexicon. This work is based on a conference paper that is less than 5 pages long, with scarcely four paragraphs devoted to the core argument. In it, he reports on two small case studies conducted in the Barbados and Niagara-on-the-Lake in Canada. The basic premise is that *"in the mind of a 'resident' anyone not a native of the destination is an outsider and consequently is a possible source of friction. What is important is the extent that, in the minds of the residents, the outsider represents a challenge to the life style of the destination"* (Doxey 1975: 195). The concluding paragraph of his work states: *"unless local residents are psychologically capable of adjusting to change, irritation levels will continue to rise with long term disastrous effects to the total environment of the tourist area, and to the touristy industry"* (Doxey 1975: 198).

He then developed his 'Irridex' on the understanding that given time, conflict between hosts and guests will lead the hosts to express their feelings progressively more overtly. The Irridex basically has the lifecycle model developed a few years later by Butler (1980), but instead of adopting a physical carrying capacity perspective, it focusses on the social carrying capacity of the destination, and how tourism can inevitably change the character of a place for the worse. The four stages and the descriptions in the original paper of causes and solutions are shown in Table 12.2.

Table 12.2: Doxey's Stages of relationship between hosts and visitors. Source: Doxey 1975

Stage	Conditions	Impact and resolution
Euphoria	Associated with the initial stages of development	Visitors and investors are welcome Usually little planning and few control mechanisms
Apathy	Visitor numbers increase Tourists taken for granted	Contact begins to become more formal Most planning will be concerned with marketing
Irritation	Saturation point is approached	Residents begin to show misgivings about the tourist industry Policy makers increase infrastructure rather than setting limits to growth
Antagonism	Tourism overwhelms places Saturation point exceeded Reputation of the destination suffers	Irritation overtly expressed verbally and physically Mutual politeness gives way to mutual antagonism Outsiders seen as the cause of all problems Remedial planning and marketing to address issues and generate new business
Missing 5th Stage?		

As can be seen, Doxey, an economics professor, focused his work on the need to plan tourism effectively, rather than engaging in a deep discussion of the social impacts on host communities. Yet, this work has been referred to thousands of times in the literature and has been used as a lens to examine a range of impacts in a variety of settings. In fact, some authors have devoted more time to a discussion of his four stages than he did.

The question must be asked, though, is whether Doxey's model is complete, or whether a fifth stage can be added? As mentioned, conflict is rarely an end state and instead tends to resolve itself as either people adapt or are displaced. Doxey's work, like many other studies of social impacts, examined attitudes to tourism during the rapid development phase of the destination lifecycle. It is during this phase that the most significant changes to local communities are likely to occur and where the whole range of negative impacts are most likely to be felt. Doxey's model seems to apply to the first three stages of the life cycle, but is incomplete in addressing latter stages. Communities adjust to tourism as it becomes part of it social, cultural and economic fabric.

In some cases, people may modify their behaviours to cope with the annoying aspects of tourism, but they begin also to appreciate what it means to the community in terms of improved livelihood, increased leisure and recreation opportunities, more and better shopping and dining opportunities and a range of other possible benefits. Certainly the experience with COVID-19 has caused some people to rethink how much tourism has been embedded as part of a community's lifestyle, even if they acknowledge adverse impacts (Kamata, 2022). Residents in tourist dependent areas, in particular, suffered severe social and economic hardship during the pandemic, leading Huser et al. (2022) to conclude the positive impacts of tourism have a significant positive correlation with residents' well-being, whereas negative tourism impacts do not have a convincing effect.

■ Budowksi – coexistence, symbiosis, conflict

Gerardo Budowski (1976) came to the same conclusion as Doxey when he examined the relationship between tourism and the environment, or more precisely human uses of natural environments. He argued that any human use of natural or near natural areas exerts a range of impacts. Tourism, through the introduction of non-traditional activities, uses and/or user groups is no different. He felt tourism could have three possible relationships with the environment. It may coexist during the early stages when the tourism industry and those promoting conservation have relatively little contact. He feels coexistence is a transitory phase especially as tourism activity increases and induces a range of mostly adverse impacts. The second type of relationship is typified by symbiosis in which tourism and conservation derive mutual benefits from each other. Natural assets continue to be conserved as far as possible or evolve towards an even more satisfactory condition while providing a range of benefits, including economic benefits to user groups. He feels that the situation occurs rarely. Instead, the most likely relationship is one is of conflict, especially when the presence of tourism is detrimental to the needs of nature.

Bukowski (976) essentially saw the relationship as existing along a continuum from coexistence through symbiosis towards conflict due to an increase in tourism and the shrinking of natural areas. He feels that the explosion of tourism in the 1970s concerned with wildlife, wildlife areas and scenic beauty based on natural resources is inimical with nature conservation and almost always leads to widespread degradation of the resource or a reduction in the assets of nature. Like one of the possible outcomes of Butler's (1980) life cycle model, once a saturation point is reached, rapid degradation seems inevitable.

He does hold out some hope, though, that tourism may not necessarily be detrimental to protected areas, provided appropriate management practices are undertaken. He qualifies this point by stating that in the 1970s, the policy of building hotels, restaurants and road systems in natural areas lead to their degradation. The solution is to build tourism infrastructure outside of natural areas, limit access and provide some kind of interpretation center at the entrance. He also illustrates that a symbiotic relationship can be fostered when quality tourism programs support conservation goals, when tourism contributes financially to park management and when an overall conservation ethos drives protected area management practices.

This work must be placed within the context of the emerging environmental movement of the 1970s when the anthropocentric and biocentric approaches to protected area management were being debated (Hendee et al., 1978). Here, it must be understood that natural areas, or wilderness areas are essentially human constructs that produce human values. There is no such thing as wilderness. It is a human concept. The preservation of 'wild' areas is not done solely for the benefit of the flora and fauna, but also for the important human benefits and values such preservation provides. In fact, to say some feature is unique and worthy of preservation is a human value judgement. As well, fundamental to the preservation ethos is that these areas should be 'used' to some extent.

Conflict emerges over how these areas should be used. The anthropocentric perspective argues that the main goal of nature is for its use and enjoyment opportunities (Pyra, 2009). People who support this approach view wilderness from a sociological, human oriented perspective and value ecosystems for their instrumental or use value (Hunter et al., 2014), or as Rottman (2014), states anthropocentric concerns for the environment are narrowly aimed at preserving the welfare of humans. The naturalness of wilderness is less important that its direct use benefits. On the other hand, biocentrics advocate the philosophy that unhindered operation of natural ecological processes is the fundamental goal of any natural area. Their concerns are oriented toward protecting non-human organisms and nature as a whole (Rottman, 2014). While human use can occur, it must only do so in a manner that is subservient to the natural ecosystem's needs.

Clearly, these two ideologies are incompatible. Natural areas managed for anthropocentric goals place management priorities on the provision of recreational services. In doing so, substantial impacts might be noted to natural processes to facilitate greater use. In its extreme form, the natural environment would be manipulated to provide tourist benefits, as has been the case in many island resort communities. On the other hand, efforts by people to restrict access or hinder use would be rejected. In reality, though, biocentrism and anthropocentrism lie along a continuum (Phillips et al., 2019). While each end represents an extreme view, a more middle ground represents a mix of use and protectionist views to a greater or lesser extent. Some people holding strong anthropocentric values may seek to preserve natural environments, for their use benefits, while others with equally strong biocentric views may be active recreationists who visit protected areas frequently.

Impacts as a function of place change

In essence, what Doxey and Budowksi document is the impact of place change on attitudes to tourism, as well as perceived conflicts. Doxey's work overtly addresses this issue, while Budowski infers place change strongly as protected areas become used increasingly for anthropocentric purposes at the expense of biocentric ideals.

Historically space and place have been theorized as bounded geographic entities (Tapsell & Tunstall, 2008). As Tuan (1975) notes, the social meaning people give to geographic space transform it into place and give it meaning. In other words, there is an emotional tie between people and place (Mahon, 2007), where sense of place is defined by place identity, place dependence, and place attachment (Jorgensen & Stedman, 2001). Place identity involves dimensions of self that define the individual's personal identity through a complex pattern of conscious and unconscious ideas, beliefs, preferences, feelings, values, goals and behavioural tendencies (Proshansky, 1978). It is is formed throughout their daily routines and experiences (Pretty et al., 2003). Place dependence is the strength of the association between an individual and a specific place, while place attachment reflects a positive bond between groups or individuals and their environment. It relates to the idea of insidedness/outsidedness, where being an 'insider' is to feel safe and attached (Turner et al., 2005) and have an emotional right to occupy place (McCabe & Stokoe, 2004). Outsiders do not have that right.

Place attachment affects attitude to tourists and tourism development (Eusébio et al., 2018; Lee, 2013), especially when tourism leads to a form of tourism gentirification. Over time as tourism activity increases, the nature of the neighbourhood may change until tourism becomes the predominant activity, inducing a degree of touristic gentrification (Liang & Bao, 2015). The concept of tourism gentrification is a relatively novel idea that occurs when the influx of tourism activities fundamentally changes the character of neighbourhoods and in doing so, transforms their social space (Lo & McKercher, 2024). When the tourism gentrification exceeds the degree of acceptability of local residents, place identity can be altered, making them feel like strangers in their home community (Wassler et al., 2018).

McKercher, Wang and Park (2015) discuss each of the elements in detail. They argue that destinations are comprised of three types of place: tourism place, non-tourism place and shared place (as noted in Chapter 11). Tourism place is signed and signaled as being for tourists, while tourists are welcome in shared space, providing they behave appropriately. Non-tourism place, as the name implies, is part of the community that is signed and signaled for use by local residents. Normally, as with most other systems, the relationship between these three types of place evolves toward some form of equilibrium, or coexistence, where each is well defined and both locals and tourists respect their boundaries (Figure 12.1).

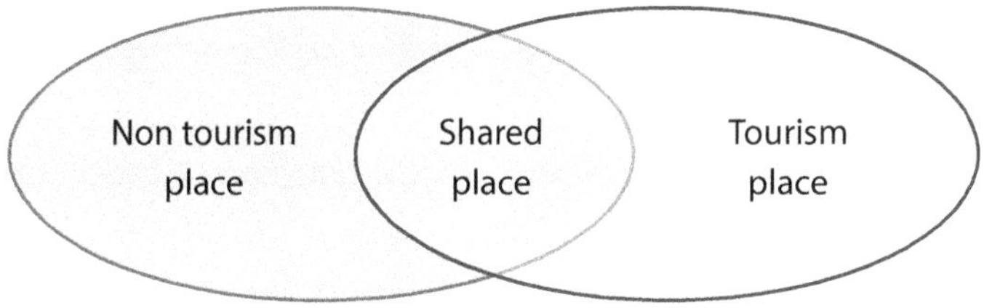

Figure 12.1: Destinations have three types of place. Source: the authors from Kim & Park, 2023.

Their meta-analysis of over 100 social impact papers examined the relationship between attitudes to tourism, lifecycle stage and place change. As shown in Table 12.3, they identified nine types of tourism place. Adverse impacts had been documented in all nine, yet positive place chance was noted in only three types. Some forms of place change affect individuals directly (personal lifestyle, congestion, dislocation), others were felt indirectly at a community level (generic community, economic, structural), while some represented a hybrid (inflation, resource, amenity).

Impacts of tourism are most likely to occur when otherwise stable systems are disrupted. Social disruption theory is closely related to place change. This theory states that communities experiencing rapid growth typically enter a period of generalized crisis that may lead to a wide array of social problems as changes penetrate throughout communities and at individual levels (Park & Stokowski, 2009). Perceived quality of life may decline initially, but will eventually stabilize and then improve as both individuals and communities adapt, for as Greider, Krannich, and Berry (1991) note, they are resilient. Non-tourism place may become shared or even exclusively tourism place and in doing so displace local residents, while shared place may evolve into tourism place. Many examples abound. 'Place' within a touristic sense has multiple social, spatial and economic connotations. Impacts are most likely to occur when that sense of belonging is threatened, but over time, stasis is re-established and a new normal of tourist, shared and non-tourism place emerges (McKercher et al. 2015).

Table 12.3: How tourism affects nine types of place. Source: McKercher et al. 2015

Indicators of negative place change	Indicators of positive place change
Generic community	
loss of traditional lifestyles; rapid cultural change; community character; moral decay; family breakdown	improved quality of life; stronger ethnic identity; cultural exchange
Personal lifestyle	
sense of 'invasion; need to curtail activities; turn locals into attractions; tourism imposed on lifestyle	
Congestion	
traffic; shortage of parking spaces; crowding at public places	
Dislocation	
physical displacement	
Cost of living	
inflation; higher costs of living; higher housing costs	
Economic	
loss of shops; low paying jobs; job loss; economic dependency on tourism	more jobs; higher standard of living; reinvigorate local economy
Resource	
increased competition for natural and recreational resources; land disputes; urban sprawl	
Amenity	
loss of recreational diversity	more leisure and recreational opportunities; better public transport and other services
Structural	
unstable residential patterns; influx of new residents; changing demographic profile; second home development	

Many of the conflicts over tourism involve place change, where once non-tourism place evolves into shared or exclusive tourism place, or where shared place becomes exclusive tourism place. In extreme cases, local residents and businesses are displaced. Here in Macau, SAR of China, this pedestrian street leading up to the Ruins of St Paul's Church used to cater to local residents and visitors from Hong Kong. It had a variety of clothing and furniture shops. The expansion of Macao's casinos, coupled with the relaxation of visa requirements for citizens of China has led to a massive influx of visitors that has completely transformed the street. Now, its shops cater almost exclusively to tourists, selling souvenirs, dried meats and range of snack items.

This and other studies applying the same method (Gössling et al., 2020; Kim & Park, 2023; Lo & McKercher, 2023; Šegota et al., 2022; Yeager et al., 2020) revealed impacts, both positive and negative, represent a change in the community's perception of place identity, place dependence or place attachment. Positive impacts are noted usually through the creation of shared place. Negative impacts, on the other hand are associated with diminished place attachment, loss of distinctiveness, continuity, self-esteem or self-efficacy, that eventually make residents feel more like outsiders in their own community. Here, non-tourism place is seen to change into either shared or tourism place.

The rise of Airbnb and other shared accommodation providers is accelerating the transition of non-tourism place into shared and possibly exclusive tourism place (Yeager et al., 2020). Tourists were once confined to hotel or hostel precincts prior to the advent of this type of lodging. Now, they can book accommodation in what were formally exclusive residential neighbourhoods or apartments buildings and can 'live like locals.' While beneficial in some areas and for some investors, others have called for regulation to preserve the integrity of some neighbourhoods (Yeager et al., 2020).

Importantly, there is a direct correlation between attitudes to tourism and the destination lifecycle stage as shown Figure 12.2. The line represents attitudes to tourism while the columns represent the perception of positive and negative impacts of tourism. As Doxey suggests, attitudes to tourism are generally positive in the early stages of development where little tourism occurs and where the perception exists that it will bring a number of positive changes to communities. But, attitudes change as tourism expands, and especially during the rapid growth stage, when a nadir is reached. Here the negative consequences of tourism are seen to far outweigh the positive benefits as place change is at its apex. But, a strange thing happens. While perceptions of negative impacts remain high, attitudes generally improve and the community begins to appreciate the positive impacts of tourism once more. What happened? The community has adjusted to the new normal and stasis has been regained.

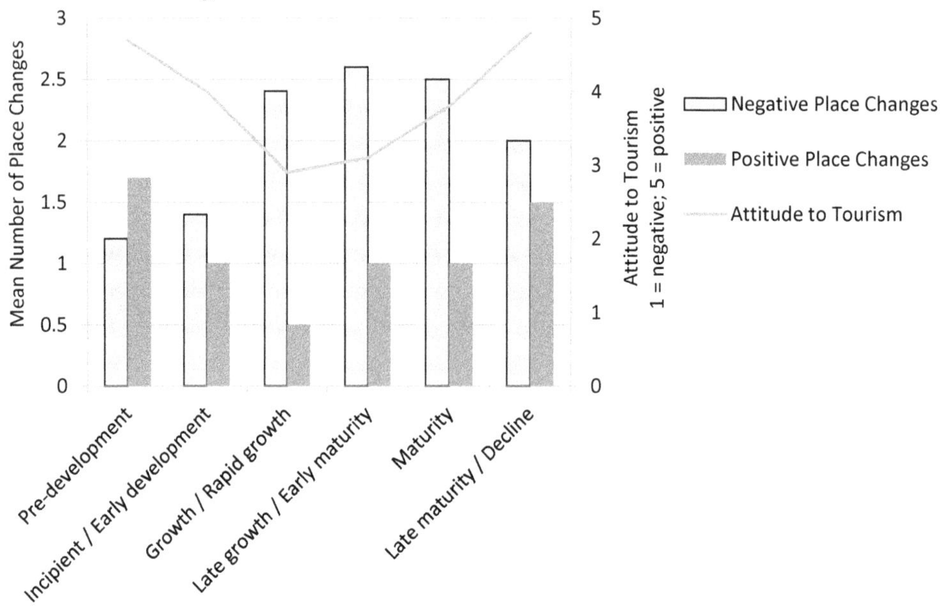

Figure 12.2: Social impacts as a function of place change. Source: McKercher et al., 2015.

Conclusion

The study of impacts of tourism, both positive and negative, is an enduring theme in tourism scholarship, for impacts have been a core study area right from the earliest days of tourism research. While research has documented well the 'what' of impacts, it is still struggling to understand the 'why'. Here a range of competing paradigms have been developed, each with its own merits but also each having its own limitations. This chapter tried to simplify much of the discussion by placing impacts within two dominant fields of study. The first is conflict literature, where the direct and indirect actions of one group of stakeholders can exert a range of impacts on another group of stakeholders. The second is the idea of space and place, whereby most adverse impacts can be summarised as a function of place change.

Discussion points / Exercises

1. Review a current issue centered on a tourist destination and determine what stage of Doxey's Irridex the destination is in. Is conflict inevitable and if not, what can be done to resolve the situation?
2. Impacts are a function of place change. Choose a destination and map out tourism, shared and non-tourism place. Look for areas where adverse impacts have been felt and document if place change has occurred, and if so, which of the nine times of place change are evident.
3. Tourism becomes part of the lifeblood of most mature destinations, and a degree of stability emerges in the spatial structure of destinations. Indeed, in many cases, local residents are also heavy users of so-called tourism places. Identify and discuss how local residents have made tourism place their own.

Points to consider for revision

- Early work on impacts of tourism was framed within a conflict paradigm.
- Multiple 'theories' have been developed that try to explore impacts, attitudes and perceptions. Each has its place, but none is ideal.
- Most impact are caused by place change, where tourism disrupts normally stable systems.

Suggested additional readings

Budowski G. (1977) Tourism and Conservation: conflict, coexistence or symbiosis? *Parks*, 1, 3-6.

Hadinejad, A., Moyle, B., Scott, N., Kralj, A. & Nunkoo, R., (2019). Residents' attitudes to tourism: A review. *Tourism Review*, 74(2), 150-165.

McKercher, B., Wang, D., & E Park (2015) Social impacts as a function of place change. *Annals of Tourism Research*. 50, 52-56.

Woosnam, K. & Ribeiro, M., (2023). Methodological and theoretical advancements in social impacts of tourism research. *Journal of Sustainable Tourism*, *31*(2), 187-203.

References

Almeida, J., Costa, C. & Da Silva, F., (2017). A framework for conflict analysis in spatial planning for tourism. *Tourism Management Perspectives*, 24, 94-106.

Andereck, K., Pachmayer, A. & Zhao, S. (2016) Resident attitude towards tourism – the state of knowledge. *Travel and Tourism Research Association: Advancing Tourism Research Globally*. 10. https://scholarworks.umass.edu/ttra/2012/Oral/10.

Ap, J (1990) Residents' Perceptions research on the social impacts of tourism. *Annals of Tourism Research* 17, 610-616.

Ap, J. & Crompton, J. (1988) Developing and testing a tourism impact scale, *Journal of Travel Research*, 37(2), 120-130.

Budowski, G. (1976) Tourism and conservation: Conflict, coexistence or symbiosis? *Parks*, 4, 3-6.

Butler R. W. (1980) The concept of a tourist area cycle of evolution: Implications for management of resources *Canadian Geographer* 24(1), 5-12

Butler, R. W. (2018). Challenges and opportunities. *Worldwide Hospitality and Tourism Themes*, 10(6), 635-461

Butler, R.W., (2020). Tourism carrying capacity research: a perspective article. *Tourism Review*, *75*(1), 207-211.

Deery, M., Jago, L., & Fredline, L. (2012). Rethinking social impacts of tourism research: A new research agenda. *Tourism Management*, 33(1), 64-73.

Doxey, G. V. (1975) A causation theory of visitor-resident irritants: methodology and research inferences. In *Travel and Tourism Research Associations Sixth Annual Conference Proceedings* (195-198), San Diego, September.

Easterling, D. (2004). The residents' perspective in tourism research. *Journal of Travel and Tourism Marketing* 17(4), 45-62.

Eusébio, C., Vieira, A. L., & Lima, S. (2018). Place attachment, host–tourist interactions, and residents' attitudes towards tourism development: the case of Boa Vista Island in Cape Verde. *Journal of Sustainable Tourism*, 26(6), 890-909. doi:10.1080/09669582.2018.1425695

Gössling, S., McCabe, S. & Chen, N.C., (2020). A socio-psychological conceptualisation of overtourism. *Annals of Tourism Research*, 84, 102976.

Greider, T., Krannich, R., & Berry, H. (1991). Local identity, solidarity, and trust in changing rural communities. *Sociological Focus*, 24(4), 263-282.

Gu, H. & Ryan, C. (2008), Constructionism and culture in research: understandings of the fourth Buddhist festival, Wutaishan, China, *Tourism Management*, 31(2), 167-178.

Hadinejad, A., Moyle, B., Scott, N., Kralj, A. & Nunkoo, R., (2019). Residents' attitudes to tourism: A review. *Tourism Review*, 74(2), 150-165.

Hendee, J., Stankey, G., & Lucas, R. (1978) *Wilderness Management*, Forest Service, US Dept. of Agriculture, Misc. Publication No 1365, Washington, pp 15 - 22.

Hunter, M., Redford, K. & Lindenmayer, D (2014) The complementary niches of anthropocentric & biocentric conservationists. *Conservation Biology* 28(3), 641-645.

Huser, L., Wiles, E. & Westerlaken, R., (2022). Impact of tourism on residents' well-being amid the Covid-19 pandemic in Bali. *Research in Hospitality Management*, 12(3), 285-293.

Jackson, E. L., & Wong, R. (1982) Perceived conflicts between urban cross country skiers and snowmobilers in Alberta. *Journal of Leisure Research* 14(1), 47-62.

Jacob, G. R., & Schreyer, R. (1980) Conflict in outdoor recreation: A theoretical perspective. *Journal of Leisure Research*. 12(4), 368-380.

Jorgensen, B. S., & Stedman, R. C. (2001). Sense of place as an attitude: Lakeshore owners attitudes toward their properties. *Journal of Environmental Psychology*, 21(3), 233-248.

Kamata, H., (2022). Tourist destination residents' attitudes towards tourism during and after the COVID-19 pandemic. *Current Issues in Tourism*, 25(1), 134-149.

Katz-Navon, T.Y. & Goldschmidt, C., (2009). Goal orientations in negotiations: The influence of goal orientations on fixed-pie perceptions and bargaining outcomes. *International Journal of Psychology*, 44(1), 60-70.

Kim, S. & Park, E., (2023). An integrated model of social impacts and resident's perceptions: From a film tourism destination. *Journal of Hospitality & Tourism Research*, 47(2), 395-421.

Lee, T. H. (2013). Influence analysis of community resident support for sustainable tourism development. *Tourism Management* 34, 37-46. doi:10.1016/j.tourman.2012.03.007

Liang, Z.-X., & Bao, J.-G. (2015). Tourism gentrification in Shenzhen, China: causes and socio-spatial consequences. *Tourism Geographies*, 17(3), 461-481. doi:10.1080/14616688.2014.1000954

Lo, J.S.K. & McKercher, B., (2023). Tourism gentrification and neighbourhood transformation. *International Journal of Tourism Cities*, 9(4). https://doi.org/10.1108/IJTC-03-2023-0038.

Mahon, M. (2007). The changing faces of rural populations. *Journal of Rural Studies*, 23(3), 345-356.

McCabe, S., & Stokoe, E. H. (2004). Place and identity in tourist accounts. *Annals of Tourism Research*, 31(3), 601-622.

McKercher, B. (1993) Some fundamental truths about tourism: understanding tourism's social and environmental impacts, *Journal of Sustainable Tourism*, 1(1), 6-16.

McKercher, B., Wang, D., & E. Park (2015) Social impacts as a function of place change. *Annals of Tourism Research*. 50, 52-56. http://dx.doi.org/10.1016/j.annals.2014.11.002.

Monterrubio, C., Sosa-Ferreira, A.P. & Osorio-García, M., (2018). Local perceptions of the impacts of second-home tourism: a qualitative approach based on social-exchange theory. *Revista LiminaR. Estudios Sociales y Humanísticos*, 16(1).

Morgan, N., Newman, P. & Wallace, G., (2007). Conflicts associated with recreational shooting on the Pawnee National Grassland. *Human Dimensions of Wildlife*, 12(3), 145-156.

Nunkoo, R. & Gursoy, D. (2019) Introduction to tourism impacts. In Gursoy, D and Nunkoo, R (Eds) *The Routledge Handbook of Tourism Impacts: Theoretical and Applied Perspectives.* Oxon: Routledge. Pp 1-20.

Nunkoo, R., Smith, S. L. J., & Ramkissoon, H. (2013). Residents' attitudes to tourism: A longitudinal study of 140 articles from 1984 to 2010. *Journal of Sustainable Tourism,* 21(1), 5-25.

Park, M., & Stokowski, P. (2009). Social disruption theory and crime in rural communities: Comparisons across three levels of tourism growth. *Tourism Management,* 30(6), 905-915.

Perdue, R., Long, P., & Kang, Y. (1999). Boomtown tourism and resident quality of life: The marketing of gaming to host community residents. *Journal of Business Research,* 44(3), 165-167.

Philips, L., Szuster, B. & Needhamd, M. (2019) Tourist value orientations and conflicts at a marine protected area in Hawaii. *International Journal of Tourism Research.* 21(6), 868-881

Pretty, G. H., Chipuer, H. M., & Bramston, P. (2003). Sense of place amongst adolescents and adults in two rural Australian towns: The discriminating features of place attachment, sense of community and place dependence in relation to place identity. *Journal of Environmental Psychology,* 23(3), 273-287. doi:10.1016/s0272-4944(02)00079-8

Proshansky, H. M. (1978). The city and self-identity. *Environment and Behavior,* 10(2), 147-169.

Pyra, L. (2009). The anthropocentric versus biocentric outlook on nature. In: Tymieniecka, AT. (eds) *Memory in the Ontopoiesis of Life.* Analecta Husserliana, vol 101. Springer, Dordrecht. https://doi.org/10.1007/978-90-481-2501-2_5

Rottman J (2014) Breaking down biocentrism: two distinct forms of moral concern for nature. *Frontiers of Psychology* 5:905. doi: 10.3389/fpsyg.2014.00905

Russell, R., & Faulkner, B. (1999). Movers and shakers: Chaos makers in tourism development. *Tourism Management, 20*(4), 411-423.

Šegota, T., Mihalič, T. & Perdue, R. (2022). Resident perceptions and responses to tourism: individual vs community level impacts. *Journal of Sustainable Tourism,* 1-24.

Tapsell, S., & Tunstall, S. (2008). "I wish I'd never heard of Banbury": The relationship between 'place' and the health impacts from flooding. *Health & Place,* 14(2), 133-154.

Tokarchuk, O., Barr, J.C. & Cozzio, C., (2022). How much is too much? Estimating tourism carrying capacity in urban context using sentiment analysis. *Tourism Management, 91,* 104522.

Tuan, Y. F. (1975). Place: An experiential perspective. *Geographical Review,* 65(2), 151-165.

Turner, P., Turner, S., & Carroll, F. (2005). The tourist gaze: Towards contextualized visual environments. In P. Turner & E. Devenport (Eds.), *Spaces, Spatiality and Technology* (pp. 281-297). Netherlands: Springer.

Wall, G., (2020). From carrying capacity to overtourism: A perspective article. *Tourism Review*, 75(1), 212-215.

Wassler, P., Schuckert, M., Hung, K., & Petrick, J. F. (2018). You're welcome? Hong Kong's attitude towards the Individual Visit Scheme. *International Journal of Tourism Research, 20*(5), 637-649.

Woosnam, K. & Ribeiro, M., (2023). Methodological and theoretical advancements in social impacts of tourism research. *Journal of Sustainable Tourism*, 31(2), 187-203.

Yeager, E.P., Boley, B.B., Woosnam, K.M. & Green, G.T., (2020). Modeling residents' attitudes toward short-term vacation rentals. *Journal of Travel Research*, 59(6), 955-974.

13 The Challenge of Sustainable Tourism

By the end of this chapter, the reader will be able to:

- Define the key concepts involved in sustainable development and sustainable tourism
- Critique the inherent contradiction in the term
- Identify and discuss models that offer some scope to assist tourism to be managed in a more sustainable manner

Introduction

From its early beginnings, travel and tourism has created a range of adverse social, cultural and environmental impacts (Lane, 2017). And, over the same period, policy makers, land use planners and more recently academics, have tried to develop protocols to optimise the economic benefits of tourism while minimising its costs. These efforts have met with only modest success, with Cave and Dredge (2020) arguing they have failed due in part to the massive growth of tourism and the political economy under which it operates. The challenge is even greater today as the world is facing problems caused by population growth, pollution, continued exploitation of non-renewable resources, growth in demand for goods and services, including tourism, and climate change. The climate crisis in particular is emerging as the key threat to long-term sustainability and will require a switch to net-zero emissions of greenhouse gases (GHG) in the very near future if the problems caused by rising global temperatures are to be averted. It may even require a transition away from the current neoliberal "take, make and dispose" approach to resource use, to one based on economy-wide recycling of resources.

The magnitude of the task of transitioning to a global tourism industry that is truly sustainable is enormous. Between 1924 and 2023 the global population quadrupled from 2 billion to 8 billion people, placing significant pressure on the Earth's fixed stock of non-renewable resources as well as its renewable resources. Growing demand for travel has added to the problem. International arrivals increased from an estimated 25 million in 1950 (Rosselló & Santana-Gallego, 2014) to reach an estimated 1.3 billion in 2023 (UNWTO, 2023) with the potential for even further growth as middle-class demand for travel expands in rapidly developing economies including China and India.

Tourism, as a sector that accounts for 8% of global carbon emissions (Lenzen et al., 2018) will be expected to contribute to strategies to deal with these problems. A number of models and frameworks have been developed over past decades aimed at managing tourism at a more sustainable level. Most, though, have met with limited success. This chapter reviews our attempts to transition to a more sustainable tourism sector.

Sustainability

The ideas of 'sustainable development' and 'sustainable tourism' are not new, with Hall et al. (2015) positing that the creation of national parks in the late 19th century represented the first primitive form of sustainability. Lane (2017) indicates that the development of the concept of sustainable tourism can be traced back to the 1970s, when the impacts of unregulated tourism in parts of Europe provoked much concern among academics.

■ The dilemma of sustainable development

The paradigm shift occurred with the formalization of the idea of 'sustainability' in 1987 when *Our Common Future* was published by the World Commission on Environment and Development (WCED, 1987). The issue crystalised further in the 1992 Rio Summit on sustainable development (Obst, 2016; Torres-Delgado & Lopez-Palomeque, 2012). Sustainable development was defined as development that meets the needs of the present without compromising the ability of future generations to meet their own needs (WCED, 1987). In essence, it represented the convergence between the economy and environmentalism (Hardy et al., 2002). This approach contains within it two key concepts: the concept of 'needs', in particular the essential needs of the world's poor, to which overriding priority should be given; and the idea of limitations imposed by the state of technology and social organization on the ability of the Earth's ecosystems to meet present and future needs.

Sustainability has a number of key principles including:

- *Inter-generational equity* - meaning that the range of activities and the scope of ecological diversity available to future generations is at least as broad as that felt by current ones;
- *Intra-generational equity*, social justice and poverty alleviation - improving the well-being of all residents in a community, and not just benefiting the powerful or the rich;
- *Public participation* – which means that we all share a role to play and that communities need to collectively make decisions rather than having them imposed by external forces;
- *Environmental protection* as an integral component of economic development – economic development without environmental conservation is no longer acceptable;
- *Dealing cautiously with risk and uncertainty* – in situations where environmental impacts of activities are not known, the preferred option is to proceed cautiously or not at all, until the likely impacts can be determined.

Almost from the beginning, though, the term 'sustainable' became contested (Pearce et al., 1989) and remains so to the present (Biely et al., 2018). Salas-Zapata and Ortiz-Munoz (2018) conclude the lack of clarity about the concept is a recurring obstacle to its implementation. Carillo and Jorge (2017) add that multiple interpretations enable it to be molded to fit the needs of different stakeholders with competing views. The ideal of inter-generational equity is possibly the most difficult principle to deal with. The ongoing emphasis on economic growth and associated depletion of renewable and non-renewable resources is likely to ensure that generations in the next century, let alone the next millennium, will not have access to the range of resources and ecological diversity enjoyed by current generations.

Weak and strong sustainability

Two competing and polar opposite approaches to 'sustainability' and sustainable development have emerged. The weak sustainability model assumes sustainability can be achieved through a constant wealth approach. The alternate view of strong sustainability is based on the premise of constant natural capital. Both models accept that the world's total asset base is comprised of both natural (ecological and social) and manufactured capital (Islam et al., 2019).

The constant wealth, or weak sustainability, philosophy assumes that natural capital and manufactured capital are essentially substitutable and considers that there are no differences between the types of well-being they create (Pelanc & Dedeurwaerdere, 2015). It sees economic sustainability as dominating, with the belief that technological advances will resolve environmental issues (Biely et al., 2018). Thus, the only thing that matters is the total value of the aggregate stock of natural and manufactured capital should be maintained and ideally added to for the sake of future generations (Pelanc & Ballet, 2015). Hence, natural capital can be traded off providing it creates sufficient manufactured capital to compensate for its loss (Islam et al., 2019). This philosophy argues for a business as usual approach. Unsurprisingly, studies of organizational sustainability schemes align closely with the weak sustainability approach (Demastus & Landrum, 2023). In a similar manner, the touristic focus at the enterprise and destination levels also tends to incorporate the ideas of flawed sustainability, forced by the need to adopt a short-term view (Peeters et al., 2024). The intellectual argument for weak sustainability is weak. Importantly, no account is given to consideration of the key differences between non-renewable resources and renewable resources.

The constant natural capital or strong sustainability model, argues that natural capital is non-substitutable, either totally or partially and needs to be maintained. As Islam et al. (2019: 160) note, the contribution of natural capital to total wealth is unique and critical, so that it cannot be measured in monetary terms and straight-forwardly compared with other types of capital, particularly in social cost–benefit analysis. Pelanc and Ballet (2015) also highlight that a distinction must be made between natural and manufactured capital. They point out that natural capital has the risk of irreversibility and thresholds. Once used, non-renewable resources are lost forever. The extinction of a species is irreversible. Moreover, above a certain threshold of pollutants, ecosystems can become overloaded and enter into decline. Importantly, although the amount of manufactured capital can be increased or decreased, natural capital can disappear if it has become stressed beyond the point of being able to replenish itself. Proponents of this view argue we need to do things differently so that the absolute stock of natural capital does not diminish (Randall, 2022), and as will be discussed in the next section, a number of people are now promoting the regenerative idea of tourism, whereby it should seek to enhance the natural capital base, rather than maintain it (Tham & Sharma, 2023).

Definitions and practices of sustainability exists along a continuum with the weak and strong positions representing the extreme ends. All consumption involves some use of natural resources, some of which are not renewable (Kuhlmann & Farringdon, 2010). DesRoches (2020) argues that there is a preservation paradox in trying to maintain or restore natural capital, for to do so will make it somehow artificial. As a consequence, the idea of *critical natural capital* emerged as a core tenet of sustainability. Here it is argued that critical natural capital highlights the need to maintain the ecological functioning of natural systems above certain thresholds of degradation to conserve the capacity of the natural capital asset to renew itself (Pelanc & Dedeurwaerdere, 2015). Moderately strong sustainability has evolved from a purist point of view to one that espouses the belief that sustainability can be seen as a series of thresholds that must not be crossed (Kuhlman & Farrington, 2010). But there is a problem in that ecosystems change over time. Thus the suite of environmental thresholds that applied in the past may not be relevant in a warmer future as species migrate into or out of ecosystems.

Two competing views of sustainable development have emerged since the concept of sustainability was first mooted. *Weak sustainability* argues that natural and manufactured capital are essentially substitutable, and as long as the net sum of both remains the same or increases, sustainable development can occur. Unfortunately, this view does not consider the adverse impacts development may have on the environment, as shown in the first picture of a heavily eroded beach in Vietnam. The ecological integrity of the beach was destroyed due to the encroachment of the resort onto the waterfront which resulted in erosion. *Strong sustainability* argues that natural capital is non-substitutable. Development can occur and is encouraged to occur, providing that the net sum of natural capital does not diminish. This model tends to be applied most easily in protected areas where a strong legislative imperative exists, as shown in the second image of Taman Negara National Park in Malaysia taken some 30 years ago. In reality, most tourism development will involve some loss of natural capital, and so the concept of *critical natural capital* has emerged as a preferred alternative path forward. It argues development can occur as long as the ecological functioning of natural ecosystems remains above a critical threshold.

The need to rethink tourism

During the early 1980s, it became apparent that major global environmental changes were occurring suddenly and silently and were only beginning to be recognized as an emerging threat by the global scientific community. Scholars and an increasing number of citizens became aware that there was a growing element of uncertainty and risk from human activities on global environments. At that time, moderate environmentalists were wrestling with the idea that we can live in a world where economic development and economic pressures for both developing and developed countries would continue. Short-term pressures to develop national economies (along with the social benefits that derive from it) were considered by policy makers to outweigh long-term potential impacts from that development process. In other words, we failed to ask the question of how can a sustainable economy be developed that integrates both industrial and ecological needs? More recently, advocates of regenerative tourism (Day, 2022; Dredge, 2022; Tham & Sharma, 2023) and alternative economic production systems (Hall et al., 2021; Ranworth, 2017; Geissdoerfer et al., 2017) have suggest that a more radical change is needed.

The fundamental problem for the tourism industry, and why it faces ongoing problems dealing with sustainability, is that there are a finite number of tourism resources that are demanded by an increasing number of tourists. Further, many of the resources that are in the greatest demand are under considerable threat from overexploitation, poor planning and climate change. Several examples of this conundrum are coral reefs that face rapid decline as sea water temperatures rise beyond 1.5^{0}C while the demand for coral experiences continue to grow, coastal resort areas are under threat from sea level rise but continue to attract investment in new tourism related infrastructure driven by increasing demand, and the loss of winter sports destinations from climate change continues even as the demand of winter sports increases. Dealing with these problems will require many difficult decisions to be made by industry and the public sector. For example, do coastal areas attempt to defend coastal infrastructure or retreat and rebuild as sea levels rise? In any discussion of theories and models currently used to deal with these issues, it is pertinent to consider how we as a global society created this problem and how it can deal with the growing demand for tourism in an era of declining availability of current and potential future resources.

The root cause of the need to balance demand for resources with the need to consumer resources was best summarised by Hardin's (1968) essay on the *tragedy of the commons*. He drew attention to the overuse of the common pool of resources that may lead to their exhaustion. No one single person to blame, yet all are collectively responsible. His concerns resonate in tourism for much of what tourists consume can be classified as common resources – rivers, beaches, natural areas, urban communities and iconic scenic views. The tourism industry enjoys the benefits of the resource without contributing directly to its management, making tourism an extractive industry.

Sustainable tourism

Sustainable tourism has been widely accepted at a conceptual level as a strategy that is able to balance tourism demand for built and natural resources with the long-term protection of these resources. At face value, tourism appears to be ideally suited to adopting sustainability as a guiding philosophy. At a conceptual level, tourism's use of resources, both natural and cultural, should be non-consumptive, making them renewable. The reality however is that tourism is a fierce competitor for resources where the provision of cultural and ecotourism opportunities may mean that local residents are displaced, and pristine ecosystems redeveloped for tourist consumption. As the demand for tourism related resources increases, the difficulty of supplying this demand for resources also increases. Moreover, since the needs of tourists are different from those of local residents, serving tourists may not suit the needs of locals. Instead, tourism is often imposed on local communities, especially rural and ethnic minority communities, at levels and speeds that cause great social disruption (McKercher, 1993).

The concept of sustainable tourism has evolved significantly since it was first mooted almost 30 years ago. Lane's (2017) reflective article illustrates that the issues have become more complex and refined as our knowledge has grown. He illustrates that the idea began as a purely reactive response to try to address tourism's adverse impacts. First generation scholars (1990 – 2010) and advocates needed to devote a great deal of time to define and refine the concept, understand residents' and communities' attitudes to tourism, strive to develop indicators, conduct small case studies and engage in a debate about the ethics of sustainability. The second generation was more focussed on governance issues, managing visitor experiences, finance and taxation instruments, seeking ways to change tourists' behaviour, focusing on indigenous tourism and positioning degrowth as a strategy. The conversation has also broadened substantially. Originally, it was thought that only small, locally owned enterprises could be sustainable. However, Weaver (2017) argues that sustainable tourism exists along a continuum, with sustainable mass tourism no longer seen as a contradiction in terms.

Significantly, we have moved beyond just looking at sustainable tourism in isolation and now realise that tourism is part of a larger debate about overall sustainability (Bramwell et al., 2017; Budeaunu et al., 2016). Earlier efforts were coordinated by tourism departments with little or no input from other government agencies. Tourism departments or ministries often are seen as junior departments that have limited influence over other government agencies and even less ability to influence policy emanating from those departments (Muangasame & McKercher, 2015). It is impossible to move towards an effective sustainable tourism sector unless a whole of government approach towards sustainability is adopted. As Budeaunu et al. (2016) observed, desirable change in the tourism system is not possible without broader economic change. The need for change will become increasingly apparent as the stock of the Earth's non-renewable resources continues to decline. Perhaps that decline will produce another version of net zero – net zero copper left to mine for example?

Tourism represents a powerful stakeholder that has the potential to out-compete other stakeholders. One of the challenges facing sustainable tourism is that much of the resource base it relies on is part of the social fabric of communities. Resources such as beaches, shopping districts, even car parking spaces are used by local residents. But, if tourism is allowed to develop in an unsustainable manner, the needs of local residents can be compromised and they ultimately may be displaced. This economic transformation may benefit local business people, but comes at a cost to local residents. Mass tourism development in places like Mallorca, Spain have displaced many residents from access to beaches and local shopping areas that used to be a core part of the community's social fabric.

The need to incorporate tourism into the broader sustainability debate received a boost with the publication of the United Nations Development Program (UNDP) 17 Sustainable Development Goals. The UNWTO endorsed this policy and developed its own policy document illustrating how and where tourism fits within these broad goals (UNWTO, 2018). The document states "*tourism's role in achieving the 17 Sustainable Development Goals (SDGs) can be significantly strengthened when sustainable development becomes a shared responsibility and moves to the core of decision-making within the tourism sector"* (UNWTO, 2018: 10). It feels tourism can contribute to 14 goals directly or indirectly (UNWTO, 2018: 16, 17), while three goals are specific to tourism matters.

The challenge however, is that reality keeps getting in the way. Hall et al. (2015) comment that the gap between sustainability, sustainable tourism and the realities of tourism, raises fundamental questions about the ability to achieve a balance between economic, social and environmental goals. A number of scholars have highlighted the inherent and apparently intractable challenges, with Ateljevic and Sheldon (2022) noting that while sustainability has generated some progress, it has fallen well short in providing the type of radical change needed in all stakeholders.

A large part of the problem lies in the political economy under which tourism operates (Cave & Dredge, 2020), and in particular, the underlying growth imperative that reinforces negative social and ecological impacts (Bellato et al., 2023). Tourism, like most economic sectors follows a model based on continuous growth for its economic well-being (Gössling & Higham, 2021; Higham et al., 2021). This approach is inherently unsustainable.

Day (2022), Bausch et al. (2021) and Weaver (2004), have highlighted more prosaic challenges. These issues include:

- The malleability of the sustainable tourism as a cultural construct that is open to ambiguous and multiple interpretations;
- The inherent complexity of tourism and its linkages with all other economic sectors making it difficult to isolate tourism sustainability from broader community or national sustainability;
- The obsession with trying to develop quantifiable indicators creates multiple problems, including debate about what indicators to include, how to weight them, how to measure them and how to monitor them;
- The emergence of indicators as an end in itself (tick the box) rather than as a means to a better end (improve performance);
- How the term has become jargonised and lost its meaning over time;
- Political baggage associated with the term and the left/right political divide over both the need to act and how to act, and;
- The feeling that the bar is set too low to maintain the status quo, rather than to improve the situation.

To these can be added climate change challenges that reduce the availability of resources for tourism use and the adoption of globally agreed strategies to deal with these actions. Despite most nations signing the Kyoto protocol to limit global warming via reducing greenhouse gas (GHG) emissions, emissions have continued to climb. In fact, in 2023, global temperatures rose to 1.45C above average for the first time in recorded history.

Attempts at limiting overuse impacts

A number of models and frameworks have been developed that aim to assist tourism be to managed in a more sustainable way. Most have met with limited success.

■ Carrying capacity and limits of acceptable change

The development of the carrying capacity concept was one of the earliest attempts to manage excessive numbers of visitors to recreational areas that were in danger of being 'loved to death' (Wall, 2020). The issue was first raised in the 1950s and gained popularity in the 1960s (Butler, 2020). Essentially, carrying capacity was defined as the maximum number of users that can be supported without an unacceptable decline in the quality of the resource or of the visitor experience (Wall, 2020). It sought to determine a 'magic number' of visitors that could be accepted in an area without causing undue harm to the physical environment they were encountering. It was soon realized that an arbitrary focus on numbers alone, without considering other variables did not work (Butler, 2020). Fennell and Cooper (2020) note that measurement depends on a range of issues including indicators, location, ethical issues and who determines the physical limits and enforces decisions.

The concept has found its way into debates on sustainable tourism. The UN World Tourism Organization (2018:3) defines tourist carrying capacity as "*the maximum number of people that may visit a tourist destination at the same time, without causing destruction of the physical, economic and sociocultural environment and an unacceptable decrease in the quality of visitors' satisfaction.*" However, echoing an earlier observation, Fennell and Cooper (2020: 121) state that in "*implementing the system as an inter-organizational instrument, the carrying capacity concept has not been particularly useful*". Earlier, Lindberg, McCool and Stankey (1997) stated that while carrying capacity has an intuitive appeal, it is not able to deal with the complexity that typifies many areas of tourism. Despite concerns of this type, carrying capacity has gained some support particularly in areas where there is a desire to practice sustainable tourism.

Carrying capacity has a number of weaknesses. The relationship between levels of use and environmental change is not linear (Wall, 2020). Instead, a threshold effect is noted, where places are able to accommodate high visitation with limited impacts until a threshold is crossed, after which catastrophic impacts can occur. Second, not all sites have an inherent capacity. Some can be sufficiently hardened to enable them to cope with additional visitors although the danger of breaching thresholds remains.

The greatest challenge though, is that carrying capacity is a social construct that relies on subjective judgements (Tokarchuk et al., 2022). It is seen as the highest tourism presence at a destination that does not interrupt the ordinary activities of residents and does not damage the touristic experience. Overtourism is said to occur when the carrying capacity is exceeded (Butler & Dodds, 2019). In Venice, concerns about overtourism led to the imposition of a peak season visitor fee in 2024 to dampen demand. But what may be too many tourists for one person may mean quite the opposite for another. Likewise, some residents may enjoy the benefits brought by tourists while others may be disturbed by them. An associated challenge is that changes in the external environment, an increasing number of hot days for example, may upset previous relationships between the factors that determine use capacity.

The concept of limits of acceptable change was introduced to overcome problems encountered with carrying capacity. It is described by McCool (2013: 286) as a framework that can be "*embedded within planning processes of protected area decision making, or can be used as a framework to assist decisions that are specific to visitor and tourism management.*" As McCool (2013) states, limits of acceptable change should not be seen as a method for calculating carrying capacity but rather a process that assists natural area managers to resolve problems that arise when attempting to mediate between overlapping goals such as preserving natural heritage values while enabling access to protected areas. In principle it should work, but again, in practice, it remains a social construct that is open to interpretation, political interference and vested interests.

The precautionary principle

The precautionary principle (PP) is often referred to in issues related to environmental and health policy as a risk management strategy. But it remains an ill-defined idea. A product of the increasing interest in environmental policy in Europe that emerged in the late 1970s, it has been adopted in many international treaties and declarations. For example, the precautionary principle was embedded in the Treaty on European Union (1992) as the basis for European environmental law. However, the precautionary principle has generated considerable controversy particularly over the interpretation of 'precautionary'. As Foster, Vecchia and Repacholi (2000) note, precautionary has been used as the basis by the EU to ban agricultural products that have been genetically engineered, potentially reengineering the principle as a protectionist tool. Sunstein (2003: 1003) suggests that "*taken in its strongest form the precautionary principle should be rejected, not because it leads to bad directions, but because it leads in no direction at all. The principle is literally paralyzing – forbidding inaction, stringent regulation and everything in between.*" From this perspective, every step, including inaction, can be seen as creating risk.

In a sustainable tourism context where timeframes are generally regarded as being long-term, Fennell and Cooper (2020) state that the precautionary principal is a useful mechanism for encouraging stakeholders, including operators and communities, to exercise precaution in actions that may have short and long-term implications. This may take the form of audits that measure long-term effects of policies and actions and assist in the maintenance of appropriate standards of practice.

■ Life Cycle Assessment (LCA)

Life Cycle Assessment (LCA) is used to evaluate the environmental impacts of services and products using a through-life approach (Ilgin & Gupta, 2010). Unlike other measures of impact, LCA seeks to assess the environmental impacts that occur cumulatively through the lifecycle of a product, service or process. Life cycle includes initial raw material extraction, processing, transport, manufacture, use, and disposal, including recycling or dumping. A typical LCA study includes developing an inventory of all materials and energy required in the value chain of the product, service or process followed by calculation of cumulative emissions.

The International Organisation for Standardisation (ISO) has approved LCA which is now used in assessments in the ISO 14040 series of standards. A number of scholars (De-Camillis et al., 2010; Filimonau et al., 2014) have demonstrated how LCA can be applied in a tourism setting. However, as Filimonau, Dickinson and Robbins (2014) caution, there remain number of problems with LCA that may reduce its accuracy. Problems include lack of regular updates of databases that LCA analysis draws on and limitations of the methodologies used to collect data.

■ Ecological footprint

One measurement technique that has attracted the attention of scholars working in the sustainable tourism area is the ecological footprint. First suggested by Rees (1992), ecological footprint is an ecological accounting approach for estimating the quantity of resources used on a range of scales and industries including tourism. On a global scale, it can be used to estimate the amount of resources used by the global society and then compared to the ability of the Earth to renew the resources used. In 2014, the Global Footprint Network estimated that humanity was using resources 1.7 times faster than they can be renewed (Lin et al., 2018) with rich countries consuming a far higher proportion of available resources than poor countries. The inequality of resource use between rich and poor countries is also reflected in tourism use of resources particularly in destinations located in less developed countries. The problems highlighted by the **ecological footprint** concept are likely to be magnified in the future unless the overconsumption that is a characteristic of many economies is reined in. Given that part of this 'overconsumption' is generated by tourism demand, future strategies to reduce the extent of the global **ecological footprint** may require rethinking the structure, availability and distribution of tourism activity.

Hunter and Shaw (2005) argued that ecological footprint analysis should be widely used as an environmental indicator. Despite significant data measurement problems, the successful development of a global measure will allow for cross country and cross sector comparisons and change management. Building on the observations of Hunter and Shaw (2005) and acknowledging that it does not capture all environmental impacts, Castellani and Sala (2012) applied and compared both ecological footprint and Life Cycle Assessment (LCA) in a tourism destination case study in northern Italy. They found that it was a useful screening tool for resource use on a tourism system wide scale but was not suitable for the more detailed assessments possible with LCA. In the future, continued refinement of the data measurement is likely to lead to ecological footprint being used to measure impacts on a tourism system scale while LCA is likely to offer a more detailed and comprehensive measurement at activity scale.

Triple and quadruple bottom line approach

Many of the models discussed in earlier chapters of this book include the idea of a threshold. What are those thresholds? The concept of the triple or quadruple bottom line provides some answers. The triple bottom line is dual edged sword. Sardianou et al. (2016), summarising the literature, remind us that sustainable tourism development must focus on the four areas of economic, environmental, social and cultural sustainability. These themes adhere to Elkington's (1994) universally accepted triple-bottom line approach to sustainability, or quadruple bottom-line if the dimension of social impacts is disaggregated to include social and cultural dimensions. An alternative version of the quadruple bottom line adds the dimension of governance. As Bac (2012) noted, strong sustainability, which had focused on natural resources, evolved to consider the social and economic aspects of the concept. Without economic viability, social and ecological sustainability are difficult to achieve. Figure 13.1 illustrates the quadruple bottom line approach.

Moscardo (2015) highlighted criticisms of the triple bottom line approach in that it encourages people to think about the dimensions as forms of capital that are separate and equal and, therefore, finding a balance amongst them is both desirable and easy. This approach is problematic on many levels. First, the timeframe to consider each dimension is variable, with economic sustainability measured in terms of daily, weekly or monthly cash flow, liquidity and profitability. Social sustainability is measured on a longer term and ecological sustainability on an even longer term. Second, the issue of what to measure is problematic and is something that has vexed the UNWTO in its attempts to develop its own set of indicators (UNWTO, 2004). Again, economic sustainability has clearly defined and measurable indicators, while social and environmental sustainability can be measured in terms of thresholds or debatable limits of acceptable change. Third, and perhaps most important, it is difficult, if not impossible, to separate values from facts (Carillo & Jorge, 2017). The final challenge is what and how to measure progress.

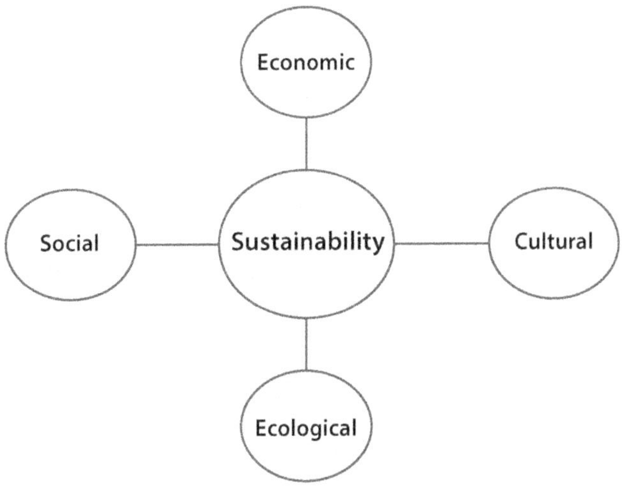

Figure 13.1: Quadruple bottom line. Source: The authors

■ Indices

The development of sustainability indicators was an attempt to place measurable thresholds on levels of tourism activity. Butler (1999) reminds us that indicators are critical in the move toward sustainability, for without indicators, the term 'sustainable' becomes little more than a meaningless hyperbole. However, a number of scholars have suggested that we are overwhelmed by indicators, rather than having too few (Marzo-Navarro et al., 2015). Part of the challenge is that tourism touches on so many aspects of society that Jovicic (2014) suggests an infinite number of indicators could be developed. The extent of the challenge is highlighted by the UNWTO's (2004) 500 plus page long publication *Indicators of Sustainable Development for Tourism Destinations: A Guidebook.*

Our efforts to be inclusive have created a classic situation of choice overload where the number of alternatives or choice options is greater than the person's ability to make effective and efficient decision. As Agyeiwaah et al. (2017: 27) observed in summarizing the literature, choice overload depends on the complexity of the choices available, how well the options align with each other and the presence or lack of a clearly preferred option. Preparing vast lists may be an impressive intellectual exercise but is largely ineffective, with the net result being a tendency to avoid acting because of choice overload (Park & Jang, 2013) or to pick and choose options that best suit one's narrow personal interests (Miller & Twining-Ward, 2005).

Regenerative and transformative tourism

The enforced stoppage of tourism during the COVID-19 pandemic provided the catalyst for a number of scholars to call for a radical transformation of tourism and a stop to return to 'business as usual' practices (Peeters et al., 2024). They called for an ontological and operational revolution to fundamentally transform tourism instead of following the hoped for, but slow and unsatisfactory sustainability path we are currently on. They illustrate that the sustainable tourism regime sees tourism as an economic activity that responds to prioritised, top-down initiatives (Bellato et al., 2023). They also argue that sustainability that seeks to do less harm is an inadequate action, and instead argue that tourism needs to actively undo the harm it has done. To support these views, Bellato et al. (2023) suggested a two-tiered framework that incorporates regenerative design dimensions to guide tourism development based on regenerative philosophy.

Regenerative and transformative tourism is seen by some as a viable option. This approach seeks to place tourism in a more holistic framework by incorporating the expectations of entire communities and stakeholders to make a destination more attractive to all (Tham & Sharma, 2023). While its definition is still evolving, Dredge (2022) summarises its core concepts as tourism that delivers net positive benefits to people, places and nature and that supports long term renewal and flourishing of social and ecological systems. In short, it seeks to create opportunities for healing of destinations while balancing social-economic-environmental impacts (Tham & Sharma, 2023). The key question however, is how can this be achieved when power imbalances between stakeholders operate within destinations and how will losers be compensated by winners?

Regenerative and transformative tourism recognise, at their heart, that the current tourism system is not sustainable (Cave & Dredge, 2020). Instead, a change in the economic model of how tourism functions and where is sits in relation to other economic activities is essential if it is to be a healing force (anon, 2022). These and other authors argue we have to move away from simply striving to be 'sustainable' to a tourism system that improves human health and wellbeing while actively encouraging ecological recovery. Doing so calls for a seismic shift in our current world view. Currently, a mechanistic and industrial paradigm dominate discussions about sustainable tourism, when an ecological worldview is needed for a regenerative sector (Bellato et al., 2023). As Dredge (2022:270) so eloquently states *"we have been lulled into a false sense of security that scientific tools like sustainability criteria and sustainable development goals will solve the challenges that confront us. However, simple tools are unlikely to solve complex dynamic challenges. The real challenge is rooted in our social-ecological consciousness and our capacity to evolve our thinking from 'me' to 'we' and to develop compassion, empathy and collaboration."*

Degrowth

Degrowth is a philosophical world view that emerged from fears that the current economic production systems will deplete non-renewable resources, accelerate industrialization, cause continuing deterioration of the global environment and lead to socio-economic collapse. Degrowth has been described as a critique of the current neoliberal economic development hegemony (Demaria et al., 2013).

The degrowth approach calls for drastic reduction in consumption and production (Weiss & Cattaneo, 2017). Until recently, the debate about degrowth as a solution to climate change has been largely ignored in the tourism literature. However, growing concern about overtourism (Fletcher, et al., 2019; Milano et al., 2019) has stimulated interest in degrowth as a potential solution. It has reinvigorated interest in earlier work by scholars including Andriotis (2014) and Scott, Gössling and Hall (2012). Hall (2009) for example stated that degrowth may be used as strategy to overcome some of the impacts of climate change. However, as Fletcher et al. (2019: 1745) note, degrowth is "*both social theory and social movement.... that has yet to engage systematically with the tourism industry*".

The central idea of degrowth is an emphasis on: quality of life rather than quantity of consumption; structural transition to service-based activities by downscaling production and consumption; investment in ecological assets; and changes in worktime as a mechanism to achieve economic stabilisation (Jackson, 2009). One outcome of this process is an increase in human well-being, reduced consumption and greater leisure time. The latter outcome has significant implications for tourism. The flaw in the degrowth view of transition is that a model able to achieve the desired economic and political outcomes has yet to be developed (Weiss & Cattaneo, 2017). As Prideaux and Pabel (2021) observed, the degrowth literature has yet to address strategies needed to convince the general public that they will need to limit their consumption of touristic experiences.

The circular economy model

A succinct definition of the circular economy (CE) is "*an industrial economy that is restorative or regenerative by intention and design*" (Ellen MacArthur Foundation, 2013:14). The Ellen MacArthur Foundation has emerged as a leading advocate of the CE approach and has been at the forefront in theory development. A more comprehensive definition describes the circular economy as a "*regenerative system in which resource input and waste, emission, and energy leakage are minimised by slowing, closing, and narrowing material and energy loops. This can be achieved through long-lasting design, maintenance, repair, reuse, remanufacturing, refurbishing, and recycling*" Geissdoerfer et al., (2017). The new relationships that emerge between use and reuse can be found at the Foundation's website (https://www.ellenmacarthurfoundation.org/circular-economy-diagram). To transition to an economic system of this type will require industrial processes and economic systems to move from an open loop system characterised by 'dispose', to a closed loop where resources are remade or recycled. It will also require extensive interventions by the public sector to ensure waste, reuse and energy reduction targets are achieved.

The impact of the CE model on tourism is not clear – a result of the lack of research into this potential model for economic organisation. If adoption of a CE model results in greater consumer surplus or reduced hours of work, there is a strong possibility that tourism will be a beneficiary.

There have been a number of critiques of the CE model. Hobson and Lynch (2016) for example view the prevailing view of the CE as a rebooting of capitalism variously described as 'cooperative capitalism', 'breakthrough capitalism' and 'naturalised capitalism'. This approach argues that the environment and capitalism can co-exist to ensure a level of sustainability that does not disrupt the status quo of politics and power (Hobson & Lynch, 2016). Other scholars view the idea of the CE as an alternative to the green growth model and for this reason an inadequate solution to ensuring long-term sustainability within the limits of the Earth's ecological systems (Schröder et al., 2019).

■ Doougnut Economy

Based on the premise that the current neoliberal economic production model has both failed to care for people and the environment, UK economist Kate Raworth (2017: e48) proposed a new economic model termed the 'doughnut economy' which she describes as:

> *"new model of human wellbeing ...... to guide humanity in the Anthropocene. In essence, it recognises that wellbeing depends on enabling every person to lead a life of dignity and opportunity, while safeguarding the integrity of Earth's life-supporting systems. The conceptual framework of social and planetary boundaries—which has come to be known as the Doughnut—contributes to this paradigm by concisely visualising its ambition, and so providing a compass for humanity's 21st century progress".*

The model establishes criteria for achieving a more just society that acknowledges the concept of social boundaries operating within a set of planetary resource boundaries. Actions that exceed planetary resource boundaries (climate change, deforestation, loss of biodiversity) are shown as overshoots, while actions that adversely affect the quality of the 'social foundations' of life (including shelter, energy, water food etc) are referred to as shortfalls. Hartman and Heslinga (2022: 279) noted that model *"offers an overarching perspective seeking wider system transformation, which we believe is crucial, as well as still underdeveloped in tourism studies, to counter the still quite dominant growth-oriented, boosterism-inspired ways of developing tourism destinations"*.

The challenge - will these actions work?

The preceding discussion highlights both the greatest opportunity and simultaneously the greatest challenge by recognizing that 'sustainability', however one defines it, involves a tripartite relationship between government, industry and the consumer (McKercher & Prideaux, 2014). Government can play a macro role through the development of policies and the enactment of legislation. Industry can provide more environmentally friendly products and services, and can engage in more sustainable business practices if it makes good business sense, as part of their broader commitment to corporate social responsibility (CSR) or if compelled to by legislation.

Changing consumer behaviour, is the third and most challenging pillar, for it involves convincing people individually to change their often long-established habits. Becken (2004) highlights the challenge that when people feel the personal benefits of engaging in an action are high, the perceived environmental costs are low. While tourists may see others as acting in an 'unsustainable' manner, they are unlikely to apply the same standard to their own actions. Anable, Lane and Kelly (2007) add further that people rarely make a link between their own actions and broader environmental problems. For example, most people are aware that driving a car produces greenhouse gases, but few people relate their own driving habits as making a direct contribution. McKercher and Prideaux (2011) also found that tourism ranked low on the list of personal environmental impact agendas, and moreover, because it represented a break from social norms, tourists could justify behaving differently from how they did at home.

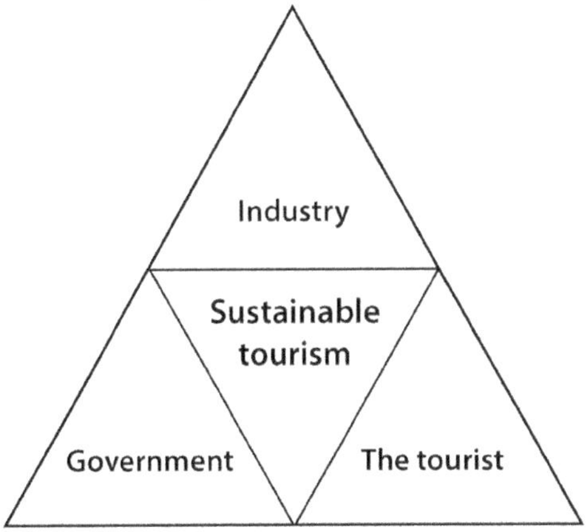

Figure 13.5: The tripartite solution to sustainable tourism

The ability to address the growth paradox remains unresolved. Tourism is expected to continue to grow at a faster rate than the global economy as countries continue to move into the realm of the developed world, incomes grow and people have the ability to travel for the first time. One of the greatest challenges facing social scientists, in general, and tourism social scientists in particular, is that they (we) can raise issues but are largely impotent in resolving them, especially those issues that require engineering or technical solutions.

Government can enact sustainability legislation, while industry may be motivated to act in a more sustainable manner. The tourist, though, represents the third and most difficult component of this tri-partite sustainability relationship, for it involves changing behaviour at an individual level. A number of studies have indicated that while people are interested in environmental issues in general and climate change in particular, these considerations play no role in their travel decision making process. Indeed, the opposite may be true, for when the perceived benefits of participation are seen to be high, the perceived impacts of such activities are seen to be low. Moreover, tourists adopt a number of neutralisation actions – verbal justifications to rationalise behaviour that they know is inappropriate. 'Everyone else is doing it', 'I am just one of many people, my actions have no real impact' and 'I am a tourist, I came all this way to participate in this activity and no one is going to stop me' are common explanations used. And so, in spite of requests to behave, some tourists, individually and collectively, will act in a way that is socially, culturally and ecologically irresponsible, as shown in these two images of a tourist souveniring a piece of the Great Pyramid at Giza, Egypt and accumulated litter at the base of the Sphinx.

Conclusion

Sustainability has been an issue for as long as people have travelled. Its importance has become critical today as we are facing a real climate crisis, to which tourism is a significant contributor. There are a large number of models and frameworks that can be used to assess environmental impacts. In most instances, no one model, concept or framework is likely to successfully address the multitude of issues that can arise when managers and decision makers need information to solve problems related to use of the environment for tourist purposes. The challenge remains of how to move from the current less sustainable practice of tourism to a more sustainable future. Will it be evolutionary or is a revolution needed? If so, are consumers prepared to make necessary changes to their lifestyles and travel habits?

Discussion points / Exercises

1. Adherents of hard sustainability argue that some activities are irreversible, such as species extinction, and as such we must adopt this approach. Proponents of soft sustainability argue that we cannot upset the global economic system without creating huge social problems. Which point of view do you endorse and why?
2. Is tourism a smokestack free industry as early proponents of sustainability argued?
3. Which model, limits to acceptable change, or carrying capacity is most appropriate for tourism?
4. We are overwhelmed with sustainability indictors. Develop a list of five core indicators that you feel are both reasonable to achieve and essential for sustainable tourism.
5. Changing tourists' behaviour is the greatest challenge in developing a more sustainable tourism sector. Identify why most tourists seem unwilling to change and what actions may be adopted to encourage change?
6. The current global economic system is not sustainable. What are the implications for tourism next century?
7. Could implementation of the 'donought' economy model solve many of the problems addressed in the UNs SDGs?

Points to consider for revision

- Sustainability is a contested issue.
- Sustainable tourism is an ideal goal for the sector which corresponds closely with many of the UNDP sustainable goals.
- A number of concepts and models have been developed to encourage, monitor and move toward a more sustainable tourism future.
- Continued overuse of the earth's resources will lead to serious deficiencies in a large range of renewable and non-renewable resources in the future.

Suggested additional readings

Agyeiwaah, E., McKercher. B., & W. Suntikul (2017) Proposing a core set of achievable sustainable tourism indicators: a meta-analysis approach. *Tourism Management Perspectives.* 24, 26-33.

Lane, B. (2017). Sustainable tourism: its evolution and its future. *Cuadernos económicos de ICE,* 93, 9-28.

Fennel, D., & C. Cooper (2020) *Sustainable Tourism: Principles, Contexts and Practices.* Channel View Publications.

References

Agyeiwaah, E., McKercher. B., & W. Suntikul (2017) Proposing a core set of achievable sustainable tourism indicators: a meta-analysis approach. *Tourism Management Perspectives.* 24, 26-33.

Anable, J., Lane, B., & T. Kelay (2006). *An evidence base review of public attitudes to climate change and transport behaviour.* London: Dept for Transport. http://www.fcrn.org.uk/sites/default/files/Evidence_of_public_attitudes_and_behaviour.pdf

Andriotis, K. (2014) Tourism development and the degrowth paradigm. *Turističko poslovanje,* 13, 37-45.

Ateljevic, I. & Sheldon, P.J., (2022). Guest editorial: Transformation and the regenerative future of tourism. *Journal of Tourism Futures,* 8(3), 266-268.

Bac, D. (2012) The emergence of sustainable tourism: A literature review. *QUAESTUS Multidisciplinary Research Journal,* 131-140. http://www.quaestus.ro/wp-content/uploads/2012/03/bac4.pdf.

Bausch, T., Schröder, T., Tauber, V. & Lane, B., (2021). Sustainable tourism: The elephant in the room. *Sustainability,* 13(15), 8376.

Becken, S. (2004). How tourists and tourism experts perceive climate change and carbon-offsetting schemes. *Journal of Sustainable Tourism,* 12(4), 332–345

Bellato, L., Frantzeskaki, N. & Nygaard, C.A. (2023). Regenerative tourism: a conceptual framework leveraging theory and practice. *Tourism Geographies,* 25(4), 1026-1046.

Biely, K., Maes, D. & Van Passel, S. (2018) The idea of weak sustainability is illegitimate *Environmental Development and Sustainability 20*(1), 223-232.

Bramwell, B., Higham, J., Lane, B., & Miller, G. (2017) Twenty-five years of sustainable tourism and the *Journal of Sustainable Tourism*: looking back and moving forward, *Journal of Sustainable Tourism*, 25(1), 1-9, DOI: 10.1080/09669582.2017.1251689

Budeaunu, A., Miller, G., Moscardo, G., & Ooi, C. (2016) Sustainable tourism, progress, challenges and opportunities: An introduction. *Journal of Cleaner Production.* 111, 285-294.

Butler, R. W. (1999). Sustainable tourism: A state-of-the-art review. *Tourism Geographies*, 1(1), 7-25.

Butler, R. W., & Dodds, R. (Eds.). (2019). *Overtourism: Issues, realities and solutions.* Berlin: De Gruyter.

Butler, R.W., (2020). Tourism carrying capacity research: a perspective article. *Tourism Review*, 75(1), 207-211.

Carillo, M., & Jorge, J. (2017) Multidimensional analysis of regional tourism sustainability in Spain. *Ecological Economics* 140, 89-98.

Castellani, V. & Sala, S. (2012), Ecological Footprint and Life Cycle Assessment in the sustainability assessment of tourism activities, *Ecological Indicators*, 16, 135-147.

Cave, J. & Dredge, D, (2020): Regenerative tourism needs diverse economic practices, *Tourism Geographies*, DOI: 10.1080/14616688.2020.1768434

Day, J. (2022) *Why regenerative tourism now… and what about sustainable tourism?* START. http://sustainabletourismandresponsibletravel.com/why-regenerative-tourism-now-and-what-about-sustainable-tourism.

De-Camillis, C., Raggi, A. & Petti, L. (2010) Tourism LCA: State-of-the art and perspectives, *International Journal of Life Cycle Assessment*, 15(2), 148-155.

Demaria, F., Schneider, F., Sekulova, F. & Martinez-Alier, J. (2013) What is degrowth? From an activist slogan to a social movement. *Environmental Values*, 22(2), 191-215.

Demastus, J. an&d Landrum, N.E., (2023). Organizational sustainability schemes align with weak sustainability. *Business Strategy and the Environment.* https://doi.org/10.1002/bse.3511

DesRoches, C.T., (2020). The preservation paradox and natural capital. *Ecosystem services*, 41, 101058.

Dredge, D., (2022). Regenerative tourism: Transforming mindsets, systems and practices. *Journal of Tourism Futures*, 8(3), 269-281.

Elkington, J. (1994). Towards the sustainable corporation: Win-win-win business strategies for sustainable development. *California Management Review*, 36(2), 90-100.

Ellen MacArthur Foundation, (2013), *Towards the Circular Economy - Economic and business rationale for an accelerated transition*. https://archive.ellenmacarthurfoundation.org/assets/downloads/publications/TCE_Report-2013.pdf

Fennell, D.A. & Cooper, C., (2020). *Sustainable Tourism: Principles, contexts and practices* (Vol. 6). Channel View Publications.

Filimonau, V., Dickinson, J. & Robbins, D. (2014) The carbon impact of short-haul tourism: a case study of UK travel to Southern France using life cycle analysis, *Journal of Cleaner Production*, 64, 628-638.

Fletcher, R., Mas, I., Blanco-Romero, A. & Blázquez-Salom, M. (2019) Tourism and degrowth: An emerging agenda for research and praxis. *Journal of Sustainable Tourism*, 27(12), 1745-1763.

Foster, K., Vecchia, P. & Repacholi, M. (2000) Science and the Precautionary Principle. *Science*, 288 (5468), 979.

Geissdoerfer, M., Savaget, P., Bocken, N. & Hultink, E. (2017). The Circular Economy – a new sustainability paradigm. *Journal of Cleaner Production*, 143, 757-768.

Gössling, S., Balas, M., Mayer, M. & Sun, Y.Y., (2023) . A review of tourism and climate change mitigation: The scales, scopes, stakeholders and strategies of carbon management. *Tourism Management*, 95, 104681.

Hall, C. M. (2009) Degrowing tourism: Décroissance, sustainable consumption and steady-state tourism. *Anatolia*, 20(1), 46-61.

Hall, C.M., Gössling, S. & Scott, D. (2015). The evolution of sustainable development and sustainable tourism. In *The Routledge Handbook of Tourism and Sustainability* (pp. 15-35). Routledge.

Hall, CM., Lundmark, L. & Zhang, J. (eds) (2021), *Degrowth and Tourism: New perspectives on tourism entrepreneurship, destinations and policy*, Routledge, pp 116-132.

Hardin, G. (1968), The tragedy of the commons, *Science*, 162, 1243–1248.

Hardy, A., Beeton, R.J. & Pearson, L., (2002). Sustainable tourism: An overview of the concept and its position in relation to conceptualisations of tourism. *Journal of sustainable tourism*, 10(6), 475-496.

Hartman, S., & Heslinga, J. H. (2022). The Doughnut Destination: applying Kate Raworth's Doughnut Economy perspective to rethink tourism destination management. *Journal of Tourism Futures*, 9(2), 279-284. https://doi.org/10.1108/JTF-01-2022-0017 .

Higham, J., Font, X. and Wu, J., (2021). Code red for sustainable tourism. *Journal of Sustainable Tourism*, *30*(1), pp.1-13.

Hobson, K. & Lynch, N. (2016) Diversifying and de-growing the circular economy: Radical social transformation in a resource-scarce world, *Futures*, 15-25.

Hunter, C. & Shaw, J (2007) The ecological footprint as a key indicator of sustainable tourism, *Tourism Management*, 28, 46-57.

Ilgin, A. & Gupta, S. (2010) Environmentally conscious manufacturing and product recovery: A review of the state of the art, *Journal of Environmental Management*, 91(3), 563-591.

Islam, M., Yamaguchi, R., Sugiawan, Y., & Managi, S. (2019) Valuing natural capital and ecosystem services: A literature review. *Sustainability Science* 14, 159-174

Jackson, T. (2009) *Prosperity Without Growth. Economics for a finite planet.* London: Earthscan.

Jovicic, D. Z. (2014). Key issues in the implementation of sustainable tourism. *Current Issues in Tourism*, 17(4), 297-302.

Kuhlman, T. & Farrington, J. (2010) What is sustainability? *Sustainability*, 2(11), 3436-3448; https://doi.org/10.3390/su2113436.

Lane, B., (2017). Sustainable tourism: its evolution and its future. *Cuadernos económicos de ICE*, 93, 9-28.

Lin, D., Hanscom, L., Murthy, A., Galli, A., Evans, M., et al. (2018). Ecological Footprint Accounting for Countries: Updates and Results of the National Footprint Accounts, 2012–2018. *Resources*, *7*, 58.

Lindberg, K., McCool, S. & Stankey, G. (1997) Rethinking carrying capacity, *Annals of Tourism Research*, 24(2): 461-465.

Marzo-Navarro, M., Pedraja-Iglesias, M., & Vinzón, L. (2015). Sustainability indicators of rural tourism from the perspective of the residents. *Tourism Geographies*, 17(4), 586-602, DOI: 10.1080/14616688.2015.1062909.

McCool, S. (2013), Limits of acceptable change in tourism, in Holden, A. & Fennel, D. (eds), *Routledge Handbook of Tourism and the Environment*, Routledge, Oxon, , pp 285-298.

McKercher, B. (1993) Some fundamental truths about tourism: understanding tourism's social and environmental impacts *Journal of Sustainable Tourism*, 1(1), 6-16.

McKercher, B., & B. Prideaux (2011) Are tourism impacts low on personal environmental agendas? *Journal of Sustainable Tourism*. 19(3), 325-345

Milano, C., Novelli, M. & Cheer, J. (2019) Overtourism and degrowth: A social movements perspective. *Journal of Sustainable Tourism*, 27(12), 1857-1875.

Miller, G., & Twining-Ward, L. (2005). *Monitoring for Sustainable Tourism Transition: The challenge of developing and using indicators*. Cambridge, MA: CABI.

Moscardo G. (2015) Tourism and sustainability: Challenges, conflict and core knowledge. In: Moscardo G., Benckendorff P. (eds) *Education for Sustainability in Tourism. CSR, Sustainability, Ethics & Governance*. Springer, Berlin.

Muangasame, K., & McKercher, B. (2014). The challenge of implementing sustainable tourism policy: A 360-degree assessment of Thailand's "7 Greens Sustainable Tourism Policy", *Journal of Sustainable Tourism*, DOI: 10.1080/09669582.2014.97878.

Obst, C. (2016) *Framing Sustainable Tourism*. Madrid: UNWTO. http://cf.cdn.unwto.org/sites/all/files/pdf/mst_issue_paper_1.pdf.

Park, J., & Jang, S. (2013). Confused by too many choices? Choice overload in tourism. *Tourism Management*, 35, 1-12.

Pearce, A. Markandya & E. B. Barbier (1989) *Blueprint for a Green Economy*. London: Earthscan Publications.

Peeters, P., Çakmak, E. & Guiver, J., (2024). Current issues in tourism: Mitigating climate change in sustainable tourism research. *Tourism Management*, 100, p.104820.

Pelanc, J. & Ballet, J. (2015) Strong sustainability, critical natural capital and the capability approach. *Ecological Economics* 112: 36 – 44.

Pelanc, J., & Dedeurwaerdere, T. (2015) Weak Sustainability versus Strong Sustainability. UN Brief for GSDR 2015. https://sustainabledevelopment.un.org/content/documents/6569122-Pelenc-Weak%20Sustainability%20versus%20Strong%20Sustainability.pdf.

Prideaux, B. & Pabel, A. (2021) Degrowth as a strategy for adjusting to the adverse impacts of climate change in a nature based destination, in Hall, CM., Lundmark, L. & Zhang, J. (eds), *Degrowth and Tourism: New Perspectives on Tourism Entrepreneurship, Destinations and Policy*, Routledge. pp 116-132.

Randall, A., (2022). How strong sustainability became safety. *Sustainability*, 14(8), 4578.

Raworth K. (2017), *Doughnut Economics: seven ways to think like a 21st century economist*. Penguin Random House.

Rees, W. E. (1992). Ecological footprints and appropriated carrying capacity: What urban economics leaves out. *Environment & Urbanization*, 4(2), 121-130. doi:10.1177/095624789200400212

Rosselló, J., & Santana-Gallego, M. (2014). Recent trends in international tourist climate preferences: a revised picture for climatic change scenarios. *Climatic Change*, 124(1-2), 119-132

Salas-Zapata, W., & Ortiz-Munoz, S. (2018) Analysis of meanings of the Concept of Sustainability. *Sustainable Development*. 27, 153-161.

Sardianou, E., Kostakis, I., Mitoula, R., Gkaragkani, V., Lalioti, E., & Theodoropoulou, E. (2016). Understanding the entrepreneurs' behavioural intentions towards sustainable tourism: A case study from Greece. *Environment, Development and Sustainability*, 18, 857-879. DOI 10.1007/s10668-015-9681-7.

Schröder, P., Bengtsson, M., Cohen, M., Dewick, P., Hofstetter, J, & Sarkis, J. (2019) Degrowth within – Aligning circular economy and strong sustainability narratives, *Resources, Conservation and Recycling*, 146, 190-191.

Scott, D., Gössling, S. & Hall, C. M. (2012) International tourism and climate change. *WIRES Climate Change*, 3(3), 213-232.

Sunstein, C. (2003), Beyond the Precautionary Principle, *University Pennsylvania Law Review*, 151(3), 1003-1-58.

Tham, A. & Sharma, B., (2023). Regenerative tourism: Opportunities and challenges. *Journal of Responsible Tourism Management*, *3*(1), 15-23.

Tokarchuk, O., Barr, J.C. & Cozzio, C., (2022). How much is too much? Estimating tourism carrying capacity in urban context using sentiment analysis. *Tourism Management*, *91*, 104522.

Torres-Delgado, A. & Lopez-Palomeque, F. (2012) The growth and spread of the concept of sustainable tourism: The contribution of institutional initiatives to tourism policy. *Tourism Management Perspectives*, 4, 1-10.

UNWTO (2004). *Indicators of Sustainable Development for Tourism Destinations: A guide book.* United Nations World Tourism Organization. Madrid, Spain

UNWTO (2018) *'Overtourism'? Understanding and Managing Urban Tourism Growth Beyond Perceptions: Executive Summary.* Madrid: UN World Tourism Organization.

UNWTO (2023), *International Tourism to Reach Pre-Pandemic Levels in 2024.* www.unwto.org/news/international-tourism-to-reach-pre-pandemic-levels-in-2024

Wall, G., (2020). From carrying capacity to overtourism: A perspective article. *Tourism Review*, 75(1), 212-215.

WCED (1987) Our *Common Future.* World Commission on Environment and Development. United Nations. https://sustainabledevelopment.un.org/content/documents/5987our-common-future.pdf.

Weaver, D.B., (2004). Tourism and the elusive paradigm of sustainable development. In A. A. Lew, C. M. Hall, & A. M. Williams *A Companion to Tourism*, pp. 510-524.

Weaver, D.B., (2017). Sustainability and mass tourism: a contradiction in terms?. In *Mass Tourism in a small world* (pp. 63-74). Wallingford UK: CABI.

Weiss, M., & Cattaneo, C. (2017) Degrowth – Taking stock and reviewing an emerging academic paradigm. *Ecological Economics*, 137, 220-230.

Weiss, M., & Cattaneo, C. (2017) Degrowth – Taking stock and reviewing an emerging academic paradigm. *Ecological Economics*, 137, 220-230.

14 Crisis and Disaster Events: COVID-19 to Climate Change

By the end of this chapter, the reader will be able to:

- Describe the key issues that need to be considered in disaster preparation, response and recovery
- Outline the stages of Faulkner's tourism disaster lifecycle model
- Discuss the role of forecasting in disaster and crisis preparation
- Outline how a mean global temperature rise above 1.5°C will affect the tourism industry
- Evaluate the strategies that can be used by the public sector to achieve net zero GHG emissions.

Introduction

Crisis and disaster events may occur with little, or no warning. During the COVID-19 pandemic, international tourism flows almost ceased for two years significantly disrupting the tourism industry at every level from local to global. Fortunately, the pandemic event had little effect on the stock of tourism resources and infrastructure and the tourism industry was able to recover quickly as the impact of the pandemic eased during late 2022 and 2023. In other events, including earthquakes, floods and tsunamis, the tourism industry has suffered considerable damage to tourism related infrastructure that hampered recovery. The 2004 Indian Ocean Tsunami is one example of a regional scale event that caused significant damage to tourism infrastructure in Thailand and Siri Lanka resulting in a slow recovery (Gurtner, 2007). The COVID-19 experience and the 2004 Tsunami highlight the need to prepare for possible future crisis and disaster events. Some events may be expected based on history and location (cyclones and hurricanes in tropical areas and earthquakes in some areas in Japan and California) while others may be unexpected. The COVID-19 pandemic and 2004 Tsunami events highlight the difficulty of preparing for events that may not occur for decades, if at all. Aside from known events, it is also necessary to consider how new types of threats may emerge to disrupt the global tourism sector. Artificial Intelligence (AI) (Jabeen et al., 2022) and cybercrime (Paraskevas, 2022) are examples of new forms of threats that may impact the tourism sector in the future.

This chapter discusses a range of issues related to crisis and disaster events with a specific focus on climate change and transition to net zero GHG emission. This approach was adopted because unlike many of the disasters of the past, climate change and transition to net zero are known crisis events, the scientific literature is able to provide extensive support for planning and preparation for these events, the types of impacts we may expect are known and the time scales of events such as sea level rise are known with some degree of certainty. This knowledge, if responded to in the near future, will enable the tourism industry to respond to this threat in a proactive rather than reactive manner, and thus minimize potential impacts on destinations and tourism flows. A proactive response will also require some knowledge of the key scientific terms and concepts used in climate change mitigation and adaptation.

Defining crisis and disaster events

The terms *crisis* and *disaster* are often used interchangeably with both terms sometimes used to describe the same event. In an attempt to provide definitional clarity, Faulker and Russel (2001) and Ritchie (2004; 2009) described disasters as events that are external to an organization and which the organization has no control over. Examples include wars, pandemics, weather events and natural disasters. Crises on the other hand are described by Faulkner and Russell (2001) and Ritchie (2004) as events that are the result of an internal organizational failure, such as a failure to adapt to change. The United Nations International Strategy for Disaster Reduction (UNISDR) guide on disaster terminology provides an inclusive definition that describes a disaster as "*a serious disruption of the functioning of a community or a society involving widespread human, material, economic or environmental losses and impacts, which exceeds the ability of the affected community or society to cope using its own resources*" (UNISDR, 2009: 9). Disaster risk management is described as "*the systematic process of using administrative directives, organizations, and operational skills and capacities to implement strategies, policies and improved coping capacities in order to lessen the adverse impacts of hazards and the possibility of disaster*" (UNISDR, 2009: 10-11).

In everyday use however, the terms crisis and disaster are often used interchangeably. For example, the turbulence that occurred in global financial markets in 2008-09 is generally referred to as the 2008-09 Global Financial Crisis even though the event was external to organizations supplying tourism services and under the UN definition can be described as a disaster. Irrespective of the term used, the essential core of a crisis and disaster event is *"they have specific cause (a war for example); are generally unexpected (and therefore difficult to prepare for in advance); create uncertainty (when will climate change culminate?); disrupt the achievement of specific goals (tourism arrival forecasts for example); generate instability (economic recovery from a natural event) and; require restoration"* (Prideaux & Bierman, 2024: 7).

During the COVID-19 pandemic, a number of scholars recognized that dealing with crises not only required drawing on tourism theory and models but also needed at least a rudimentary understanding of epidemiology as a pathway for understanding the medical aspects of a pandemic and how these may affect tourism flows (Mertzanis & Papastathopoulos, 2021). The need to draw on knowledge that resides in other disciplines to deal with tourism crises is clearly demonstrated with climate change. Gaining even basic knowledge about key scientific terms, such as tipping points and feedback loops, will help researchers and tourism industry stakeholders gain a deeper understanding of what has happened, what is happening, what is likely to happen in the future and what the likely outcomes may be at various time scales. With this understanding, tourism planners and tourism academics will be better equipped to respond to climate change before, rather than after, a specific event. Similar comments can be made about other types of crisis and disaster events that may have an impact on the tourism sector. Examples include extreme weather events, earthquakes, disruption to ecosystem stability and the implications of the growing adoption of artificial intelligence (AI).

In comparison to the COVID-19 pandemic, the climate change crisis that the global community is now beginning to experience will disrupt tourism over a much longer period of time and degrade or destroy a significant amount of the physical and environmental infrastructure that currently supports the global tourism sector (IPCC, 2023). Given the likely long running nature of the climate crisis, the global tourism industry will need to respond to diverse anticipated and unanticipated changes over a time scale that far surpasses previous major crisis including the two world wars of the 20th Century and the seemingly never-ending series of regional and global financial crises.

Changes in the availability of tourism resources, the need to relocate major infrastructure, particularly in low lying coastal destinations, and the manner in which tourists respond to these changes will present challenges to the tourism industry, downstream supply chains, consumers and the public sector. The academic community will also face challenges with the need to modify, discard or develop new models and theories to meet this challenge. Previous chapters have discussed a range of academic responses to the challenges posed by the growth in tourism demand and how these responses have provided a toolbox of options and solutions. The challenge is to continually update this knowledge to provide answers and explanations to deal with future changes posed by climate change, policies designed to deal with climate change, and a myriad of other future natural and human crises and disasters.

The key issues that must be considered in crisis and disaster preparation, response, recovery and restoration include:

- Knowledge of threats which may be known or unknown
- Geographic scale, ranging from local to global
- Time scales which may be measured in minutes for an earthquake, decades for climate change or generations for social impacts
- Impacts which include scale of damage, length of recovery time, financial costs, environmental costs and human health costs
- Planning that recognises that essential differences between adaptation and mitigation and the need to consider timescales and impacts in terms of winners and losers and industry and community resilience
- Preparations for dealing with potential crisis and disaster events commencing in the pre-event stage followed by the event itself and post-event recovery and restoration stages
- Impact on consumers' future travel intentions
- Resilience of human, physical and environmental resources
- Research undertaken by multiple social science and scientific disciplines, UN agencies, NGOs and government funded research organisations.

Collectively, these terms provide a scaffold around which crisis and disaster preparation, response, recovery and restoration strategies can be formulated, tested and when necessary, implemented and identifies areas where many of the theories and models previously discussed in this book can be applied.

Models and theories related to crisis and disaster events

While global scale events such as the COVID-19 pandemic illustrate the need for ongoing research, the crisis and disaster literature is relatively sparce, with a dedicated journal in the tourism field yet to emerge. A review of the academic study of crisis and disaster research indicates a rapid increase in academic papers during and immediately after a major event, followed by the rapid decline in interest until the next event. Most of the previous academic research in crisis and disaster events was published after the event with the exception of the COVID-19 pandemic where its long running nature facilitated an enormous amount of research published during and after the event. Papers that focus on strategies to deal with the event (Kuščer et al., 2022) and recovery (Jarumaneerat et al., 2023) are particularly important and provide insights that can be used in developing strategies to deal with future crisis events.

In a review of COVID-19 related research Yang, Zhang and Rickly (2021) identified 249 academic works published during the first twelve months of the pandemic. However, many gaps in understanding remain as a number of review papers have highlighted (Duan et al. 2022; Zhang et al. 2023). To some extent the lack of academic papers has been addressed by books including Glaesser (2004), Laws and Prideaux (2005), Laws, Prideaux and Chon (2007), Henderson (2007), Ritchie (2009), Visser & Ferreira, (2013), Ritchie & Campiranon, (2014), Neef & Grayman (2018), Hosie & Pforr (2016), Beirman (2020), Lewis-Cameron, Jordan & Roberts (2021), Wilks, Pendergast, Leggat & Morgan (2021), Beirman (2022) and Kennell, Mohanty, Sharma & Hassan (2022) and Prideaux and Bierman (2024).

There also remains limited crisis and disaster theory and model development (Broshi-Chen & Mansfeld, 2021). Concern about the lack of conceptual and theoretical foundations was expressed by Ritchie and Jiang (2019) and echoed in later review papers by Duan et al. (2022) and Zhang et al. (2023). The most well known models contributing to crisis and disaster management include Faulkner's (2001) six stage disaster response framework and later variations suggested by Ritchie (2004), McKercher's (1999) approach based on chaos theory and Speakman and Sharpley's (2012) life-cycle-based framework. More recently, Broshi-Chen and Mansfeld (2021) suggested a new approach they termed the Quadratic Creativity & Innovation Model (QC&IM) that can be employed for mapping, development and implementation of creativity and innovation-oriented crisis management and mitigation solutions.

Faulkner (2001), adding to previous research into disaster responses (Roberts, 1994; Fink, 2000) suggested a six-stage tourism disaster lifecycle type model commencing at the pre-event stage, moving though to a pre-modal stage (a disaster is imminent), the emergency stage (the event), an intermediate stage (immediate response to the event), long-term recovery stage and resolution. Ritchie (2004) suggested collapsing the six stages of Faulkner's framework into a three-stage model referred to as the Strategic Management Framework. Stages 1 and elements of stage 2 of Faulkner's framework were incorporated into a crisis/disaster prevention and planning stage, elements of stage 2 as well as stages 3, 4 and 5 of framework were collapsed into a strategic implementation stage and stage 6 revised to form a resolution, evaluation and feedback stage. The key difference to earlier frameworks is its flexibility that enables ongoing and rapid evaluation, reevaluation and modification during an evolving crisis or disaster situation.

While the framework approach suggested by Faulkner (2001) has been critiqued extensively, extended and revised, the core elements remain robust and continue to provide insights into crisis and disaster management at each stage. In practice, the staged approach illustrated in Faulkner (2001) and Ritchie (2004) as well as Roberts (1994) and Fink (2000) reflect the approaches typically employed by disaster response organizations. Jiang and Ritchie (2024) provide a useful analysis of collaborative planning and response at destination level in their case study of Cyclone Marcia in Queensland, Australia in 2015.

While frameworks are useful for pre-event planning and managing crisis and disaster events, the nature of the events that the frameworks are applied are generally limited in terms of time and scale. The COVID-19 pandemic demonstrated the need for understanding longer term impacts at a global scale. As Prideaux and Bierman (2024) observed, climate change is a crisis event that is likely to be long running, perhaps stretching into the next century and beyond, and of a magnitude that is likely to remain unknow until greenhouse gas (GHG) emissions reach net zero and global temperatures stabilize. This suggests that within the timeframe of a long running crisis and disaster event, there will be numerous small scale discrete events that may occur simultaneously, intermittently or concurrently with each contributing to the overall event and each requiring a specific response.

In a climate change context, mitigation strategies to deal with the known aspects of the problem as well as unknowns will create numerous difficulties for tourism industry stakeholders. Some of them include long-term planning, return on investment and impact on demand. Collectively, the unknown nature of many of the issues will need to be dealt with by destinations, tourism firms, the public sector and supply chains denote a so-called wicked problem described as a problem that is difficult to solve because its parameters are contradictory, often incomplete and where no single solution is possible. In coastal destinations, for example, uncertainty about the eventual level and timing of sea rise is beginning to create planning problems affecting zoning and determination of whether it is more cost effective to defend tourism infrastructure or relocate it. Similar issues arise in responding to other potential crisis and disaster events that may arise from new technology such as AI.

Forecasting

Given the impact that crises and disasters can have on tourism, both academic and non-academic commentators have devoted considerable resources to try to understand how the future may unfold. Forecasting is at best a blunt but necessary tool for planning and policy development but rarely accurate (Prideaux et al., 2003). Yet forecasting is widely employed in crisis and disaster research. Where forecasting is based on the impacts of known previous events, forecasting can provide valuable insights for preparation, response and recovery particularly in planning to deal with natural events such as windstorms (cyclones, hurricanes and typhoons), earthquakes, floods and heatwaves and other events such as pandemics and terrorist attacks. Forecasting has less utility in dealing with events that have not previously occurred such as concerns about AI.

Climate change scenarios produced by the Intergovernmental Panel on Climate Change (IPCC) are based on known and relatively predictable relationships but even here, a high level of uncertainly as to actual outcomes exists. The impacts that decisions made by human actors will have on eventual outcomes and how the breaching of tipping points will cascade through to other feedback loops are often unpredictable and may create new and unexpected climate impact pathways. Despite these problems, the IPCC scenarios provide a useful guide for actors within the tourism industry in developing responses to climate change and provide a form of forecasting that is useful for long term tourism planning.

Climate change

The tourism industry is in the unique and unenviable position of being simultaneously a major contributor of greenhouse gases and a significant victim of a changing climate. Historically, research on the relationship between tourism and climate lagged significantly behind that of other sectors (Pang et al., 2012) and to some extent remains on the periphery of mainstream tourism research when measured by the volume of published research, the number of active tourism researchers involved in climate change research and the number of climate change subjects taught in tertiary level tourism programs. Much of the tourism related research has focused on social science aspects of climate change, without necessarily incorporating science-based models. The following discussion reviews a range of models and frameworks that are used in climate change in a tourism context.

The Sixth Assessment Report (AR6) of the United Nations Intergovernmental Panel on Climate Change published as a series of Working Group reports between 2001 and 2023 stated unequivocally that surface temperatures are continuing to rise and without rapid reduction of GHG emissions will soon exceed 1.5°C and continue to 2°C and beyond by the end of the century. The 1.5°C threshold is significant with temperatures above that level leading to significant and in some instances irreversible damage to many ecosystems as well as sea level rise and extreme weather events.

Despite actions agreed to at the 2015 Paris Agreement to wind back GHG emissions to limit global temperature to below the 1.5°C threshold, global temperatures rose to 1.48°C above the 1850-1900 pre-industrial level in 2023 with several months recording average temperatures well above 1.5°C (Copernicus Climate Change Service, 2023). Indeed, 2023 saw mean global temperature exceed 1.5°C for the first time.

Climate change will have far greater cumulative impacts over a far longer period of time. As the Earth warms, many of the resources that tourism relies on will be irreversibly damaged or even lost. The IPCC (2023) stated that rising global temperatures will lead to an increasing incidence of heatwaves, an increase in climate refugees fleeing rising sea levels, prolonged drought and intolerable temperatures, changes in perception, more extreme weather, wildfires, sea level rise and ecosystem change. Based on data published by the IPCC (2023), Boehm and Schumer (2023) identified the key factors that demonstrate that climate change is taking place include:

- The atmosphere has the highest level of CO_2 for 2 million years
- The decade 2010-2020 was warmest decade than any period ~ 125,000 years
- Sea level rise has been faster than in any prior century for 3,000 years
- Ocean warming is occurring at a faster rate than any period since the last ice age
- Ocean acidification has reached the highest level for last for 26,000 years.

The ultimate level of impact will be determined by the success of mitigation and adaptation strategies. Compared to the past where forecasts of the future were often guesstimates based on limited knowledge, we now have a reasonably accurate understanding of how quickly global temperatures are rising, why they are rising, how to halt temperature rise and the likely impacts on human settlements and the environment as any given level of temperature increase. Information of this type is critical for assessing likely impacts on existing tourism resources, tourism product and experience development, government policy responses, tourism demand, adaptation and potential impacts in the future.

The methodology used in compiling IPCC climate change reports is based on a synthesis of peer reviewed publications as well as appropriately documented industry publications by a panel of acknowledged academic experts. The first working group report published as part of AR6 *Climate Change 2021: The Physical Science Basis* (IPCC, 2021) involved a panel of 234 scientists from 66 countries who reviewed over 14,000 scientific papers to write a report that after two rounds of review, was approved by the 195 governments that are part of the IPCC process.

Scenarios that illustrate possible impacts on global ecosystems and human systems of various combinations of GHG emission and mitigation strategies are key features of IPCC assessment reports that should be of interest to the tourism industry. Knowledge of this type provides opportunities to see into the future, identify strategies to minimize disruption, undertake timely adaptation and adopt informed decision making. In AR6 (IPCC, 2023), five scenarios were outlined. The low scenario (SSP1 -1.9) is based on low GHG emissions falling to net zero around 2050 with global warming limited to about 1.6 ^{0}C. The middle scenario (SSP 1-2.6) assumes CO_2 emissions remaining at current levels until around 2050 and falling to net zero by 2100. Under this scenario global warming will reach about 2^0 C by mid-century and rising to 2.7^0 C by 2100. Under the high scenario (SSP5- 8.5) CO_2 emission triple by 2075 leading to global warming of 2.4^0 C mid-century rising to 4.4^0 C by 2100. Growth in emissions will affect all major climate system components leading to more extreme weather events such as heatwaves, droughts, and windstorms (cyclones, typhoons, hurricanes). Impacts will include rising seas levels, flooding in coastal and low-lying areas, loss of biodiversity, vector borne disease and the like. The scope and scale of these impacts will become more complex with every tenth of a degree of temperature rise. More than that, we have entered a period where the likelihood of abrupt and/or irreversible changes in the climate system can occur once a tipping point is reached, resulting in species extinction, irreversible loss of biodiversity, coral reef die-off, sea level rise and other impacts.

A number of social science-based models have been developed to demonstrate the links between science and social science research and show why each side of the research community needs to build bridges to facilitate a deeper understanding of the feedback loops generated by human impacts on the environment and subsequent environmental impacts on humans. One example is the five-stage climate change impact model developed by Prideaux, Coghlan and McNamara (2010) and modified by Ramis and Prideaux (2013) to investigate how coral bleaching may affect tourism demand. Using data derived from a survey of visitors to Australia's Great Barrier Reef and observations that at temperatures of 2^0 C and above coral bleaching is anticipated to become an annual event, Ramis and Prideaux (2013) found that tourists respond negatively to a decline in the quality of their reef experience. The modified model outlined in Figure 14.1 includes a feedback loop that recognizes the need for the scientific community to incorporate human impacts into their modelling as is the norm for IPCC scenario building. The factors that affect coral reef systems can also be applied in other terrestrial and marine environments to understand the impact of climate change.

The need for new approaches is highlighted by the realities of climate change and the imperative of moving to a net carbon zero world as quickly as possible. It is no longer debatable whether tourism needs to decarbonise (Gössling & Higham, 2021; Higham et al., 2021). This obligation was spelt out in the 2015 Paris Climate Agreement where leaders from 195 countries pledged to substantially reduce emissions of greenhouse gases and the more specific Glasgow Declaration on Climate Action in Tourism which supports policy actions to cut tourism's emissions in half by 2030 and reach net zero by 2050 (Scott & Gössling, 2021; UNWTO, 2021).

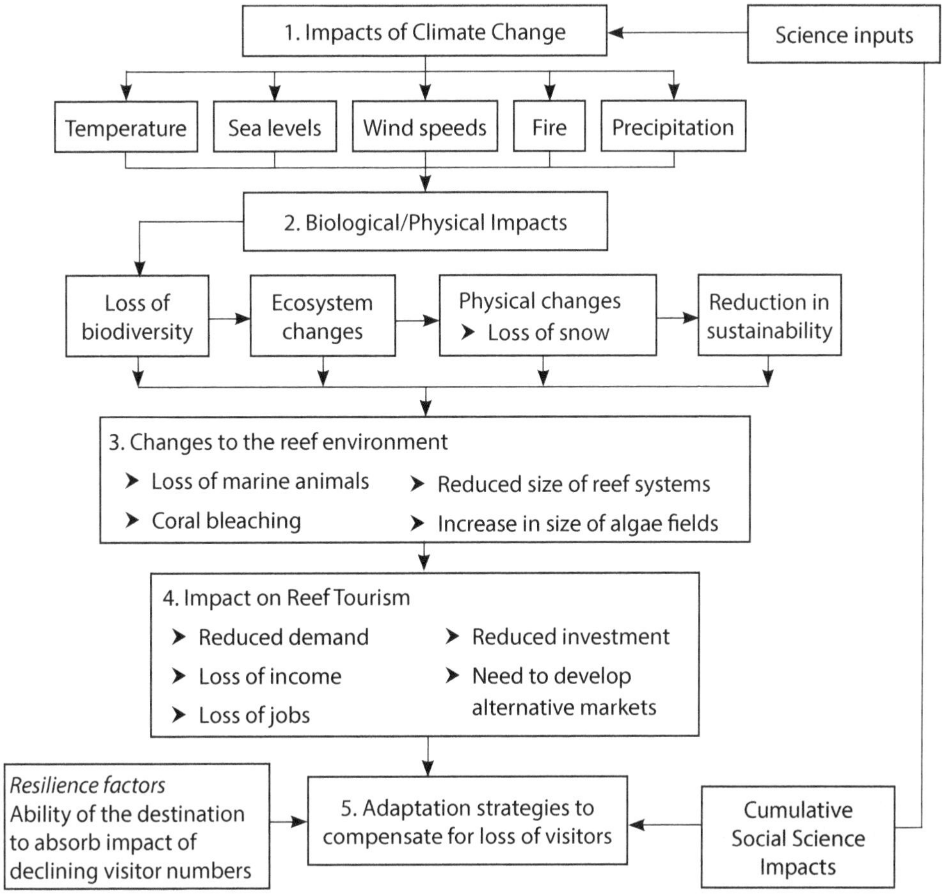

Figure 14.1: Climate change impact model applied to the Great Barrier Reef. Revised from Ramis and Prideaux (2013)

Other models developed to illustrate the impact of climate change on aspects of tourism include destination scale models such as the Beach Comfort Index (BCI) (Morgan et al., 2000), regional ecosystem models of the type produced by Williams, Bolitho and Fox (2003), national scale models such as the United Kingdom Climate Projections models (Robinson et al., 2023) and global scale scenario models produced by the IPCC. Indices of the type illustrated by the Tourism Climate Index (TCI) (Scott et al., 2016) also provide useful input into model development. The data provided by modeling is able to provide destinations, the private sector and government agencies with critical inputs that can be used for a range of planning needs. Planning needs include information on potential loss of infrastructure, impact on environmental resources, community and industry resilience, future tourist flows, the impact of mitigation strategies such as net zero, and the cost of adaptation.

Beyond the models and theories discussed above, modelling discussed in the scientific literature can also be used to inform tourism industry stakeholders about the future impacts of climate change. The IPCC literature refers to a number of scientific terms including tipping points and feedback loops that are used to describe the potential impacts of climate change. The impact of permafrost melting driven by increasing temperatures in the Arctic illustrates the role that tipping points and positive feedback loops can play in accelerating adverse climate change impacts. As permafrost melts, methane is released into the atmosphere creating a positive feedback loop. A positive feedback loop is a process that amplifies GHG emissions while negative feedback loops diminish or dampen GHG emissions. Ripple et al. (2021) identified 41 climate change feedback loops, of which only seven were classed as dampening, seven were uncertain and 27 were positive. Positive feedback loops can be reversed unless a tipping point is crossed. A tipping point refers to a critical threshold that when crossed can lead to accelerating and often irreversible change in the global climate system (Lenton et al., 2019). The interconnective nature of feedback loops can create a situation where crossing the tipping point of one positive feedback loop can lead to the breaching of tipping points in other positive feedback loops resulting in a domino like effect of a cascading chain of reactions that will only cease when the chain reaction concludes. The impact of cascading tipping points is illustrated by Wunderling et al., (2021) who postulated that rising sea water temperatures can lead to increased melting of the Antarctic and Greenland ice sheets which in turn leads to slowing of the Atlantic meridional overturning circulation (AMOC) and in turn drying and dieback of the Amazon rainforest.

Transition to net zero

Net zero emissions is a mitigation strategy that emerged out of the 2015 Paris Agreement. The aim of net zero is to wind back global GHG emissions to a state where total emissions of GHGs due to human activity are balanced by the removal of these gasses over a given period of time. Achieving net zero requires reduction in GHG emissions by switching from non-renewable to renewable energy production, using hydrogen as an alternative energy source, phasing out fossil fuel powered vehicles, ships and aircraft, developing green production, cessation of deforestation, industrial scale recycling, new methods for steel and concrete production and so on. Carbon dioxide can be removed from the atmosphere by new farming methods, banning deforestation, revegetation of unused land, tree planting and CO_2 removal by yet to be proven carbon capture and storage technologies (Mintz-Woo & Lane, 2021). By the end of 2023, 145 countries had either adopted or were planning to adopt net zero targets. Collectively, these countries account for nearly 90% of global emissions (Climatetrackerorganisation, 2023). An understanding of the magnitude of the task can be gained by using simulators such as the free online EN-ROADS climate solutions simulator (EN-ROADS, n.d.) that allows users to explore how changes in sectors such as transport and energy can increase or reduce global temperatures.

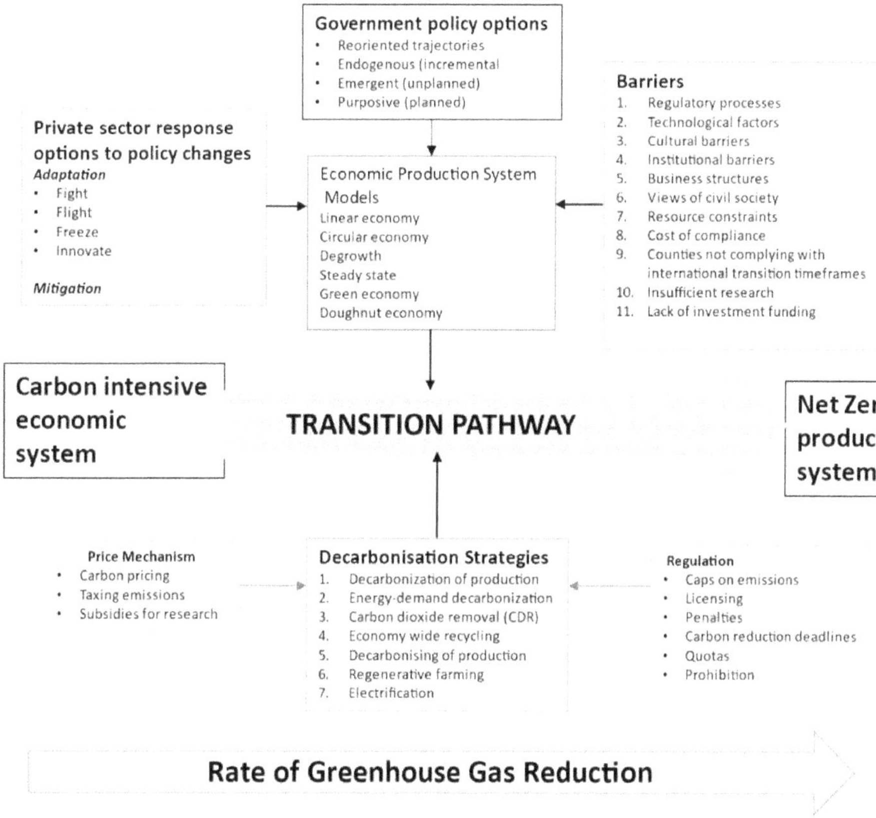

Figure 14.2: Factors that influence transition to net zero GHG emissions. Adapted from Prideaux (2022)

Achieving net zero will require a massive shift in how energy is produced and consumed and entail significant government intervention to establish mechanisms for achieving targets. These changes will undoubtedly flow on to most if not all areas of the tourism industry. Four types of transition strategies that may be used by the public sector were identified by Seyfi and Hall (2020) and are described as endogenous (incremental), reorientation of trajectories, purposive (planned) and emergent transformation (unplanned). Governments and businesses often prefer endogenous rather than purposive pathways because of the lower cost of asset conversion and staff retraining and because they are less likely to create external and internal opposition. However, rapid increase in adverse climate impacts may force the adoption of the purposive approach rather than relying on endogenous or emergent approaches. Figure 14.2, adapted from Prideaux (2022), illustrates how

14

some of these relationships can be incorporated into a descriptive model where the price mechanism and regulations can be used to support a range of decarbonization strategies and how the barriers to decarbonization may impact on the transition strategies that the public sector may adopt. The model also highlights potential industry responses to new policy settings from fighting changes to adopting innovation. The model, in an appropriately modified form, can be used to identify how various sectors may respond to policies aimed at crisis and disaster preparation other disaster and crisis events.

Implications for tourism

Scott, Hall and Gössling (2012) noted that initial interest in the impact of climate on tourism activity can be traced back to the 1960s followed by a surge of interest in the mid- to late 1970s. There was a further resurgence of interest in the late 1980s, with a noticeable increase in recent years. Early research recognised that impacts would be uneven, potentially benefiting some destinations while harming others. This 'Victim, Winner, Loser' theme has continued to dominate much of the research agenda. For example, under the victim theme low-lying islands may become inundated (IPCC, 2023; Storlazzi et al., 2015), the viability of much of the North American tourism sector may be compromised (IPCC, 2023), coral reef dependent destinations will also come under pressure as coral bleaching reduces their appeal (IPCC, 2023); golfing destinations that rely on plentiful water supplies may be affected by a decline in rainfall (UNWTO-UNEP, 2008); the alpine and cross country ski industry is particularly at risk (Pröbstl-Haider et al., 2021, Steiger et al., 2022); and the list goes on. However, there may be a number of winners. Rebetez (2011) found a correlation between the number of nights spent in mountain resorts in the summer and hot temperatures at lower elevations. The loss of ski seasons may offset by longer summer seasons enabling more hiking and mountain biking (Pröbstl-Haider et al., 2021), seasonality may shift and extend the seasons into traditional shoulder seasons (Amelung et al., 2007) and other benefits. Overall, however, the number of winners will be much smaller than the number of losers.

Tourism needs to reduce its emissions to net zero in line with other economic sectors if global warming is to be kept to 2.0° C or less. Achieving the goal of net zero is a challenge under current growth scenarios (Gössling et al., 2023) as tourism emissions could triple rather than reduce (Gössling & Scott, 2018). Conefrey and Hanrahan (2022) concur, noting that not only has the carbon footprint of tourism not been assessed definitively, it lacks serious discussion and planning, due in large part because baseline data for tourism emissions are not available. Indeed, Becken (2019) raised the specter that the challenge was 'mission impossible'. The issue is even more challenging given that the aviation sector has been excluded from the

Paris Agreement (Higham et al., 2021), even though it is the single greatest source of tourism-related carbon emissions. Air travel is simply not compatible with low carbon tourism (Scott & Gössling, 2022), especially given the popularity of low-cost airlines (Gössling & Higham, 2021). Yet people are unwilling to substitute its convenience and low cost for more time-consuming rail or drive tourism.

Many types of tourism will be victims of climate change, while some other forms may benefit. Low altitude ski fields, such as this one shown in the summer, as well as those in lower latitudes are likely to see less snowfall and shorter ski seasons, placing many ski resorts and the ancillary goods and services at risk, ranging from equipment rental through to accommodation and food and beverage outlets. This scenario may extend well into the traditional ski season.

A number of other structural issues need to be addressed, while many of the hoped for changes will need to come from outside the tourism sector. As an example, Peeters and Papp (2023) cite the electric car revolution as a key step in decarbonising drive tourism, even though it is not a tourism initiative. But even here there is substantial reluctance to embrace electric vehicles out of range anxiety concerns, the practical issues of finding charging stations and the time involved in recharging vehicles (Fitt, 2022).

As the global economy adjusts to the policies designed to combat climate change, a number of issues will need to be addressed. We are still lacking in some basic tools, such as clear and common definitions, agreed systems boundaries, compatible data collection about tourism and transport and the availability of fine enough, tourism-specific information. The issue of metrics and measurements is highlighted by Scott and Gössling (2021) who examined the International Energy Agency's road map to net zero. The Agency has set out 400 (yes 400) specific emission reduction strategies across different emitting sectors. Some of these recommendations sound good in principle, but are unlikely to be adopted in practice. For example, the road map calls for a freeze on international long-haul flights at 2019 levels and a modal shift of shorter flights to high speed rail. Likewise, they call for hotels to make a rapid transition to zero-carbon-ready buildings by converting to renewable energy or sourcing their energy from fully decarbonized suppliers.

Jones (2023), highlighted four other issues facing the tourism sector:

- Circumstances where the concept of net zero is becoming a synonym for inaction as companies adopt 'targets' rather than action
- The preponderance of businesses to look to buy in natural capital solutions from others, such as carbon-offsetting and tree planting, rather than reducing their own emissions
- An over-reliance on waiting for technology to provide the solution
- The challenge of the growth paradox of how to reduce emissions when the entire economic system is predicated on growth.

The challenges of future crises and disasters

As history has demonstrated, the tourism industry will continue to face an enormous range of future crisis and disaster events, some of which can be forecast and others that will come as a complete surprise. While the scale of the COVID-19 pandemic was a complete surprise to many actors in 2020, there is now an expectation that there may be future global scale pandemics, but the timing, scale and location remain matters for speculation. In comparison, the potential impacts of the climate change crisis are now reasonably well known and based on IPCC modelling, there is a growing understanding of the policies that will need to be implemented to deal with this crisis. Given this knowledge, the global tourism industry is much better placed to use the models, theories and tools, some of which have been discussed in this chapter, to deal with the problems that will arise in future decades.

Aside from climate change and mitigations strategies, such as net zero, to deal with the problem of rising GHG emissions, the tourism industry will need to respond to an enormous range of other crisis and disaster events in coming decades. Models such as those suggested by Faulkner (2001) and Ritchie (2004) continue to provide valuable guidance to actors within the tourism industry who must respond to these events. Newer models such as the Quadratic Creativity & Innovation Model (Broshi-Chen & Mansfeld, 2021) will also assist. Ultimately however, the responsibility of actors within the tourism industry to deal with crisis and disaster events will initially fall on individual firms and organisations. This indicates the need for preparation as outlined in Faulkner's (2001) six stage and should include training, adequate insurance cover, design of infrastructure that can withstand anticipated local threats and so on. Depending on the scale of the event, external assistance may also be given by the public sector. As part of the process of planning for, responding to, and recovery from crisis and disaster events there is considerable scope to incorporate other models discussed in the book.

Beyond the current crisis events that have defined the past two decades of the twenty first century problems are the even larger issues of how does humanity ensure its long-term survival, measured not in decades or even centuries but in millennia? From this perspective should we view the climate crisis as part of the even larger centuries long crisis of sustainability? What implications does an approach to crisis on this almost unimaginable scale have for the tourism industry in the near, medium and long term?

Conclusion

The global tourism industry has a demonstrated ability to recover from major crises and disaster events. In the first decades of the twenty first century the tourism industry recovered from a global pandemic, a global financial crisis, numerous wars, natural disasters and political crises. Unfortunately, research has to date tended to focus on how these events were responded to and the success or otherwise of recovery, with relatively scarce theoretical advances. From a literature perspective this may reflect the ongoing lack of a focal point, a specialist journal for example, where these issues can be debated.

It is also apparent that the global community, and the tourism industry, faces two other immense challenges. The most immediate is climate change which will have a significant impact on the availability of many important tourism resources, including coastal areas subject to sea level rise and degraded ecosystems, over a yet undetermined time period. The second is the long-term sustainability of the Earth as its current stock of non-renewable and renewable resources are consumed to meet the needs of the present with little regard for the future. Both challenges will create enormous problems for all tourism stakeholders.

■ Discussion points / Exercises

1. Assess how Faulkner's crisis impact model could be employed to prepare for a natural disaster such as a flood in the area you live in.
2. Give examples of how the tourism industry can prepare for the loss of coastal areas due to sea level rise.
3. Consider how the climate change impact model could be applied in a climate change threatened ecosystem in the area where you live.
4. Will new technology, including AI and the metaverse, create new crisis events in the future?

Points to consider for revision

- Although crisis and disaster events can have an enormous impact on the global tourism system, research in this area is limited and often tied to specific crisis events.
- Preparation is a key element in planning to deal with future crisis and disaster events but is often ignored.
- Many of the crisis and disaster events that have affected the tourism industry were anticipated but the timing and extent of impacts are often unknown.
- Identifying the potential impacts of climate change and planning to minimize impacts must be considered a high priority for all destinations.
- Understanding the potential impacts of the climate change crisis requires at least a basis understanding of key scientific terms such as feedback loops and tipping points.

Suggested additional readings

Prideaux, B. & Bierman, D. (eds) (2024) *Handbook on Crisis and Disaster Management in Tourism*, Edward Elgar.

Jianf, Y. & Ritchie, B. (2024), Collaborative structures and actions in tourism disaster management: The case of cyclone Marcia in Central Queensland, Australia, in Hall, M and Prayang, G. (eds) *Tourism, Cyclones, Hurricanes and Flooding*, Channel View.

Ritchie, B. W. (2004). Chaos, crises and disasters: A strategic approach to crisis management in the tourism industry. *Tourism Management*, 25, 669-683

References

Becken, S. (2019). Decarbonising tourism: mission impossible? *Tourism Recreation Research*, *44*(4), 419-433.

Beirman, D. (2020). *Restoring Tourism Destinations in Crisis: A strategic marketing approach*. Routledge.

Beirman, D. (2022). *Tourism Crises and Destination Recovery*, Sage.

Boehm. S. && Schumer, C. (2023), *10 Big Findings from the 2023 IPCC Report on Climate Change*, World Resources Institute, https://www.wri.org/insights/2023-ipcc-ar6-synthesis-report-climate-change-findings

Broshi-Chen, O., & Mansfeld, Y. (2021). A wasted invitation to innovate? Creativity and innovation in tourism crisis management: A QC&IM approach. *Journal of Hospitality and Tourism Management*, 46, 272-283.

Climatetrackerorganisation, (2023) CAT Net Zero Evaluations, https://climateactiontracker.org/global/cat-net-zero-target-evaluations/

Conefrey, A. & Hanrahan, J. (2022). Comparative analysis of national tourism decarbonisation plans. *European Journal of Tourism Research*, 31, 3105-3105.

Copernicus Climate Change Service (2023), *Copernicus: 2023 is the hottest year on record, with global temperatures close to the 1.5°C limit*, https://climate.copernicus.eu/copernicus-2023-hottest-year-record

Duan, J., Xie, C., & Morrison, A. M. (2022). Tourism crises and impacts on destinations: A systematic review of the tourism and hospitality literature. *Journal of Hospitality & Tourism Research*, 46(4), 667-695. doi.org/10.1177/1096348021994194

EN-ROADS, (n.d.), The En-ROADS Climate Solutions Simulator, EN-ROADS. https://www.climateinteractive.org/en-roads/

Faulkner B. (2001). Towards a framework for tourism disaster management. *Tourism Management*, 22(2), 135-147. doi.org/10.1016/S0261-5177(00)00048-0

Faulkner, B. and Russell, R. (2001) Turbulence, chaos and complexity in tourism systems: A research direction for the new millennium, in Faulkner, B., Moscardo, G. and Laws, E. (eds), *Tourism in the Twenty-First Century: Reflections on experience*, pp. 328-349, Cabi.

Fink, S. (2000) *Crisis Management: Planning for the Inevitable*, iUniverse.

Fitt, H. (2022). Boring and inadequate? A literature review considering the use of electric vehicles in drive tourism. *Current Issues in Tourism*, 25(12), 1920-1946.

Glaesser, D. (2004). *Crisis Management in the Tourism Industry*. Routledge.

Gössling, S., & Scott, D. (2018). The decarbonisation impasse: global tourism leaders' views on climate change mitigation. *Journal of Sustainable Tourism*, 26(12), 2071-2086.

Gössling, S., & Higham, J. (2021). The low-carbon imperative: Destination management under urgent climate change. *Journal of Travel Research*, 60(6), 1167-1179.

Gössling, S., Balas, M., Mayer, M. & Sun, Y. (2023). A review of tourism and climate change mitigation: the scales, scopes, stakeholders and strategies of carbon management. *Tourism Management*, 95, 104681.

Gurtner, Y. (2007) Phuket: tsunami and tourism - a preliminary investigation. In Laws, E., Prideaux, B. & Chon, K., (eds) *Crisis Management in Tourism.* Cabi Oxon. pp.217-233.

Henderson, J. (2007). *Tourism Crises: Causes, Consequence and Management*. Elsevier.

Higham, J., Font, X. & Wu, J., (2021). Code red for sustainable tourism. *Journal of Sustainable Tourism*, 30(1), 1-13.

Hosie, P., & Pforr, C. (2016). *Crisis Management in the Tourism Industry: Beating the odds?* Routledge.

IPCC (2021), *Climate Change 2021: The Physical Science Basis. Contribution of Working Group I to the Sixth Assessment Report of the Intergovernmental Panel on Climate Change.* Cambridge University Press, Cambridge, UK, doi:10.1017/9781009157896.

IPCC (2023), *Climate Change 2023: Synthesis Report. Contribution of Working Groups I, II and III to the Sixth Assessment Report of the Intergovernmental Panel on Climate Change.* IPCC, Geneva, Switzerland, pp. 35-115, doi: 10.59327/IPCC/AR6-9789291691647

Jabeen, F., Al Zaidi, S., & Al Dhaheri, M. H. (2022). Automation and artificial intelligence in hospitality and tourism. *Tourism Review*, 77(4), 1043-1061.

Jarumaneerat, T., Rabiul, M. K., Prideaux, B., & Promsivapallop, P. (2023). Pandemic travel fear, travel intention, and self-protection intention of Chinese residents: an application of protection motivation theory. *Asia Pacific Journal of Tourism Research*, 28(9), 1019-1030.

Jiang, Y. & Ritchie, B. (2024), Collaborative structures and actions in tourism disaster management: The case of cyclone Marcia in Central Queensland, Australia, in Hall, M. and Prayang, G. (eds) *Tourism, Cyclones, Hurricanes and Flooding*, Channel View.

Jones, P., (2023). Towards net-zero emissions in the global tourism industry. *Athens Journal of Tourism*, 10(2), 125-136.

Kuščer, K., Eichelberger, S., & Peters, M. (2022). Tourism organizations' responses to the COVID-19 pandemic: An investigation of the lockdown period. *Current Issues in Tourism*, 25(2), 247-260.

Kennell, J., Mohanty, P., Sharma, A., & Hassan, A. (Eds.). (2022). *Crisis Management, Destination Recovery and Sustainability: Tourism at a Crossroads.* Taylor & Francis.

Laws, E., & Prideaux, B. (2005). *Tourism Crises: Management responses and theoretical insight.* Psychology Press

Laws, E., Prideaux, B. & Chon, K., (eds) (2007) *Crisis Management in Tourism.* Cabi Oxon.

Lenton, T. M., Rockström, J., Gaffney, O., Rahmstorf, S., Richardson, K., Steffen, W., & Schellnhuber, H. J. (2019). Climate tipping points—too risky to bet against. *Nature*, 575(7784), 592-595.

Lewis-Cameron, A., Jordan, L. A., & Roberts, S. (2021). *Managing Crises in Tourism.* Springer International Publishing.

McKercher, B. (1999). A chaos approach to tourism. *Tourism Management*, 20(4), 425-434. doi.org/10.1016/S0261-5177(99)00008-4

Mertzanis, C., & Papastathopoulos, A. (2021). Epidemiological susceptibility risk and tourist flows around the world. *Annals of Tourism Research*, 86, 103095.

Mintz-Woo, K., & Lane, J. (2021). Why and where to fund carbon capture and storage. *Science and Engineering Ethics*, 27(6), 70.

Morgan, R., Gatell, E., Junyent, R., Micallef, A., Özhan, E., & Williams, A. T. (2000). An improved user-based beach climate index. *Journal of Coastal Conservation*, 6, 41-50.

Neef, A., & Grayman, J. H. (Eds.). (2018). *The Tourism-Disaster-Conflict Nexus.* Emerald Group Publishing.

Pang, S. F., McKercher, B., & Prideaux, B. (2013). Climate change and tourism: An overview. *Asia Pacific Journal of Tourism Research*, 18(1-2), 4-20.

Paraskevas, A. (2022). Cybersecurity in travel and tourism: a risk-based approach. In *Handbook of e-Tourism* (pp. 1605-1628). Cham: Springer International.

Peeters, P., & Papp, B. (2023). *Envisioning Tourism in 2030 and Beyond. The changing shape of tourism in a decarbonising world.* Travel Foundation. https://www.

thetravelfoundation.org.uk/wp-content/uploads/2023/02/Envision2030_SummaryFINAL.pdf

Prideaux, B & Bierman, D. (eds) (2024) *Handbook on Crisis and Disaster Management in Tourism,* Edward Elgar

Prideaux, B. (2022). Transition to a carbon-neutral economy. In *Encyclopedia of Tourism Management and Marketing.,* Edward Elgar Publishing. pp. 564-567.

Prideaux, B. & Bierman, D. (2024) Keys issues in crisis and disaster management and research in Prideaux, B. & Bierman, D. (eds), *Handbook on Crisis and Disaster Management in Tourism,* Edward Elgar Publishing.

Prideaux, B. & Bierman, D. (eds), *Handbook on Crisis and Disaster Management in Tourism,* Edward Elgar Publishing.

Prideaux, B., Coghlan, A. & McNamara, K.E. (2010) Assessing the impacts of climate change on mountain tourism destination using the climate change impact model, *Tourism Recreation Research,* 35, 187-200.

Prideaux, B., Laws, E., & Faulkner, B. (2003). Events in Indonesia: Exploring the limits to formal tourism trends forecasting methods in complex crisis situations. *Tourism Management,* 24(4), 475-487.

Pröbstl-Haider, U., Hödl, C., Ginner, K., & Borgwardt, F. (2021). Climate change: Impacts on outdoor activities in the summer and shoulder seasons. *Journal of Outdoor Recreation and Tourism,* 34, 100344.

Ramis, M. & Prideaux, B. (2013) The importance of visitor perceptions in estimating how climate change will affect future tourists flows on the Great Barrier Reef, in Reddy, M. & Wilkes, K. (eds), *Tourism, Climate Change and Sustainability,* Routledge, London, pp. 173-188.

Ripple, J., Wolf, C., Newsome, T.M., Gregg J.W., Lenton, T.M., Palomo, I., Eikelboom, J.A., Law, B.E., Huq, S., Duffy, P.B. & Rockström, J. (2021) World scientists' warning of a climate emergency *Bioscience,* 71(9), pp. 894-8. https://doi.org/10.1093/biosci/biab079

Ritchie B. W., Jiang Y. (2019). A review of research on tourism risk, crisis and disaster management. *Annals of Tourism Research,* 79(November), 102812. doi.org/10.1016/j.annals.2019.102812

Ritchie, B. W. (2004). Chaos, crises and disasters: A strategic approach to crisis management in the tourism industry. *Tourism Management,* 25, 669-683. doi.org/10.1016/j.tourman.2003.09.004

Ritchie, B. W., & Campiranon, K. (Eds.). (2014). *Tourism Crisis and Disaster Management in the Asia-Pacific* (Vol. 1). CABI.

Ritchie, B.W., (2009). *Crisis and Disaster Management for Tourism.* Channel View Publications. Bristol. UK

Roberts, V. (1994), Flood management: Bradford paper, *Disaster Prevention and Management,* 3(2), 44-60.

Robinson, E. L., Huntingford, C., Semeena, V. S., & Bullock, J. M. (2023). CHESS-SCAPE: high-resolution future projections of multiple climate scenarios for the United Kingdom derived from downscaled UK Climate Projections 2018 regional climate model output. *Earth System Science Data,* 15(12), 5371-5401.

Scott, D., Hall, C. M., & Gössling, (2012). *Tourism and Climate Change: Impacts, adaptation and mitigation*. Routledge.

Scott, D. & Gössling, S. (2021). Destination net-zero: what does the international energy agency roadmap mean for tourism? *Journal of Sustainable Tourism*, 30(1), 14-31.

Scott, D., & Gössling, S., (2022). A review of research into tourism and climate change. *Annals of Tourism Research*, 95, 103409.

Scott, D., Hall, C. M., & Gössling, S. (2016). A report on the Paris Climate Change Agreement and its implications for tourism: Why we will always have Paris. *Journal of Sustainable Tourism*, 24(7), 933-948

Seyfi, S. & Hall, CM. (2020) Political transitions and transition events in a tourism destination, *International Journal of Tourism Research*, 22:493–506. https://doi.org/10.1002/jtr.2351

Speakman, M., & Sharpley, R. (2012). A chaos theory perspective on destination crisis management: Evidence from Mexico. *Journal of Destination Marketing & Management*, *1*(1-2), 67-77. https://doi.org/10.1016/j.jdmm.2012.05.003

Storlazzi, C. D., Elias, E. P., & Berkowitz, P. (2015). Many atolls may be uninhabitable within decades due to climate change. *Scientific Reports*, *5*(1), 14546.

United Nations International Strategy for Disaster Reduction (UNISDR) (2009) *2009 UNISDR United Terminology on Disaster Risk Reduction*. https://www.unisdr.org/files/7817_UNISDRTerminologyEnglish.pdf

UNWTO. (2021). *The Glasgow Declaration on Climate Action in Tourism*. United Nations World Tourism Organization. www.unwto.org/the-glasgow-declaration-on-climate-action-in-tourism.

UNWTO-UNEP-UNWTO (2008) Climate Change and Tourism: Responding to Global Challenges, (authors Scott, D., Amelung, B. Becken, S. Ceron, J-P., Dubois, Gossling, S., Peeters, P. & Simpson, M.) Madrid: United Nations World Tourism Organisation (UNWTO), United Nations Environment Program (UNEP) and World Meteorological Organisation (WMO).

Visser, G., & Ferreira, S. (Eds.). (2013). *Tourism and Crisis*. Routledge.

Wilks, J., Pendergast, D., Leggat. P., & Morgan, D. (2021). *Tourism Safety and Wellbeing in the New Normal*. Springer.

Williams, S. E., Bolitho, E. E., & Fox, S. (2003). Climate change in Australian tropical rainforests: an impending environmental catastrophe. *Proceedings of the Royal Society of London. Series B: Biological Sciences*, 270(1527), 1887-1892.

Wunderling, N., Donges, J. F., Kurths, J., & Winkelmann, R. (2021). Interacting tipping elements increase risk of climate domino effects under global warming. *Earth System Dynamics*, 12(2), 601-619. https://doi.org/10.5194/esd-12-601-2021

Yang, Y., Zhang, C. X., & Rickly, J. M. (2021). A review of early COVID-19 research in tourism. *Annals of Tourism Research*, 91, 103313.

Zhang, Y., Moyle, B., Dupré, K., Lohmann, G., Desha, C., & MacKenzie, I. (2023). Tourism and natural disaster management: a systematic narrative review. *Tourism Review*. DOI 10.1108/TR-08-2022-0377.

15 Tourism Planning

By the end of this chapter, the reader will be able to:

- Describe the difference between planning and policy
- Identify and describe the scales of planning
- Explain the need to consider new approaches to planning in the future

Introduction

The last few decades of the 20th Century, saw a shift from modernist forms of public administration where government bureaucracies were given licence to 'define and act for the public good' (Dredge & Jamal, 2015: 287) to the current globalist/neoliberal perspective of governance, where governments have downsized bureaucracies and outsourced many functions. The outcome has seen many governments move into an arbiter/enabler role, shifting from public administration to public management. This change significantly influenced thinking about the role that governments play in planning. The consequence has been that while there is apparent public sector support for development that adheres to the principles of sustainability, the reality is that neoliberal style policy continues to place growth rather than sustainability at the heart of policy formulation. This approach is no longer appropriate in a global economy facing numerous changes generated by climate change. Addressing these challenges, and policies to combat climate change, will require a more interventionist approach to planning in the future.

Previous chapters have observed how contemporary systems of economic production and consumption have created a situation where global society faces a future that will be significantly different from the present. Characteristics of this 'future' will include rising sea levels, degraded ecosystems and higher temperatures. Future planning challenges will include responding to the impact of climate change, the depletion of non-renewal resources, ecosystem decline and the need to transition into a new carbon-neutral economy. From a planning context these changes will require rethinking of how tourism planning is undertaken and the objectives that future planning may wish to achieve compared to the current objectives of planning which are often based on a desire for continuing growth but with a cursory nod to a need for development that is 'sustainable'.

This chapter briefly reviews the role of planning, the relationship between planning and policy, and issues that planning will need to address in the future. Underlying this discussion is the view that a lack of planning, particularly at the strategic level, but also at the destination and enterprise level, will result in ad hoc, unregulated, inefficient and reactive responses to calls to supply tourism infrastructure and services.

The literature on planning is vast and growing with an increasing interest in specific forms and aspects of tourism. Readers are referred to Saarinen, Rogerson and Hall (2017), Dredge and Jamal (2015) and Rahmafitria et al. (2020), Costa (2020) and Lew (2007) for detailed discussions on key trends in tourism planning. Hall (2015) provides a framework that illustrates the relationships between planning traditions, key themes, global development milestones and humanity's global footprint. He illustrates, for example, that planning has evolved from boosterism, through to a focus on economic development, community tourism, sustainable planning and now to a greater concern on public tourism planning.

One outcome of the interest in planning has been the emergence of special interest areas in the planning literature. Examples include the role of planning in sustainable development, (Simão & Partidárion, 2012), power exercised through social representation in planning (Moscardo, 2011), community participation (Tosun, 2005), cultural heritage planning (du Cros, 2001) and destination development (Haughland et al., 2010). Commenting on the apparent gap between academic model building and their application in practice, Ruhanen (2004) observed that there is a need to move the 'wealth of academic knowledge' into the real world of planners and decision makers. This observation by Ruhanen (2004) continues to have merit and will become even more important in the future as human society begins to face the sustainability crisis that has resulted from its over exploitation of scare renewable and non-renewable resources.

Planning traditions

Planning has a long tradition stretching back to the time when the first urban settlements emerged in the middle east. It is now undertaken in a number of forms and at various scales. The forms of planning include public and private sector, development, infrastructure provision to support tourism activity, land and resource use, and promotion and marketing (Hall, 2005). Scales include sectoral, local, regional, national and international. Over time, planning has emerged as a systematic process that assists various actors to engage in dialogue. This process acknowledges that there are usually overlapping and often competing interests that need to be considered in a policy environment that is becoming increasingly complex and contested, as factors such as social justice, sustainability and impacts of climate change and other events are considered. The outcome of this process is generally expressed as a plan that incorporates strategies and targets designed to meet future objectives.

Tourism planning emerged as a specialised field in the 1960s in response to growing demand (Tosun & Jenkins, 1998). Governments began to support tourism at destination, regional and national levels, generally following a focus on investment incentives, industry organisation and spatial considerations (Yang & Wall, 2008). As Gunn (1988) noted, planning in this period was essentially an exercise in matching demand with tourism products. Planning failures highlighted the need for a more participatory approach (Harrill & Potts, 2003), supporting earlier calls by Gunn (1979) for participation by multiple stakeholders in tourism planning.

In recent decades tourism planners have been confronted by the need to deal with the contradictions inherent in responding to increasing demand for tourism in a global context where there are increasing concerns about the ability of the global environment to support growth. In practice, a gap has emerged between what scholars suggest should be done and the factors that should be considered, and what actually happens (Yang & Wall 2008). Despite a growing volume of calls to preserve what remains of nature, development of new tourism related infrastructure continues to sanction the destruction of pristine environments. One of many possible examples is Mexico's Mayan Train, a 1554 kilometre railway track build in the Yucatán Peninsula to connect tourist destinations on the Caribbean coast with historic Mayan archaeological sites within the peninsula. The project has been criticised on environmental grounds and concerns about further marginalization of indigenous groups living in the area (Camargo & Vázquez-Maguirre, 2021).

Although Hall (2000: 6) observed that "*planning is an extremely ambiguous and difficult word to define*", the need for a definition cannot be denied. Dredge and Jennings (2007: 22) for example, define planning as a "*strategic activity comprising a number of stages that lead to the determination of a course of action to meet predetermined goals. Planning is concerned with the future; is devoted to acquiring knowledge and identifying appropriate courses of action; is about anticipating change, developing a strategic vision and facilitating political decision making; and is value-laden and political.*" Other definitions generally include similar relationships that should be embraced by planners, but are not always done. The specific methodology used in planning will depend on a large number of considerations including the aim of a particular plan, the environment that planning occurs in, the objectives of the organisation sponsoring the plan, the level of consideration given to environmental, key stakeholders, social and economic cost and benefits of the plan, and distribution of benefits between winners and losers of the planning process.

Planning is not a standalone activity. It takes place in a policy environment. However, planning and policy are dialectical concepts that have socially constructed meanings and are often defined by the context in which they are applied (Dredge & Jamal, 2015). That context may be 'fuzzy' or have a number of meanings depending on the context in which the terms are used. Dredge and Jenkins (2007: 7) for example define policy as being "*a position, strategy, action or product adopted by government and arising from contests between different ideas, values and interests*". From a slightly different perspective, Hall (2000) observed that tourism policy is whatever governments choose to do or not do, including actions and inactions, decisions and non-decisions.

In practice however, the ability of governments to balance community welfare and environmental conservation with the need for profit has created conflicts in policy making and planning (Rahmafitria et al., 2020). The need to balance tourism development with social aspects of community has also created other areas of tension. For example, in a neoliberal policy context the market is given primacy, while in a socialist context, the public sector sets the goals that policy and planning should achieve. Planning occurs in a political environment that determines how agendas are set, what issues are important, the roles that are assigned to the market and the state, and the degree to which winners compensate losers. From these perspectives it is difficult to disengage planning from the political context that it occurs in and from the form of political economy that determines policy formation. While acknowledging the difficulty inherent in treating planning in isolation from policy and political economy the discussion in this chapter will focus primarily on planning.

Planning approaches, typologies and frameworks

Getz (1986), in a study of over 150 tourism planning models identified a number of problems associated with them including; limited analysis and evaluation of the costs and benefits of tourism development; the failure of many models to integrate into other development processes; failure to take note of non-economic factors; and a narrow focus on specific development planning at project levels. To these can be added failure to consider the impact on stakeholders in the supply chains that support the projects that planning is concerned with. Planning (and policy) can be broadly categorised as either substantiative ideas that articulate the direction and underlying values of tourism policy and planning or procedural ideas that suggest how planning should be undertaken (Dredge & Jenkins, 2007). The typology suggested by Getz (1987, cited in Hall, 2005) provides an example of the substantive approach to planning. Getz identified four approaches:

- Boosterism
- Economic
- Physical/spatial
- Community.

Getz observed that these approaches were not mutually exclusive, and planning may incorporate elements of two or more approaches applied in tandem or sequentially. Hall (2000) later added sustainable planning in recognition of the growing interest in sustainability that had begun to occur in the literature and in global forums such as the United Nations Conference on Environment and Development (also referred to as the Rio de Janeiro Earth Summit). Soon after, Faulkner (2001) responding to the need to develop planning instruments that could be used for crisis management and suggested a six-stage community response plan for managing and responding to disasters in a tourism setting.

Commenting on the approaches suggested by Getz and Hall, Dredge and Jenkins (2007) cautioned that easily identifiable planning categories run the danger of oversimplifying and obscuring the richness and diversity of the transcendent shifts that take place in social science and public administration ideology and the often conflicting ideas that determine policy development and decision-making. As an alternative, Dredge and Jenkins (2007) suggested a typology based on six substantive approaches able to be identified in tourism planning (and policy formation):

- *Industry development planning and policy* – use of financial measures and tools such as incentives and tax breaks to strengthen industry capacity
- *Market planning and policy* – the promote the development and expansion of markets with the aim of product development and diversification
- *Spatio-physical destination planning and policy* – spatial destination planning, modelling and relationships between attractions, services, origins, transport routes and marketing
- *Conflict management planning and policy* – measures to mitigate the social, cultural, environmental and economic effects of tourism on local destinations.
- *Communicative tourism planning and policy* – establishing collaborative dialogue where multiple stakeholders are involved
- *Crisis-responsive planning and policy* – addressing crises as they emerge and drawing where appropriate from the previous approaches.

Costa (2018) also suggested a typology with planning models grouped into

- Rational tourism models
- Systematic tourism models
- Production tourism models, and
- Strategic management tourism models.

The typology suggested by Dredge and Jenkins (2007) exhibits many similarities to those suggested by Getz (1987) and Hall (2000) with the exception the crisis-response planning and policy approach. The contributions by both Hall (2000) and Faulkner (2001) led to significant advances in ideas about planning. For example, growing concerns about ability of tourism planning to ensure that development is sustainable has generated considerable debate and more recently has begun to merge with concerns over climate change and how tourism has contributed to it and suffered from it. More recently, the need to respond to the downturn in global tourism activity caused by the COVID-19 pandemic s generated an enormous uptick in interest in crisis and disaster management.

From a procedural perspective, spatial planning models based on a 'stagist' approach have been widely adopted by planning bodies. Popular models following the procedural approach include the Recreation Opportunity Spectrum (ROS), Zero Based Budgeting (ZBB), Management by Objectives (MBO) and Environmental Assessments (EIA). Lew (2007) stated that many of these models have been subjected to considerable criticism because of underlying assumptions including predictable behaviour, unlimited knowledge and problem-solving capabilities, a failure to list all possible scenarios and objectivity in a closed system. In most cases planners have limited knowledge and time horizons and are subject to unpredictable events and influences. Examples of models based on sequential stages include Dredge and Jenkins (2008) who suggested eight stages and Lew (2007) who recommended 10 stages. Lew's stages are:

1. Formulation of goals and objectives
2. Identify a goal (to solve the problem)
3. Collect background data
4. Identify guidelines for assessing alternative plan scenarios
5. Identify alternative plan scenarios, including policies and guidelines to achieve the goals
6. Assess alternative plans scenarios using pre-defined assessment guidelines
7. Select the preferred alternative
8. Implement the plan
9. Monitor, evaluate, and revise the Implementation
10. Identify new problems and begin the process again. Development of alternative solutions or scenarios

Bittar-Rodrigues and Prideaux (2018), provide a typical example of this type of 'stagist' approach but with a strong emphasis on community empowerment in a Community Based Tourism (CBT) model. In their model (Figure 15.1), Column 2 identifies nine stages which are influenced by factors identified as community issues (column 1), non-community actors (column 3) and external factors, such as national governance issues (column 4).

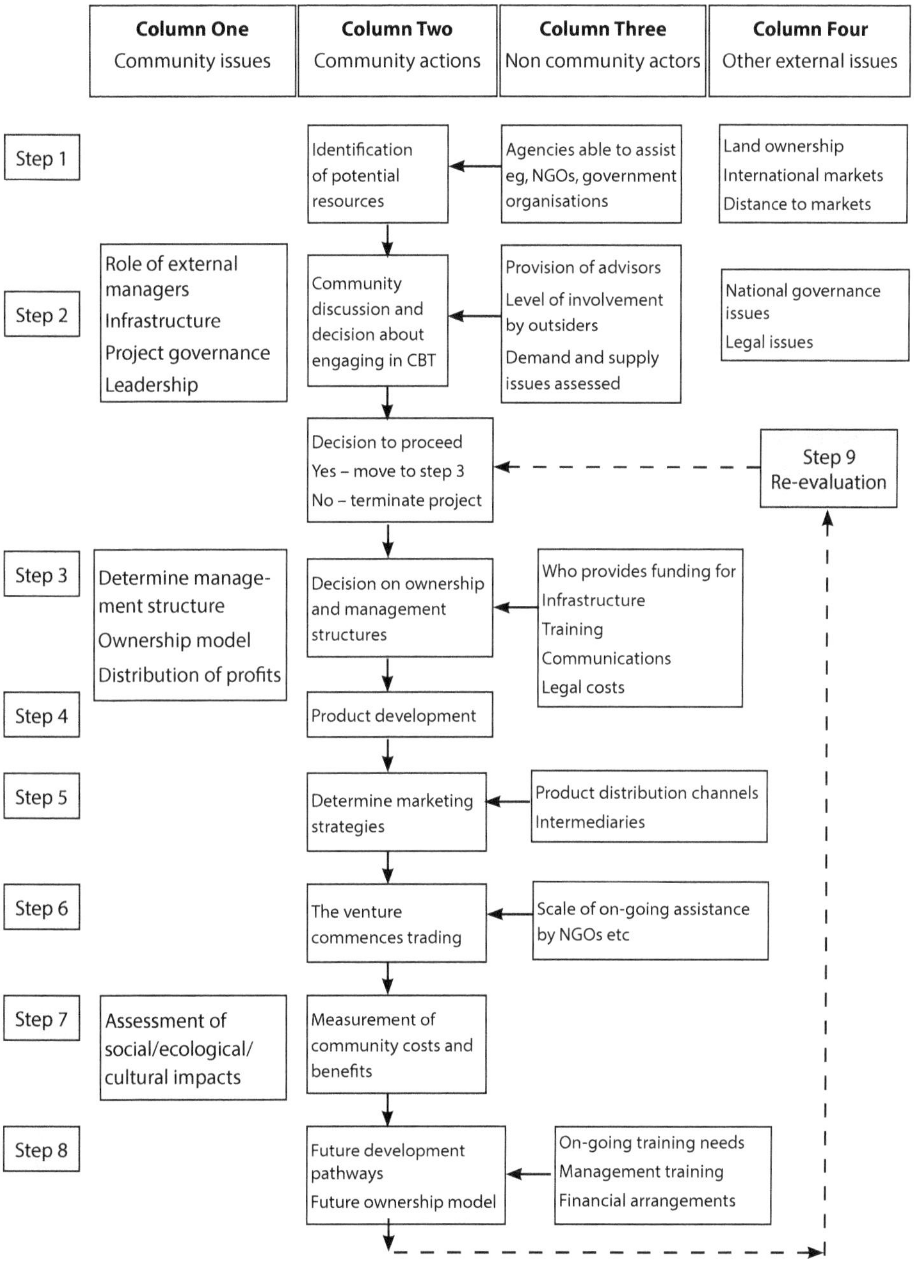

Figure 15.1: CBT Implementation model. Source: Bittar-Rodrigues and Prideaux (2018)

A new perspective

Many of the models discussed in this chapter have been applied on a 'one fit suits all' solution to various issues including planning. In reality, one model is often insufficient to deal with the range of issues that may be encountered when looking for answers. Approaching the question of which model is the most appropriate, Prideaux and Thompson (2021) suggested an open architecture approach described as a process that allows various models and theories to be 'plugged in' on an 'as required' basis as planning moves through a series of stages such as those identified by Lew (2007). The opposite to the open architecture approach is a 'closed architecture' approach where only one model is used to inform on planning issues. The open architecture approach facilities flexibility allowing models to be used in parallel or sequentially where required to solve various issues in the planning process. For example, supply considerations such as location and resources typically considered in stage 3 of the planning model suggested by Lew (2007) can be addressed by a variety of non-planning models including economic base, periphery-core and destination pull.

Impending sea level rise will create enormous difficulties for coastal destinations. Coastal located airports, including at Phuket, Thailand and on many Pacific islands including Tuvalu, face the threat of flooding as sea levels rise. Similarly, many beachfront hotels and other tourist infrastructure will be faced with the need to either relocate or defend by building seawalls. However, the timing and eventual height of sea level rise is unknown and will depend on how quickly annual rises in greenhouse gas emissions can be halted.

On a different scale, future planning will also need to consider the implications of a rapidly urbanising global population. The United Nations Department of Economic and Social Affairs (2018) estimates that by 2050 about 68% of the global population will live in urban areas. Changes in technologies including a possible switch to autonomous vehicles, increased use of new technologies including facial recognition and robots, Artificial Intelligence (AI), further growth of the sharing economy (with products such as Airbnb and Uber) will call for new planning solutions. On the climate change front, the impending loss of many coastal areas to sea level rise will generate a need to relocate some urban areas or parts thereof. This will require a significant re-evaluation of spatial and scale elements of tourism planning as well mobility requirements. Emphasis will also need to be given to the changing moral and ethical aspects of tourism development particularly in the Global South (Saarinen et al., 2017).

Conclusion

This chapter has highlighted a number of models, typologies and frameworks that have emerged in the tourism planning literature. The authors noted that in the future, planning will need to take a path different from the present one. For example, government, rather than market led, intervention may be required. In the future it is reasonable to expect that governments and their bureaucracies will take a more interventionist role in public policy administration than has been the practice in recent decades, particularly as the need to transition to a net zero GHG economy becomes more urgent.

Discussion points / Exercises

1. What are the major planning factors that should be considered when deciding which planning method should be used?
2. What is the difference between the open architecture and closed architecture approach to planning?
3. What are the key differences between subjective and procedural approaches to planning?

Points to consider for revision

- Planning takes place in a policy environment. Thus a range of non-spatial issues including community welfare, the environment and economic return need to be considered.
- The problems associated with climate change will require new planning approaches in future decades.
- Why will consideration need to be given to replacing neoliberalist approaches to planning in the future?

Additional readings

Costa, C. & Brandao, F. (2018) Tourism planning: evolution and future trends, in Cooper, C., Volo, S., Gatner, W. & Scott, N. (eds), *The SAGE Handbook of Tourism Management*, Sage London, pp. 538 -555.

Dredge, D. & Jamal, T. (2015), Progress in tourism planning and policy: A post-structural perspective on knowledge production, *Tourism Management*, 51, 285-297.

Rahmafitria, F., Pearce, P., Oktadiana, H. & Putro, H. (2020) Tourism planning and planning theory: Historical roots and contemporary alignment, *Tourism Management Perspectives*, 35. doi.org/10.1016/j.tmp.2020.100703.

References

Camargo, B. A., & Vázquez-Maguirre, M. (2021). Humanism, dignity and indigenous justice: The Mayan train megaproject, Mexico. *Journal of Sustainable Tourism*, 29(2-3), 372-391.

Bittar Rodrigues, C. & Prideaux, B. (2018), A management model to assist local communities developing Community Based Tourism ventures: A case study from the Brazilian Amazon, *Journal of Ecotourism*, 17(1), 1-19.

Costa, C. (2020), Tourism planning: a perspective paper, *Tourism Review*, 75(1), 198-202.

Dredge, D. & Jamal, T. (2015), Progress in tourism planning and policy: A post-structural perspective on knowledge production, *Tourism Management*, 51, 285-297.

Dredge, D. & Jenkins, J. (2007), *Tourism Planning and Policy*, Wiley, Milton, Queensland.

Du Cros, (2001), A new model to assist in planning for sustainable cultural heritage tourism, *International Journal of Tourism Research*, 3(2), 165-179

Faulkner, B. (2001), Towards a framework for tourism disaster management, *Tourism Management*, 22, 135-147.

Getz, D. (1986), Models in tourism planning: Towards integration of theory and practice, *Tourism Management*, 7, 21-32.

Getz, D. (1987), *Tourism planning and research: Traditions, models and futures*, paper presented at the Australian Travel Workshop, Bunbury, Western Australia, 5-6 November.

Gunn, C. (1979), *Tourism Planning*, Crane Rusak, New York.

Gunn, C. (1988), *Tourism Planning: Basics, Concepts, Cases*, Taylor and Francis, New York.

Hall, C. M. (2000). *Tourism Planning: Policies, processes and relationships*. Harlow: Pearson.

Hall, C. M. (2005) *Tourism, Rethinking the Social Science of Mobility*, Pearson, Harlow.

Hall, C. M. (2015) Economic greenwash: On the absurdity of tourism and green growth, in V. Reddy and K. Wilks (eds.), *Tourism in the Green Economy*, Earthscan, London, pp. 339-358.

Harrill, R. & Potts, T. (2003), Tourism planning in historic districts: Attitudes towards tourism development, *Journal of the American Planning Association*, 63(3), 233-244.

Haughland, S., Ness, H., Grønseth, B., & Aarstad, J. (2010), Development of tourism destinations: an integrated multilevel perspective, *Annals of Tourism Research*, 38(1), 286-290.

Lew, A. (2007), Invited commentary: Tourism planning and traditional urban planning theory – the planner as an agent of social change, *Leisure/Loirir*, 31(2), 383-391.

Moscardo, G. (2011), Exploring social representations of tourism planning: issues for governance, *Journal of Sustainable Tourism*, 19 (4-5), 423-436.

Prideaux, B. & Thompson, M. (2021), developing tourism in remote communities: an open architecture approach, *Advances in Hospitality and Leisure*, 16, 3-21. Doi 10.1108/S1745-354220200000016003

Rahmafitria, F., Pearce, P., Oktadiana, H. & Putro, H. (2020) Tourism planning and planning theory: Historical roots and contemporary alignment, *Tourism Management Perspectives*, 35, doi.org/10.1016/j.tmp.2020.100703.

Ruhanen, L. (2004), Strategic planning for local tourism destinations: an analysis of tourism plans, *Tourism and Hospitality Planning & Development*, 1(3), 239-253.

Saarinen, J., Rogerson, C. & Hall, CM. (2017), Geographies of tourism development and planning, *Tourism Geographies*, 19(3), 307-317.

Simão, J. & Partidário, R. (2012) How does tourism planning contribute to sustainable development? *Sustainable Development*, 20, 372-385.

Tosun, C. (2000), Limits to community participation in the tourism development process in developing countries, *Tourism Management*, 21(6), 613-633

Tosun, C. & Jenkins, C. (1998) The evolution of tourism planning in third world countries: A critique, *Progress in Tourism and Hospitality Research*, 4, 101-114.

United Nations Department of Economic and Social Affairs (2018) 68% of the world population to live in urban areas by 2050, says UN, United Nations Department of Economic and Social Affairs, https://www.un.org/development/desa/en/news/population/2018-revision-of-world-urbanization-prospects.html

Yang, L. & Wall, G. (2008), The evolution and status of tourism planning: Xishuangbanna, Yunnan, China, *Tourism and Hospitality Planning and Development*, 5(2), 165-182.

Epilogue

This book explored a range of theories, concepts and ideas that shape how we think about tourism the way we do. In doing so, it revealed that tourism is a true multi-discipline. It is informed by a range of core and peripheral disciplines as identified in Chapter 2. Historically, though, tourism studies has been beset by a high degree of silofication – a varied field of study examined strictly within the confines of individual disciplinary silos. Even when attempts have been made to be multi-disciplinary, the results have often been less than satisfactory, for usually one school of thought dominates, while others are placed in subservient roles. Add to this Tribe's (2006) treatise on the force field of tourism, and it is not surprising that tourism studies have been labelled as fragmented and disjointed (Benckendorff and Zehrer 2013, Laws and Scott 2015), typified by multiple communities of discourse with historically little cross fertilization between communities (Gibson 1998).

Yet, at its most basic, we are talking about two simple things, regardless of the disciplinary focus, methodology adopted or personal bias of the author. They are: tourism as a phenomenon and; the tourist as the person participating in that phenomenon. The former relates to its structure, functioning, evolution, interaction with the broader community and its impacts, while the latter examines people who travel, why they travel, what they do and what impacts they exert. Indeed, different disciplines often examine the same event, but from different sides of the same coin. Destination evolution is a classic case in point. Butler (1980) adopted a supply-side, geographer's perspective, Plog (1974) adopted a demand side, social psychologist's perspective, while the advent of complexity theory added a third dimension. The same situation occurs with motivation studies where sociologists and psychologists have vied for dominance. Geographers, anthropologists, economists and environmentalists have examined impacts. And the list goes on. Each approach has its merits and each has its limitations. None is perfect and certainly none is more 'right' or valid than others.

To date, though, the task of trying to present a holistic overview of tourism, what it is and how it works, has proven to be daunting. This book has attempted to do so by discussing the contributions made by many of the foundation scholars who legitimized tourism as a field of study and who explored this fascinating topic from the unique perspective of their own disciplines. How does it all fit together? The figure attached to the epilogue lists about 50 major themes that formed the basis of the book and shows the interactions between and among them. Each of the themes is developed throughout the book, but because of the intricacy of interactions, it has been impossible to show the links within each chapter. Instead, the final task of integration rests with this Epilogue.

The chart highlights just what a multi-discipline tourism is and why it is such a fascinating field to examine. It also highlights the innate challenges we face when trying to teach it in a simple manner, for it is impossible to understand tourism without having at least a basic knowledge of the many areas, fields and disciplines that have contributed to our collective knowledge, and how they interact.

We have written previously how we are now in the third generation of tourism studies. The first generation of tourism scholars studied it from within their own single disciplinary lens in order to inform that discipline. Few if any pure tourism programmes existed, and instead tourism scholars often worked on the fringes of their core disciplines, sometimes at the derision of their more 'serious' colleagues who scoffed at the validity of tourism as a research topic. These people had a strong core grounding and were then able to build a strong conceptual foundation that has informed the field for over 50 years.

The second generation came from industry or government, had a reasonably strong single disciplinary focus and were usually employed in burgeoning tourism departments, typically located in business faculties. Importantly, they usually completed their doctoral studies within a traditional discipline, as few of the nascent tourism programmes had reached the level of maturity to offer stand-alone doctorates. They were able to bring a combination of applied business skills that complemented their core disciplines. They are responsible for progressing the field to where it is today, for the work of the second generation of academics led to the proliferation of tourism journals and the ultimate acceptance of the field as a legitimate area of academic enquiry.

The third generation of tourism scholars emerged over the last 20 years and is now dominating the field as the first and second generations enter or are close to retirement. They make take the field to new heights for they possess a unique skill set, having graduated from dedicated tourism programmes. But, the jury is still out to some extent, for they often lack strong background in any single discipline to inform their studies. They may produce impressive empirical studies using sophisticated statistical techniques and have access to unprecedented big data sources. But a risk exists that their work will be atheoretical and add little real new knowledge to the field, for most have not had the chance to explore tourism as an integrated multi-disciplinary phenomenon. This book is written in part for them, in order to give them a chance to appreciate where our collective body knowledge came from and the various sources of these fonts of knowledge. In doing so, we hope that the reader can gain a deeper appreciation of tourism as a dynamic field of study.

Figure 1: Integration of Tourism

	Tourism as a multidiscipline	Tourism as a journey away and back	Heuristic vs technical definitions	Pleasure tourism as an extreme form of leisure	Tourism as a system	Individual tourism systems	Concept of a destination	Attractions	Tourist attractions system	Attractions taxonomy	Valuation of time (commodity or resource?)	Distance decay	Market access	Effective tourism exclusion zones	Segment transformation	Itinerary models	Visa requirements	How tourists consume a destination	Tourism as trade in services	Tourism as a democratic activity	Butler's lifecycle	Plog's lifecycle	Complexity theory	Prideaux's evolutionary model	Risk	Carrying capacity	Barriers to travel	Tourism as a life priority	The leisure paradox	Generational cohort theory	Boorstin's contrived reality	MacCanell's quest for authenticity	Cohen - strangeness vs familiarity	Cohen's tourist typology	Cohen's centre and the other	Tourism as a liminal experience	Dann's push pull model	Maslow's hierarchy of needs	Crompton's motives	Pearce's travel career pattern
Force field of knowledge	x		x	x	x		x	x	x	x		x	x	x		x				x	x	x	x	x	x	x	x	x	x	x	x	x	x	x	x	x	x	x	x	x
Tourism as a multidiscipline			x	x	x	x	x	x	x	x	x	x	x	x	x	x	x	x	x	x	x	x	x	x	x	x	x	x	x	x	x	x	x	x	x	x	x	x	x	x
Tourism as a journey away and back			x	x	x	x					x	x	x			x	x	x	x	x		x					x	x				x	x	x	x	x	x		x	x
Heuristic vs. technical definitions				x			x				x	x	x					x		x		x			x		x	x	x	x	x	x	x	x	x	x	x	x	x	x
Pleasure tourism as an extreme form of leisure					x	x		x	x	x	x	x	x	x	x	x		x		x					x	x	x	x	x	x	x	x	x	x	x	x	x	x	x	x
Tourism as a system						x	x	x	x	x	x	x	x	x	x	x	x	x	x		x	x	x	x	x	x											x	x	x	x
Individual tourism systems							x	x	x	x	x	x	x	x	x	x	x	x		x			x		x	x	x	x	x	x	x	x	x	x	x	x	x	x	x	x
Concept of a destination								x	x	x				x		x		x			x	x	x	x		x											x			
Attractions									x	x	x	x	x		x	x		x	x		x	x	x	x	x	x					x	x	x	x	x	x	x	x		x
Tourist attractions system										x		x	x			x		x	x						x	x											x	x		
Attractions taxonomy											x	x	x		x	x		x		x	x	x	x	x	x	x		x	x		x	x	x	x	x		x	x		x
Valuation of time												x	x	x	x	x	x	x		x	x	x	x	x	x		x	x	x	x							x	x	x	x
Distance decay													x	x	x	x	x	x		x			x	x	x	x			x	x			x			x	x		x	x
Market access														x	x	x	x	x				x	x	x	x	x					x	x	x	x	x	x	x	x	x	x
Effective tourism exclusion zones															x	x	x						x	x	x									x	x	x	x			x
Segment transformation																x	x	x				x	x	x	x	x	x	x	x	x							x			x
Itinerary models																	x	x	x				x		x	x	x	x									x			

Visa requirements																		x	x	x		x			x	x	x	x		x				x	x	x	x		x	x
How tourists consume a destination																			x	x	x	x	x		x	x	x		x	x	x	x	x	x	x	x				
Tourism as trade in services																				x			x	x	x	x														
Tourism as a democratic activity																					x	x	x	x	x	x	x	x	x	x	x	x	x	x	x	x	x	x	x	x
Butler's lifecycle																						x	x	x		x	x						x	x	x	x	x		x	x
Plog's lifecycle																							x	x	x	x	x	x	x	x	x	x	x	x	x	x	x	x	x	x
Complexity theory																								x	x	x	x	x		x						x	x			x
Prideaux's evolutionary model																									x	x	x			x		x		x		x	x	x	x	x
Risk																										x	x	x	x	x	x	x	x	x	x	x	x	x	x	x
Carrying capacity																																								
Barriers to travel																												x	x	x			x	x	x	x	x	x	x	x
Tourism as a life priority																														x			x	x	x	x	x	x	x	x
The leisure paradox																														x			x	x	x	x	x	x	x	x
Generational cohort theory																																	x	x	x	x	x	x	x	x
Boorstin's contrived reality																																x	x	x	x	x	x			
MacCanell's quest for authenticity																																	x	x	x	x	x	x	x	x
Cohen - strangeness vs familiarity																																		x	x	x	x	x	x	x
Cohen's tourist typology																																			x	x	x			x
Cohen's centre and the other																																				x	x	x	x	x
Tourism as a liminal experience																																					x	x	x	x
Dann's push pull model																																						x	x	x
Maslow's hierarchy of needs																																							x	x
Crompton's motives																																								x
Pearce's travel career pattern																																								

Index

www.ingramcontent.com/pod-product-compliance
Ingram Content Group UK Ltd.
Pitfield, Milton Keynes, MK11 3LW, UK
UKHW051847100926
8556IPUK00005B/502